An Introduction to Evolutionary Ecology

Andrew Cockburn

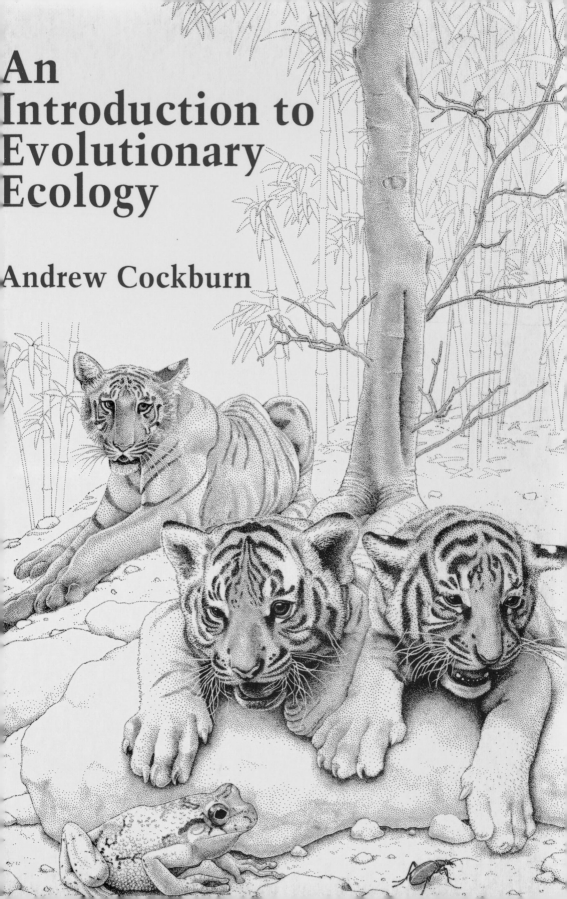

AN INTRODUCTION TO
EVOLUTIONARY ECOLOGY

An Introduction to Evolutionary Ecology

ANDREW COCKBURN

Professor of Botany and Zoology
School of Life Sciences
Australian National University
Canberra, Australia

ILLUSTRATED BY
KARINA HANSEN

OXFORD
BLACKWELL SCIENTIFIC PUBLICATIONS
LONDON EDINBURGH BOSTON
MELBOURNE PARIS BERLIN VIENNA

© 1991 by
Blackwell Scientific Publications
Editorial offices:
Osney Mead, Oxford OX2 0EL
25 John Street, London WC1N 2BL
23 Ainslie Place, Edinburgh EH3 6AJ
3 Cambridge Center, Cambridge,
  Massachusetts 02142, USA
54 University Street, Carlton
  Victoria 3053, Australia

Other Editorial Offices:
Arnette SA
2, rue Casimir-Delavigne
75006 Paris
France

Blackwell Wissenschafts-Verlag
Meinekestrasse 4
D-1000 Berlin 15
Germany

Blackwell MZV
Feldgasse 13
A-1238 Wien
Austria

First published 1991
Reprinted 1992

Set by Setrite Typesetters, Hong Kong
Printed and bound in Great Britain by
Hartnolls Ltd, Bodmin, Cornwall

DISTRIBUTORS

Marston Book Services Ltd
PO Box 87
Oxford OX2 0DT
(*Orders:* Tel: 0865 791155
         Fax: 0865 791927
         Telex: 837515)

USA
  Blackwell Scientific Publications, Inc.
  3 Cambridge Center
  Cambridge, MA 02142
  (*Orders:* Tel: 800 759-6102
             617 225-0401)

Canada
  Oxford University Press
  70 Wynford Drive
  Don Mills
  Ontario M3C 1J9
  (*Orders:* Tel: 416 441-2941)

Australia
  Blackwell Scientific Publications
  (Australia) Pty Ltd
  54 University Street
  Carlton, Victoria 3053
  (*Orders:* Tel: 03 347-0300)

British Library
Cataloguing in Publication Data

Cockburn, Andrew *1954*–
  An introduction to evolutionary ecology.
  1. Ecology
  I. Title
  574.5

  ISBN 0-632-02729-0

Library of Congress
Cataloging-in-Publication Data

Cockburn, Andrew, 1954–
  An introduction to evolutionary ecology
  Andrew Cockburn
  illustrated by Karina Hansen.
        p.   cm.
  Includes bibliographical references
  and index.
  ISBN 0–632–02729–0
  1. Biological diversity.   2. Ecology.
  3. Species.   4. Evolution.
  5. Genetics.   I. Title.
  QH313.C74   1991
  574.5 – dc20

Contents

[v]

Preface

I can remember no more exhilarating experience than the first time I snorkelled over the edge of a coral platform into a kaleidoscopic array of shelf corals and fish. On the same trip to the Great Barrier Reef of northeastern Australia, there were many moments of great power, from diving with sharks and manta rays to watching the intricate beauty of tiny nudibranchs and polychaetes. Deriving great joy from the observation of nature is common, and takes many forms. In time such joy is often replaced by curiosity—a desire to understand the diversity of organisms; what shapes their form and behaviour, and how they interact. This curiosity lies at the heart of evolutionary ecology.

The aim of the book is to encourage students to see how conclusions can be drawn about patterns in nature, rather than to provide a series of spectacles through which the world can be viewed. In order to understand how problems can be resolved, it is best to be faced with real examples, and much of the material in this book is an analysis of areas that are very controversial, subject to active research, and which will undoubtedly require a radically different treatment in later editions of this book. I hope that I have chosen the organisms which illustrate how to test controversy best, but inevitably my choice of examples is influenced by personal biases. I am a terrestrial ecologist, and know most about mammals, birds, angiosperms and insects in approximately that order.

This book aims to provide a course suitable for later year undergraduates, or graduate students early in their careers. Deciding how much knowledge should be assumed is always a difficult task, and in this case is based largely on my own experience of the interests and expertise of senior undergraduate students in Australia. I have expected some familiarity with the systematics of plants and animals, and have often described the organisms using their Linnaean classification (e.g. poriferan rather than sponge, *Solidago* rather than goldenrod), except where common names are used extremely widely (house mouse, cow, etc.). In order to provide easy access for students lacking training in this area I have closely followed the systematic arrangements offered in widely used undergraduate textbooks. I have not assumed great mathematical competence, though where algebraic or geometric analyses illustrate ideas succinctly I have included this material. For the mathematically adventurous, I include directions to more sophisticated treatments throughout the text. The most troublesome area is genetics, and this text includes a much more detailed discussion of genetics at all levels than is usual in an ecology text.

Preprints of papers helped shape my ideas at critical times. Thanks to Annie Collie, Steve Frank, Alan Grafen, Paul Harvey, Robert May,

Mark Pagel and Monte Slatkin. Grafen (1988a) convinced me that I should completely rewrite Chapter 1. Idiot questions about groups which had somehow escaped my attention in my previous education were patiently answered by my colleagues Warwick Nicholas (nematodes), Julian Ash (plants in general), Mike Howell (parasites in general), John Oakeshott (fruitflies and molecules) and Penny Gullan (arthropods). The students of my Zoology C13, C15, C32 and C34 classes at the Australian National University were the main sounding board for my ideas. Thanks also to the critical readers of the text and the original outline, both anonymous and those brave enough to face my dark scowls. John Jaenike provided a thorough critique of the whole book, and John Endler told me the original chapter organisation was 'boring' some 2 years before I decided that I agreed with him. The book was finished in the peaceful atmosphere at the Department of Zoology at Gothenburg University, where Malte Andersson was the finest of hosts. Karina Hansen performed the herculean task of converting my vague arm-waving into illustrations of great beauty. Last, an enormous thanks for the encouragement from Mark Robertson at Blackwell.

Andrew Cockburn
Göteborg

Chapter 1
The Scope of
Evolutionary Ecology

Evolutionary ecologists represent the intellectual descendants of that vast proportion of humanity who have been fascinated and puzzled by the living world, and thought deeply enough about our world to ask the question 'why?'. In boundless curiosity about nature and life, evolutionary ecologists differ very little from the shamans, gurus and clerics in their quest for answers. The difference lies in knowing how to ask questions about nature, using techniques and a philosophy that has emerged in just over a century of extraordinary leaps of under-standing. The great geneticist Theodosius Dobzhansky entitled one of his final papers 'Nothing in biology makes sense except in the light of evolution' (Dobzhansky 1973). This book is a celebration of that view, and explores the role that evolution plays in ecology, which is usually defined as the study of the distribution and abundance of organisms.

The chief aim of the book is to improve your ability to ask ques-tions about the living world and the pattern that resides within it. For make no mistake, the majority of problems in evolutionary ecology remain to be solved; what I hope to offer is the techniques and philos-ophy allowing their solution, and some sense of the joy and excitement involved in their pursuit.

PATTERNS IN THE LIVING WORLD

When faced with the complexity of nature we are challenged in two ways. The first is recognition of pattern, the second is its explanation. Some patterns in nature are clear, and contribute much of the beauty and majesty of the natural world. Some are subtle, and demand more careful analysis. In either case, explaining these patterns is always complex and challenging, and is the real warp and weft around which evolutionary ecology is woven. This Chapter describes some obvious patterns whose explanation has proved challenging for ecologists, and introduces the range of topics which will be covered in this book.

Why was 1983 so bad for giant pandas?

Few creatures are as well known to the public throughout the world as the giant panda, *Ailuropoda melanoleuca*, despite its narrow geographic range and rarity. This large relative of the bears has an unusual diet for a member of the Order Carnivora, as more than 99 per cent of its diet consists of bamboo branches, stems and leaves. Bamboos are a type of grass, but are unique among that family because of their generally large size and long lives. Despite their grandeur and longevity, flower-ing is infrequent and synchronous for a species within a region. At the conclusion of flowering, the plants set seed and die (Janzen 1976).

Death immediately after reproduction is widespread among plants and animals and has attracted a variety of names in various taxa, though the long delay before reproduction which occurs in bamboo is much more unusual. Botanists usually use the terms monocarpic and polycarpic, though I prefer the terms semelparity and iteroparity as general descriptors of the dichotomy between reproducing once and dying, and repeated reproduction throughout a season or lifetime. Explaining the advantages of semelparity and iteroparity is one of the concerns of life history theory, to which Chapters 5 and 6 are devoted. Semelparity illustrates one of the important elements of this theory; heavy investment in reproduction may be valuable to the organism but it also incurs some costs, in this case the ultimate price—death.

But what does semelparity of its food source mean to the panda? The bamboo forests inhabited by pandas have been subjected to clearing and occupation by humans, fragmenting habitat and reducing its complexity (Schaller *et al.* 1985). Pandas usually consume a variety of species of bamboos, so the prospect of starvation following the death of their food source is low, provided chance synchrony does not occur between the flowering of all the bamboo species in a particular region, and provided that the forest is not so fragmented that only one food species is available. Unfortunately, these circumstances coincided in the spring of 1983, when there was a simultaneous mass flowering of the two principal food species, *Fargasia spathacea* and *Sinarundinaria fangiana*, in the main reserve in which pandas persist. Despite the commendable response of the Chinese government to this problem, seedlings take some time to grow tall enough to provide adequate food for pandas, and many pandas died. In Chapter 5 the importance of habitat selection in evolutionary ecology is considered, and in Chapter 10 the developing science of conservation biology is discussed.

Different ways of growing up

One of the more sadistic exercises used in the comparative zoology practical examinations at our university is to ask students to identify a bug (Hemiptera) and beetle (Coleoptera) which have converged on the typical adult form of each other. Such morphological convergence is common in nature, but in this case is especially interesting given the radically different developmental patterns of these two groups of insects (Fig. 1.1). Our own view of the natural world is distorted by our resemblance to our offspring—human babies look more or less the same as adults.

However, like the beetle depicted in Fig. 1.1, most animals and plants have a complex life cycle comprising very different stages, with the young often having different habitat, diet and morphology to their parents. As their name reflects, in the mayflies (Ephemeroptera), the adults are temporary machines incapable of feeding at all, though they do mate! In many marine animals, the adults are sessile or move very little in their lifetimes, while the larvae disperse passively or actively. In other species, like the monarch butterfly, the young move very little while the adults undertake spectacular migrations. Parasites often require more than one host to complete their life cycle, and the forms which lives in each host may differ greatly.

Not only do the stages leading to adulthood differ greatly between

(a)

Egg   Young nymph   Fourth instar   Full-size nymph   Adult
                      nymph

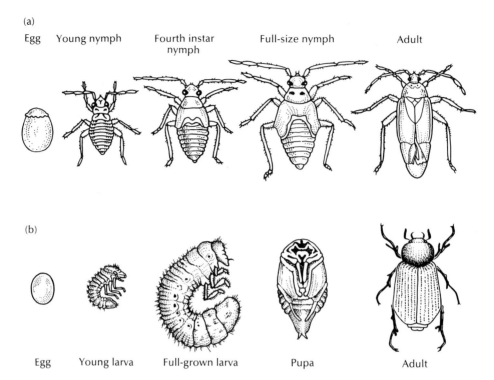

(b)

Egg   Young larva   Full-grown larva   Pupa   Adult

Fig. 1.1 Morphological similarity in adult form of an (a) exopterygote hemipteran
and an (b) endopterygote coleopteran. After Atkins (1978).

species and higher taxa, but they may be polymorphic within species.
The evolution of complex life cycles and developmental polymorphism
is the topic of Chapters 5 and 6. Most obvious examples of developmental polymorphism in animals are associated with gender. My pre-
ferred example for the most bizarre case of sexual dimorphism comes
from *Cystococcus*, a genus of Australian scale insects (Hemiptera), in
which the female induces the formation of large woody galls on the
leaves and stems of the bloodwood group of *Eucalyptus*. The adult
female is very large but degenerate, having no appendages other than
genitalia for copulation, and mouthparts which are used to tap the
phloem of the plants on which she feeds (Fig. 1.2). After the gall is
fully formed, she gives birth to the first part of her clutch, which are
exclusively males. The males are reared in the gall, feeding on a layer
of white nutritive tissue. They moult through several instars until
adulthood. Adult males are fully winged hemipterans, unusual only in
an elongation of their abdomen, but are very short-lived, probably
surviving less than 48 hours. The female gives birth to the second part
of her clutch, this time exclusively females, about the time the males
moult into their adult form. Females never possess wings, but first
instars (crawlers) do possess short stubby legs which are lost at the
first moult. They use these legs to cling to their brothers' abdomens,
and are carried to nearby host plants when the adult male leaves the
gall (Gullan & Cockburn 1986).

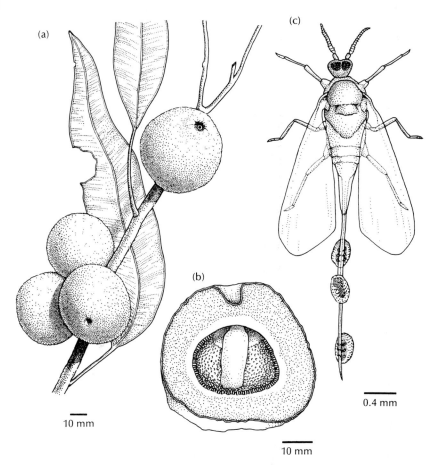

Fig. 1.2 Cystococcus gall-formers. (a) *Cystococcus* on a bloodwood eucalypt; (b) a female inside her gall, and her male offspring feeding from the white nutritive layer inside the gall; (c) a winged adult male carrying his sisters on his elongate abdomen. Females have no appendages after the first instar (Gullan & Cockburn 1986).

In most animals, males are smaller than females, as in *Cystococcus* (Greenwood & Wheeler 1985). In mammals and birds the opposite is much more likely to be true. When Portuguese seamen returned from Papua and New Guinea to Europe with the first specimens of birds-of-paradise they created a sensation. Not only were the birds immediately acknowledged as among the world's most beautiful, but their habits became a source of speculation and legend. One of the factors contributing to the myth was the way the birds were prepared by the Aru islanders from whom the specimens derived (Stresemann 1954; Forshaw & Cooper 1977). In order to accentuate the gorgeous plumes, the islanders maximised shrinkage by removing the legs. The naturalists of the day concluded that the birds lived high in the sky and never visited the ground. The female was supposed to incubate its egg in a hollowed depression in the back of the male! Like many tropical bird species, the male is inevitably more gorgeous than the female, and each species appears to display a different feature of their plumage (Fig.

1.3). For example, some have showy head plumes, others have large bald patches. The nature of gender differences and the processes generating them will be discussed in the latter half of Chapter 7.

Why have sex in the sea?

Gender is intricately linked to sexual reproduction, yet not all animals and plants reproduce sexually, though sexual reproduction arose very early in the history of life. Some plants and animals are parthenogenetic,

Astrapia
mayeri ♂

Cicinnurus regius
♀
♂

Diphyllodes
respublica

Pteridophora
alberti ♂

Fig. 1.3 Sexual differences in the male birds-of-paradise. Drawings after Forshaw & Cooper (1977).

and some alternate between parthenogenesis and sex. Some are exclusively self-fertilising, and some are parthenogenetic but still require sperm to penetrate the egg in order for reproduction to proceed. The various modes of reproduction are summarised in Table 1.1, using a rather more simplified set of terms than is occasionally applied to this problem. The distribution of this bewildering diversity of reproductive habits illustrates the first problem of an evolutionary ecologist: nature is often clearly patterned, but the explanation of those patterns is anything but clear. The best example of pattern comes from aquatic invertebrates. Across a very large range of taxa, animals living in marine habitats are much more likely to reproduce sexually than their freshwater relatives (Table 1.2). Bell (1982), who identified this pattern, calls this result '... one of the soundest generalizations in evolutionary biology'.

Confidence that this result reflects strong pressures associated with living in marine or freshwater habitats is enhanced by the repeatability through at least eleven phyla. Although the majority of these taxa were primitively marine organisms from which freshwater fauna are descended, the tendency towards increased sexual reproduction in marine habitats still occurs where a predominantly freshwater group has a few marine representatives (for example, chaetonotid gastrotrichs). These data therefore overcome one of the problems inherent in the comparative method which underpins so much of evolutionary ecology. This difficulty concerns the independence of data points. All organisms are related, and share much of their phylogenetic history. Therefore all descendants may share a common trait that originated in their ancestors. The trait may or may not have developed in the habitat they now occupy, and has only evolved once, not in each of the myriad descendants of a particular ancestor. Because repeated evolution of a trait strengthens our conclusions, we need to know how many times that trait has evolved, and not how many taxa retain a trait because of shared ancestry (Ridley 1983). In a sample of one hundred rodent species, and twenty antelopes, we might conclude that small body size is associated with continuously growing incisors, though examination of other small-bodied taxa would soon dispel this myth.

Why does each fig have its own fig wasp?

Figs (*Ficus* spp.; Moraceae) are an extremely diverse and common genus of plants with a distribution centred in the tropics but drifting into the temperate zone and, of course, into human diets. Individual fig trees produce a large crop of fruits, and while these fruits or synconia are still very small, they attract thousands of agaonid wasps; a different species of wasp visits each of the hundreds of species of fig. The female wasps crawl into the fig through overlapping scales which cover the ostiole, or hole in the apex of the fig (Fig. 1.4). It is a tight fit, and the female loses her antennae, wings, and possibly also contaminants, such as dirt laden with bacteria and fungal spores. One important part

Table 1.1 Modes of reproduction. For a more elaborate distinction of animal modes of reproduction see Bell (1982), and for plants see Stebbins (1950) and Briggs & Walters (1984). The lists of examples are not intended to be exhaustive. Further, many groups (e.g. different angiosperms and coccoids) exhibit virtually every form of reproduction. In addition, plants tend to vary facultatively in their expression of both gender and sex.

Class of reproduction	Characteristics	Common synonyms	Plant examples	Animal examples
Asexual modes				
Vegetative proliferation	Reproduction by growth, budding, or fragmentation	Apomixis	Many ferns and grasses	Some bryozoans, sponges, oligochaetes
Parthenogenesis	Reproduction by seeds, eggs, etc. without fertilisation	Agamospermy, amictogamety	Many angiosperms, e.g. dandelions	Many species, e.g. from digenean trematodes, tardigrades, cynipid wasps, teid lizards, but bdelloid rotifers are the only exclusively parthenogenetic major taxon
Combinations				
Haplodiploidy	Males are haploid and produced from fertilised eggs; females are diploid from syngamy of meiotic gametes from different individuals		None	Most Hymenoptera and various other insects; some mites and rotifers
Cyclical parthenogenesis	Parthenogenesis alternates with sexual reproduction predictably		None	Cladocerans; some rotifers; digenean trematodes; aphids and some other insects
Sexual modes				
Self-fertilisation	Syngamy of meiotic gametes from the same individual	Automixis	Wheat and many other angiosperms. May have a variety of forms	Some triclads, planarians, gastrotrichs, rhabdocoel turbellaria, notostrocans
Sex with outcrossing	Syngamy of gametes derived from different individuals	Amphimixis	Many angiosperms	All mammals and birds; many other groups

Table 1.2 Modes of reproduction in invertebrates occurring in both freshwater and marine environments. Modified from Bell (1982).

Phylum	Class	Freshwater	Marine
Porifera	Demospongiae	Usually cyclical parthenogenesis	Usually sexual
Cnidaria	Hydrozoa	Budding gives new polyps; often sexual	Budding leads to colony formation; often sexual
Platyhelminthes	Turbellaria		
	Catenulida	Asexual, by fission; sex very rare	Exclusively sexual
	Tricladida	Often by fission; one genus by obligate selfing	Exclusively sexual; regular selfing unknown
Rhynchocoela		*Prostoma* is sexual but may self	Sexual; some *Lineus* spp. may self
Gastrotricha		Parthenogenesis	Exclusively sexual
Rotifera	Digononta	Parthenogenesis	Exclusively sexual
	Monogonta	Parthenogenesis or cyclical parthenogenesis	Cyclical parthenogenesis; sex more pronounced
Nematoda		Males unknown in more than half the species; males often very rare	Males unknown in about 20% of species; males are usually as frequent as females
Mollusca	Gastropoda	Parthenogenesis in two genera; selfing may be common in opisthobranchs	Exclusively sexual; selfing probably rare
Annelida	Polychaeta	*Nereis limnicola* is selfed	Usually sexual, fission widespread
	Oligochaeta	Fission usual in Aelosomatidae and Naididae; Tubificidae sexual	Exclusively sexual
Arthropoda	Ostracoda	Many parthenogenetic	Probably exclusively sexual
	Branchiopoda		
	Cladocera	Parthenogenesis or cyclical parthenogenesis	Cyclical parthenogenesis; males common and sex frequent
	Tardigrada	Parthenogenesis common	Exlusively sexual
Bryozoa	Gymnolaemata	Cyclical parthenogenesis	Exclusively sexual
	Phylactolaemata	Cyclical parthenogenesis	No marine forms
	Stenolaemata	No freshwater forms	Exclusively sexual
Entoprocta		Sexual and budding	Sexual and budding

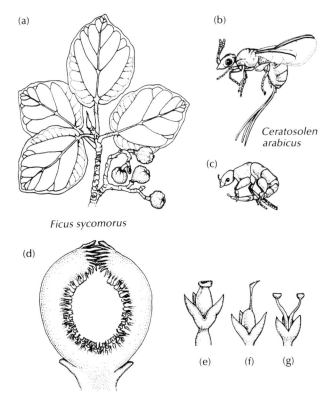

Fig. 1.4 Association between the fig wasp (*Ceratosolen arabicus*) and the fig it
pollinates (*Ficus sycomorus*); (a) the plant; (b) female wasp (note the ovipositor); (c)
the flightless male; (d) the synconium (note the ostiole); (e) the short-styled female
flower, in which wasps can develop; (f) the long-styled female flower, which
develops normally; (g) the male flower. After Palmer & Pitman (1972); Meeuse &
Morris (1984). (Individual drawings not to scale.)

of the female's body which is protected from this stripping is a pair of
thoracic sacs which are usually full of *Ficus* pollen of the appropriate
species. The interior of the fig is lined with florets with receptive
stigmata. The female moves from stigma to stigma and pollinates each
of them. The female also probes each style with her ovipositor. In
about half of all fig species, the styles vary in length. If the style is
short enough for her ovipositor to reach the ovule, the female lays an
egg at the side of the ovule. If the style is too long, an egg is not
usually laid, so pollination can occur and the ovule may develop
normally. The pattern sometimes breaks down if several females
manage to break into the synconium, as intense reproductive competition forces some of the females to lay in the long-styled flowers.
Once the female has finished oviposition she dies inside the synconium.
In the other half of the figs, there are two kinds of trees (Kjellberg *et al.*
1987). On some trees, all female flowers have short styles, and the
ovaries can be parasitised by the wasps. Such trees are functionally
male. The other trees have long styles which the wasps cannot parasitise. They do not produce pollen and so are functionally female.

Some pollination must be successful, or the fruit will be aborted. One factor which reduces the incidence of pollination is visitation by other wasp species which lay the eggs, but contribute nothing to pollination. The success of these intruders is contingent on the present of at least one pollinating wasp. Usually from 20 to 80 per cent of the ovules fail to develop.

This extraordinary association illustrates a variety of questions which preoccupy evolutionary ecologists. The close relationship between figs and fig wasps is apparently obligatory in both partners. Yet the relationship is subject to disruption by unwelcome intruders. Because the figs have presumably diversified since the evolution of the association, how is it that speciation in figs appears to be accompanied by speciation of fig wasps (Chapter 8)? Figs are often among the most diverse genera of trees in the forests in which they occur. Why do several species of figs co-occur, while in other genera only a single species occurs in each habitat (Chapter 9)? How easy is it for the invading 'cheats' to break up or exploit this close association (Chapter 5)? Why in some species are the sexes of the trees separate, while in other species the sexes are combined on a single tree (Chapter 7)?

Interactions between species: a periodic table?

Interactions like the relationships between figs and fig wasps form the basis of a discipline within ecology called community ecology, which is discussed in Chapter 5 and Chapter 9. Much of community ecology is concerned with the interactions between pairs of species, which may take a great diversity of forms (Table 1.3), several of which are illustrated in the association between figs and the wasps which enter their fruits. Classification of these interactions is considerably more difficult than it seems. This is because the mechanism of interaction may be different from its effect, and because many of the effects of interactions are indirect, and may also depend to a great extent upon context (Abrams 1987a). For example, in a classic experiment, Paine (1966) removed the starfish from a stretch of rocky shore. *Pisaster*, the

Table 1.3 Classification of participants in direct interactions.

		Effect of individual of species A on an individual of species B		
		Positive	Neutral	Negative
Effect of individual of species B on individual of species A	Positive	Mutualism	Commensalism	Predation, herbivory, parasitism
	Neutral			Amensalism
	Negative			Competitor

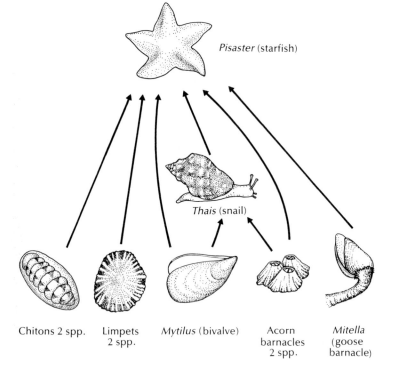

Fig. 1.5 Predation by the starfish *Pisaster* facilitates coexistence of other intertidal invertebrates by preventing competition between them. After Begon *et al.* (1986).

starfish, ate all the smaller members of the rocky shore community, including barnacles, snails and limpets (Fig. 1.5). The effect of the removal was striking, as the previously diverse group of grazing and filter-feeding invertebrates was reduced greatly, with one or two species coming to dominate the community. So although the predatory starfish ate all of the members of the community, and its effect on the fitness of individual prey was by definition deleterious, its effect on populations of many of its prey was beneficial. The benefit arises from the ability of the predator to keep the density of the superior competitors in check, preventing competition, and permitting the persistence of inferior competitors.

Why are there so many species of beetles?

Not all groups of organisms have fared equally well through the history of life. Indeed, extinction appears to be the fate of most species. At a higher taxonomic level, all the metazoan phyla appear to have evolved by the end of the Cambrian, about 500 million years ago. Many of the phyla that are recognisable in the fossil record have no descendants (Conway Morris 1979, 1989). It is also curious that many of the extant phyla contain very few species (Fig. 1.6; Strathmann & Slatkin 1983). The arthropods stand in dramatic contrast to this pattern, with estimates of the number of species ranging from just over a million to at

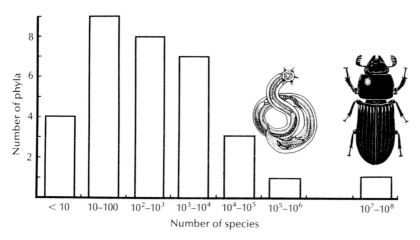

Fig. 1.6 Frequency distribution of species richness in the metazoan phyla. Note the logarithmic scale. Data from Barnes (1987), with adjustments to the two most speciose phyla (Arthropoda and Nematoda) to account for the incomplete descriptions of marine and tropical free-living nematodes, mites and canopy-dwelling beetles (Erwin 1982; May 1988; though see Thomas 1990).

least thirty million. The latter estimate was derived from sampling the canopy of a tropical tree by spraying with a 'knock-down' insecticide, which revealed an unsuspected diversity of specialised beetles in the upper canopy (Erwin 1982). Erwin's data confirm the observation by J.B.S. Haldane who, when asked what the study of biology had taught him about the creator, replied that it indicated 'an inordinate fondness for beetles'. May (1986) has suggested that to a good statistical approximation, all species are arthropods.

These data draw our attention to a variety of questions, largely unanswered, which nonetheless have come to fill an increasingly central role in evolutionary ecology. What sets the tempo of speciation and extinction (Chapters 8, 9)? How can some of the smaller taxa persist with little evidence of pronounced radiation, while others boom-and-bust (Chapter 9)? To what extent are contemporary patterns of diversity a reflection of historical influences (Chapter 9)? How can many species of beetles coexist and specialise on the same tree species (Chapter 9)?

DARWIN'S SOLUTION TO ASKING WHY?

I hope that this discussion introduces some problems worthy of explanation. Consensus has not always been reached, but the major aim of this book is to convince you, the reader, that the appropriate philosophical and technical background to solve all these problems is now available, and that the use of this approach and techniques is one of the most stimulating and aesthetically pleasing endeavours in modern science. The solution to these problems is a recent one, and was not possible until the traditional view of the world was shattered by the realisation that living things are not static entities or 'ideals' in the sense of Plato and Aristotle, but instead were mutable and had changed and diversified enormously through time. The father of evolutionary ecology, and the man credited with this revolution in our view of the world, was Charles Darwin (1809–1882). His accomplishments were manifold. First, he marshalled an extraordinary body of evidence that confirmed that animals had evolved, or changed through time, putting into a general perspective the growing recognition among nineteenth century scientists that the world was of great antiquity and that the fossil record depicted great change among its inhabitants. Second, he proposed the Principle of Natural Selection, a plausible and frighteningly simple model of the mechanics of evolution. Third, he showed with extraordinary clarity the power of the comparative method in detecting patterns in nature, including the patterns of behaviour and relationships of plants and animals in the wild.

EVOLUTION IS NOT A SYNONYM OF NATURAL SELECTION

That living organisms have evolved, or changed through time, is now an indisputable fact, excellently illustrated by geological, biochemical and biogeographical evidence, as well as abundant direct observation (Box 1.1). The mechanisms underlying evolution are a source of greater controversy. A major aim of this book is to demonstrate that Darwin was correct in claiming that much of the pattern that we observe in nature is a consequence of one of the mechanisms through which evolution can occur, the process of natural selection. Before I can proceed with this demonstration, it is important that it is absolutely clear what natural selection is, why selection is not synonymous with evolution, the levels in the biological hierarchy at which selection can operate, and the meanings of two important but confusing terms, adaptation and fitness.

What is natural selection?

The definition I offer is simplified from that given by Endler (1986a). He describes natural selection as a process, in which if a population has:

(a) (*variation*) among individuals in some attribute or trait;

(b) a consistent relation between that trait and some measure of reproductive success or survivorship, e.g. prowess in acquiring mates, fecundity (*fitness differences*);

(c) *inheritance*, or a consistent relationship, for that trait, between parents and their offspring, which is at least partly independent of common environmental effects, *then*:

1   there will be a within-generation effect; individuals of a given age will differ *predictably* from the individuals which do not survive to that age;

2   if the population is not at equilibrium, there will also be a between-generation effect; the offspring generation will differ *predictably* from their parental generation.

Box 1.1 Some of the incontrovertible evidence for evolution. All of this evidence was available during Darwin's time, and a lot of it proves evolution without requiring any use of the Principle of Natural Selection. The history of biology since Darwin has seen the repeated triumph of evolutionary biology in explaining the history of life.

1   The fossil record shows that biota of great antiquity was very different from that living today. This is a consequence of both extinction (many or most living organisms have disappeared) and radiation (there appear to be more species living today than ever before, or there were more species before the destruction of rainforests gathered pace).

2   Fragments of the earth's surface which have been isolated from large landmasses for a very long period tend to have biotas recalling more ancient fossil biotas on the large landmasses. For example, New Caledonia is the living repository of a once extant radiation of southern gymnosperms, and Madagascar has preserved a cargo of once widespread mammals which are now localised to that island.

3   Functionally complex characters are not used, demonstrating their abandonment during evolution. For example, females of bedbugs and some nematodes have complex sperm storage organs near their genitals, yet males bypass these and copulate directly through the body wall of the female (Fig. 1.7; Thornhill & Alcock 1983).

4   Functionally complex characters often developed from different sources in different taxa, with 'good' design in one group not utilised in another, demonstrating their separate origin. The most famous example is the Panda's sixth digit, or thumb (Gould 1980a).

5   Island biotas are often depauperate in taxa but rich in morphological diversity within those taxa, showing that all morphological solutions to problems posed from the environment must have come from a given source fauna. What is more, the biota of these islands often resemble most closely the biota of the nearest area of mainland. This conclusion can be applied at any scale, from the fish in African lakes to the mammals of Australia. The most famous examples are Darwin's finches (Grant 1986).

6   Members of often completely different taxa have repeatedly converged on the same morphological solution, with a precision with often beggars belief. Convergence demonstrates that problems posed by the environment may be more important than taxonomic affinity in determining morphology, a conclusion unimaginable without acceptance of evolution. Examples are legion, but my favourite is the convergence between the lemurs of Madagascar and the phalangerids, petaurids and pseudocheirids of Australia.

Types of selection

Evolution can occur in the absence of selection, by a variety of mechanisms which will be introduced fully in Chapter 2. The predictability of evolution by selection is vitally important. Richard Dawkins (1986)

Fig. 1.7 Traumatic insemination in bedbugs. After Thornhill & Alcock (1983).

eloquently summarises the confusion and indignation which surrounds the supposedly random nature of evolutionary change, and points out that it is cumulative and predictable selection which is capable of generating the breathtaking wonder of the patterns which can be observed in nature. Selection can take many forms, and has been classified by a number of overlapping dichotomies or trichotomies which are defined in the following discussion.

Directional, stabilising and disruptive selection

First, as the definition implies, selection can operate on both the mean and variance of the distribution of a trait (Fig. 1.8). Probably the commonest form of selection is elimination of extreme phenotypes at both ends of the trait distribution (stabilising selection), which has the effect of reducing the variance of traits in a population. It is important to note that intense stabilising selection can occur without change in the average phenotype. Its converse is disruptive or diversifying selection, which increases the variance in a trait. Once again, the mean of the trait need not change. Selection which affects the mean value of the trait is called directional selection, though it is important to realise that the variance may also change with directional selection.

Frequency-dependent selection

The fitness of an organism may depend on its frequency in the population. Frequency-dependent selection occurs when the fitness of a phenotype depends upon the frequency of that phenotype in the population. Selection of this sort could favour either rare phenotypes or common phenotypes. For example, consider the passion flower butterflies *Heliconius melpomene* and *H. erato*. Each species consist of a bewildering diversity of races throughout South America (Fig. 1.9). The

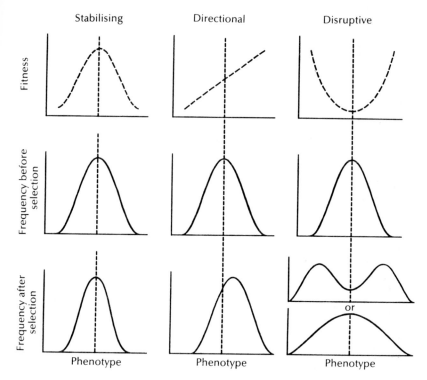

Stabilising   Directional   Disruptive

Fitness

Frequency before selection

Frequency after selection

Phenotype   Phenotype   Phenotype

or

Fig. 1.8 One classification of types of selection. After Futuyma (1986).

races have ranges which overlap (sympatry). The most astonishing feature of this diversity is that at any location, the local race of *H. melpomene* resembles sympatric *H. erato* more closely than it does the geographically adjacent race of its own species (Turner 1981). Similarity in sympatry has evolved because of Muellerian mimicry. Both species have brilliant warning or aposematic coloration displayed against a black background, and also have exceptionally long wings. The butterflies are distasteful to predators, which learn to avoid butterflies with the warning pattern. It seems clear that this example can be explained because the most important characteristic influencing coloration in *Heliconius* is resemblance to the most common model, as it is with this model that predators will be most familiar. This effect may be fairly general, though some small passerine birds prefer rare prey when density is high, for reasons that remain obscure (Allen & Anderson 1984).

Negative frequency-dependent selection favours phenotypes so long as they remain uncommon. The most obvious circumstance favouring negative frequency-dependence is the interaction between one species and another species which preys upon or parasitises it. Selection should favour specialisation in the predator or parasite on the commonest phenotype of prey or host, as this phenotype will be encountered most frequently. This should allow rare host phenotypes to evade attack by virtue of their rarity, but any advantage will disappear as they increase

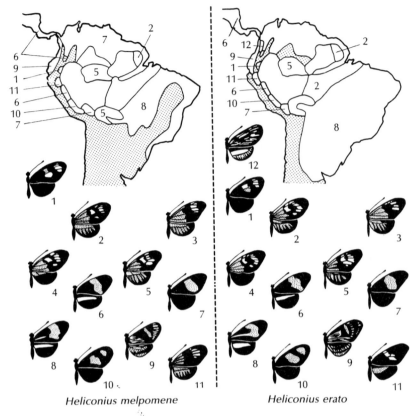

Fig. 1.9 Convergence between races of *Heliconius melpomene* and *H. erato* in the Neotropics. After Turner (1986).

in frequency in the population. A rather extraordinary example concerns predation of the seeds of *Scheelea* palms by the squirrel *Sciurus granatensis*. The fruits of the palm contain from one to three seeds. When one-seeded fruits are common, the squirrels tend to abandon the fruit after they have consumed a single seed, so the second and third seeds remain unmolested (Bradford & Smith 1977). By contrast, when the frequency of single seeds declines and the squirrels learn that greater rewards are available within each fruit, the rate of predation on second and third seeds increases. There is also density-independent selection against fruits with more than one seed, as they are smaller and provide fewer reserves for establishment of the seedling.

Frequency-dependent selection is not strictly synonymous with either disruptive or stabilising selection, as it can lead to oscillations in the mean value of the trait.

Density-dependent and density-independent selection

Selection may only emerge when competition between individuals becomes pronounced, for example when density is high. For example, food shortages and the consequent selection for competitive ability may be density-dependent. By contrast, the effects of extreme weather may be density-independent, so the survival or death of organism with a tolerance occurs regardless of density.

Hard and soft selection

Wallace (1981) points out that the outcome of selection will differ according to whether competition between individuals or direct inter-action with the environment is the most important agent of selection. When individuals are in competition for some limited resource, so that only a certain proportion will be able to survive, that proportion will be determined within a subpopulation purely by relative prowess, but some individuals will always survive (soft selection). Alternatively, if some external condition such as temporary extreme cold is important, there is no reason why all individuals might not perish, as the survival of one individual need not be conditional on the survival or death of another individual. The number of individuals which survive the selective episode will therefore be largely independent of density or relative prowess (hard selection). This dichotomy cannot be simply related to either frequency- or density-dependence.

Natural and artificial selection

A large part of the evidence marshalled by Darwin (1859) in support of his Principle of Natural Selection came from his observations of the effects on animals and plants of selective breeding by humans. These effects ranged from the progressive increase in the productivity of plants and animals used in agriculture to the bizarre plumages and shapes admired by fanciers of pigeons and dogs. This progressive improvement relies on the properties of variation and inheritance, but the fitness differences are imposed by humans. The central difference between selection of this sort (artificial selection) is not so much its human agency, as the way it is conducted with a goal in mind (more meat, more fruit, a tomato which is cuboid so that it packs into crates more neatly, longer leg feathers, ears that droop or are erect, etc.). By contrast, natural selection is blind, and operates without a goal. Humans can certainly impose natural selection. For example, we have favoured the development of resistance to DDT in insects.

Natural and sexual selection

The last dichotomy was also identified by Darwin (1871). Most of the emphasis of *The Origin of the Species* was on traits that augmented survival to breeding age, as Darwin correctly recognised that the ma-jority of organisms fall by the wayside in the 'struggle for existence'. However, Darwin was greatly puzzled by the occurrence of traits that seemed more of a hindrance than a help. How could the plumage of the male birds-of-paradise (see Fig. 1.3) aid them in survival to breed.

Darwin (1871) therefore distinguished between traits which aid in survival or viability, which were shaped by natural selection, and those which assisted individuals gain access to mates, which were shaped by a process called sexual selection. Sexual selection can produce traits inimical to the survival of the organism, so long as the disadvantage is counterbalanced by the mating advantage.

Sexual selection can arise in two ways. First, through competition between members of one sex for mates (intrasexual selection) which could favour antlers for fighting or the development of loud calls which might be more easily noticed by mates. Second, through preference for particular traits in the other sex (intersexual selection). Such preferences will be considered in some detail in Chapter 7. While in this text I will retain Darwin's terminology, I prefer to regard sexual selection as an interesting subset of natural selection. Indeed, the distinction between the two is sometimes rather obscure. For example, the antlers and other weaponry of red deer (*Cervus elephas*) stags simultaneously allow the males access to good quality feeding grounds, which provides them with both a source of food, and access to hinds which congregate at these high quality habitats (Clutton-Brock *et al.* 1982).

WHAT ARE FITNESS AND ADAPTATION?

Most discussions of the meaning of fitness begin with an apology about the number of meanings biologists have attributed to this inoffensive word (e.g. Dawkins 1981). This confusion is unfortunate, because naive definitions of natural selection (survival of the fittest) and of fitness (the fittest survive) have been juxtaposed by some critics of Darwinism to argue that the theory of evolution by natural selection is a tautology, unfalsifiable, and inadmissible as science (e.g. Popper 1974; Peters 1976; Gish 1978).

The theory of natural selection is not tautologous

The tedious arguments of tautology and unfalsifiability have been rebutted repeatedly (e.g. Stebbins 1977; Kitcher 1982; Sober 1984), yet they continue to trouble students, and warrant brief discussion. The confusion arises partly as a result of unfair attribution (the survival of the fittest—the fittest survive juxtaposition has never been considered an adequate all embracing description of natural selection), and partly because of the confusion between tautology and a set of logical consequences of accepted premisses. Tautology can be defined as a proposition that will always be true regardless of whether its component terms are true or false. An extreme example might be 'a female bird will have millions of offspring or it will not'. Although the first part of the argument is never satisfied, the caveat prevents the falsification of the sentence. Caplan (1977) offers the following example of an argument which is not tautologous:

1 All biologists are immortal.
2 All philosophers are biologists.
Therefore
3 All philosophers are immortal.

Like the definition of natural selection, the argument is a syllogism, depending on major premises (1 and 2) which lead inevitably to a logically valid argument or minor premiss (3). The prediction that all philosophers are immortal is novel, but subject to falsification, as it is easy to show that the major premises are wrong. By contrast, the syllogism of natural selection has not been generally falsified, as the major premises are repeatedly demonstrable facts. Indeed, we can use the occasional falsification of the elements of the natural selection syllogism to identify the circumstances under which selection should be most prevalent. For example, the failure of the premiss of fitness differences is the heart of the Neutral Theory of Molecular Evolution (Chapter 2), and the failure to demonstrate fitness differences and inheritance the flaw in some Darwinian models of human cultural evolution.

We can make deductions from the laws of gravity, yet few argue that predictions of the laws of gravity are tautologous and unhelpful. In the remainder of this text, we shall see just how helpful the syllogism of natural selection has been.

Absolute versus relative prowess

Almost as much confusion surrounds the definition of the term adaptation, at least in part because discussion has confused the end result with the process. The cumulative product of natural selection will be adaptedness (Dobzhansky 1970), or an ability to survive and reproduce in a given environment. The process through which adaptedness is acquired is called adaptation. Unlike fitness, adaptedness is an absolute concept.

In order to understand this important distinction, consider a population which is expanding and one which is declining to extinction. If all members of the population enjoy reproductive success during the expansion because there is little competition for resources, they all exhibit adaptedness. In turn, if there is little variation in reproductive success because conditions are benign, there may be few fitness differences and limited opportunity for selection. By contrast, as the population declines, there may be strong fitness differences and intense selection, but low adaptedness. Therefore to understand the processes of selection and adaptation we need to understand both total reproductive success and its variance. This requires a reorientation of our attitude to variance, a property of data which is often regarded as something which is annoying and to be suppressed by the appropriate statistical straightjacket. To an evolutionary biologist, variance is often a property of central interest.

Sexual selection illustrates how there may be great differences in the opportunity for selection between the sexes, even though there is no average difference in reproductive success. Kittiwakes (*Rissa tridactyla*) are long-lived cliff-dwelling seabirds which form pair bonds to reproduce. Their lifetime reproductive success has been followed by Coulson & Thomas (1985) in a splendid study spanning three decades. Because each young has a father and mother, and equal numbers of males and females are born, the average lifetime reproductive success of males and females is identical. Because the reproductive fortunes of males and females within a pair bond are closely linked, variances in lifetime reproductive success are also very similar (Table 1.4).

By contrast, red deer stags form no long-term associations with hinds, but instead attempt to attract and defend groups of hinds during a short mating season, the rut, after which they contribute nothing to the rearing of their offspring. In another elegant long-term study, Clutton-Brock *et al.* (1982) have followed the lives of a population of red deer on the island of Rhum, off the coast of Scotland. Within a season, one stag may defend a harem containing many hinds, and many males go without a mate. Only males in the prime of life stand much chance of successful copulations; both young stags and old males debilitated by past years of glory find themselves on the sidelines. Reproduction is also costly for hinds, and many are forced to skip years of reproduction. As hinds produce a single calf each time they breed, their lifetime reproductive success is generally lower than the number of years over which they reproduce (thirteen calves was the maximum in the Rhum population). However, a male which ruts successfully for several years may father as many as twenty-four offspring, but half the males do not survive to breed young (Table 1.4). The greater variance in male reproductive performance is fertile ground for sexual selection.

HOW DO WE DETECT SELECTION AND ADAPTATION?

The living world is the product of hundreds of millions of years of evolution, involving diversification and extinction, elaboration and simplification. Much of evolutionary history can therefore only be inferred retrospectively, and the measurement of evolution depends on correct assessment of the likelihood of unique historical events. Indeed, the variance upon which natural selection depends may have disappeared millions of years ago. This does not mean that the study of evolution is hopeless, but it certainly makes this rich and rewarding field particularly demanding. Before tackling specific topics of interest it is necessary to be familiar with some of the techniques which have been developed to overcome these demands, and to detect and describe selection and adaptation. The next three Chapters are devoted to the problem of method. In Chapter 2 the genetic basis for variation is briefly introduced, and the interpretation of genetic data in a selectionist framework discussed. In Chapter 3 the factors which uncouple the

Table 1.4 Coefficient of variation (standard deviation/mean) in lifetime reproductive success (LRS) in the monogamous kittiwake (*Rissa tridactyla*) and the polygynous red deer (*Cervus elephas*), measured as number of offspring fledged (kittiwake) or surviving to one year (deer). Note that variation in success in any one season depicts variation in lifetime reproductive success poorly. Data from Clutton-Brock (1983). See also Clutton-Brock *et al* (1982) and Coulson & Thomas (1985).

	Red deer		Kittiwake	
	Male	Female	Male	Female
Seasonal success of breeding animals	1.49	1.43	0.83	0.83
Lifetime reproductive success				
Individuals reaching breeding age	1.10	0.59	0.91	0.83
All individuals born	1.29	1.03	1.41	1.38

phenotype and genotype, and the limits and constraints to selection are discussed. I also address the philosophically important topic of the units to which selection is directed. In Chapter 4 the techniques of adaptationist analysis which I will use throughout the text are critically discussed.

SUMMARY

The living world is highly patterned. Explanation of this pattern has greatly preoccupied humans throughout their history. Our understanding of pattern has been transformed by the repeated demonstration that organisms are not static entities, but have changed and diversified through time. Darwin's Principle of Natural Selection is a simple mechanism of evolution which has the capacity to produce pattern as a consequence of adaptation. It depends on three features of living organisms: variation, heritability and fitness differences associated with that heritable variation. Given these properties, certain other properties are inevitable, including evolution when populations are not at equilibrium. The Principle of Natural Selection is not tautologous, but a syllogism which is logically consistent but falsifiable if the fitness differences associated with heritable variation do not occur. Confusion has surrounded the terms adaptation and fitness. Fitness is a relative measure of survival and reproductive prowess. Adaptedness is an absolute measure of the capacity to survive and reproduce. Adaptation is a process leading to adaptedness. All three terms have a strict environmental context. Adaptedness of an organism may be high at one place and time but low in another. The Principle of Natural Selection is therefore deeply rooted in ecology, and its application to problems in ecology is the aim of this book.

FURTHER READING

Endler (1986a) and Manly (1985) provide very good reviews of the evidence that is required to detect natural selection, and abundant examples where selection has been demonstrated. Futuyma (1986) is the best general text on evolution. Mayr (1982a) is a superb survey of the development of evolutionary thought.

TOPICS FOR DISCUSSION

1 Describe some additional examples of interactions between species which cannot be included in the classification in Table 1.3.

2 Discuss the explanations for the change in the fossil record which had been suggested before the discovery of the Principle of Natural Selection.

3 Evolution is a demonstrable fact, yet some people appear to have great difficulty in accepting that it has occurred. Discuss the basis for this reticence.

Chapter 2
The Genetic Basis of
Evolutionary Change

Darwin's extraordinary intellectual achievement in deducing the Principle of Natural Selection is even more staggering when we consider that he had no knowledge of the mechanism of heredity. Today, evolutionary biologists face a different challenge, to accommodate the extraordinary advances that are made daily in the study of genetics. This progress has unfortunately followed two distinct paths, generating a schism within evolutionary and ecological genetics where both undergraduate courses and research programs focus exclusively on only one approach. The first improvement of understanding flows from the study of transmission genetics, which has generated formidable mathematical description of the consequences of heredity which enable predictions of the path of evolution. The second revolution in understanding comes from molecular genetics, which studies the molecular and chemical properties of genetic material and the nature of interactions within the genome.

In the following discussion I aim to provide sufficient background for a student to understand the implications of both transmission and molecular genetics for the study of ecology. I deliberately avoid both mathematical formalism and biochemical detail. First, I consider the way that genes influence the phenotype, and the significance of the recent discovery that much of the internal dynamics of genetic material has little direct connection with the phenotype. Second, I consider the processes that introduce intrapopulation and intraorganismal variation in genomes. Third, I describe the processes that lead to changes in the frequencies of genes and genomes within populations. Last, I tackle the special problem of how variation can be maintained in the face of processes which should cause it to disappear.

GENOTYPE AND PHENOTYPE

A gene is the physical unit of heredity. When expressed it has a specific effect on the observable properties of an organism (phenotype). The phenotypic effects of genes are recognisable in both organisms and their descendants. Except in some viruses, genes are built from deoxyribonucleic acid or DNA. In prokaryotes, DNA is normally organised in the form of a circle. Eukaryotes also carry DNA in circular form in mitochondria and chloroplasts, one of the many pieces of evidence which suggest that eukaryotic cells might have originated as a symbiotic association between prokaryotes, whose separate identity has been lost through evolutionary history (Margulis 1981).

Most of the DNA in eukaryotes is combined with histone proteins into linear chromosomes. Chromosome morphology is comparatively easily observed, once leading to excitement that it might unlock the secrets of the genome. However, connections between chromosome organisation and phenotype remain extremely obscure. For example, the small deer *Muntjacus reevesii* has 64 chromosomes while its phenotypically very similar congener *M. muntjac* has either 6 or 8, depending on the population (White 1978). Overall, chromosome number may vary from one (for example, males of the ant *Myrmecia pilosula*, Crosland & Crozier 1986), to 1260 in the sporophyte of the fern *Ophioglossum reticulatum* (Soltis & Soltis 1987).

The normal mode of expression of a gene is transcription of DNA via ribonucleic acid (RNA) into a protein. However, not all DNA codes for proteins. Some genetic material has a functional role in regulation of the expression of genetic material but is not itself ever transcribed into gene products. The DNA complement and sequence in each cell of an individual is usually approximately the same, yet cells and tissues differ greatly in which proteins are expressed. In addition, expression of a particular protein is often restricted to a particular stage of the life cycle of the organism. Each gene has a regulatory region upstream from the sequence which determines the structure of the protein (Fig. 2.1), but our understanding of the biochemistry of regulation remains rudimentary. Tissue- and temporal-specificity of gene action is therefore one of the greatest challenges facing molecular geneticists, and also of profound interest in understanding the connection between phenotype and genotype.

In addition, much of the DNA in a cell often has no function either in production of proteins or in the regulation of gene action. This non-coding DNA falls into several classes. First, genes often replicate themselves within the genome, and occur as multiple copies (gene families). For example, the human genome contains many hundreds of copies of the ribosomal RNA gene. Once duplication has taken place it is possible for the gene copies to diverge, and also for some copies (pseudogenes) to lose their function or regulatory sequence without affecting the biology of the cell.

Second, the coding regions of a gene (exons) are often punctuated by regions of DNA (introns) which are excised during production of the messenger RNA which in turn codes for the proteins. Introns contribute nothing to the phenotypic expression of the gene. The origin and evolutionary significance of introns remain a source of great controversy (Doolittle 1987). Last, much of the genome of eukaryotes comprises simple sequences repeated many times (satellite or

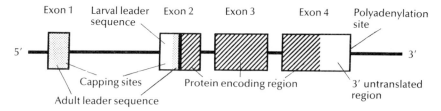

Fig. 2.1 The form of the alcohol dehydrogenase gene in *Drosophila melanogaster*. The protein-encoding sequence is split into three exons separated by two small introns. Two transcripts are produced, a larval and an adult form, which differ only in non-coding leader sequences. The larval leader is at the head of exon 2, and is unspliced in the mature form. The adult leader is some distance upstream. After Kreitman (1983).

highly repetitive DNA). The contribution that these types of DNA make to the total genome varies greatly. Some amphibians and lungfish are especially notable for the presence of very large quantities of DNA per cell. Indeed, it appears that the volume of repetitive DNA is so great that cell mechanics are affected, and there is an inverse correlation between the rate of cell differentiation and DNA content (Sessions & Larson 1987).

Alleles of genes, and dominance

Although by no means universal, most somatic cells of animals and sporophytes of plants contain more than one set of chromosomes, so that the region of DNA coding for each gene is represented two or more times. Genes can vary in their sequence, and alternative forms of each gene are called alleles. The use of electrophoresis to resolve mobility of different forms of proteins in an electric current has provided abundant evidence for multiallelism within cells and populations (heterozygosity), and this evidence has been confirmed and extended with the use of new techniques for direct sequencing of proteins, or more commonly and easily, the DNA which codes for them.

The possession of two alleles at a locus does not mean each allele will contribute equally to the phenotype. One allele may obscure the expression of the other completely (complete dominance), partly (partial dominance) or not at all. An allele whose phenotypic expression is masked is called recessive relative to the dominant allele. Under complete dominance, the fitness of individuals homozygous for the dominant allele and heterozygous at the locus are the same, even if the recessive allele is extremely harmful when homozygous. This insulates rare recessives from selection because they are only rarely expressed. In contrast to this pattern, two alleles may interact synergistically so that in combination the fitness of the heterozygote is greater (heterosis or overdominance) or lower (underdominance) than the fitness of either homozygote.

Gene interactions

The area of genetics which has been dealt with most poorly by the new molecular techniques is arguably the most important to ecologists. Most phenotypic traits are polygenic, or under the control of many separate genes. Instances where major phenotypic changes during evolution can be attributed to single genetic changes of large effect are very rare, at least among animals (Gottlieb 1984). One exception appears to be some cases of pesticide resistance (Brattsten *et al.* 1986). Conversely, some genes influence many apparently unrelated aspects of the phenotype, and are said to be pleiotropic. For example, hormones like ecdysone in insects or prolactin in mammals have many phenotypic influences, so the genes coding for these proteins or the sequences influencing their regulation are presumably subject to diverse selection pressures associated with those various influences. Genes may also interact directly. For example, an allele of one gene may influence the expression of another (epistasis), in the same way that two alleles of the same gene may interact along an axis of dominance and recessiveness.

The last important interaction between genes arises because alleles which are in close physical proximity are less likely to be separated by recombination at meiosis, unless the genetic material at a site specifically promotes recombination (so-called hotspots). This close association is called linkage, and the lower than random probability that closely associated genes will be separated by recombination is called linkage disequilibrium. For example, the flowers of the primrose *Primula vulgaris* are of two forms. The pin form has a long style and anthers set deep in the flower, and the thrum form has a short style and anthers placed round the rim of the flower (Fig. 2.2). This heterostyly promotes outcrossing because pollinators which are dusted with pollen from thrum anthers are likely to deposit that pollen on a pin style, and vice versa. This abrupt morphological discontinuity is suggestive of two alleles at a single locus, with thrum being produced by a homozygous recessive (tt), and thrum by a heterozygote (Tt), as thrums are self-incompatible, so TT is not usually produced. However occasionally both short and long homostylous flowers are produced, suggesting that more than one allele is being broken up by crossing-over. Closer genetic analysis suggests that there are three genes which involved. Thrum plants are produced by $T_1 T_2 T_3$ and pin flowers by $t_1 t_2 t_3$ (Maynard Smith 1989). Heterozygotes are presumably strongly selected against because of the disadvantage of self-fertilisation, and linkage disequilibrium is strongly selected. Groups of genes can become so closely linked that they function as a single gene, and such groups are called supergenes.

Nature versus nurture and heritability

Just as the expression of an allele depends on the other alleles present at the locus, interactions with other loci, and the regulatory processes

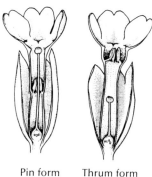

Pin form   Thrum form

Fig. 2.2 Heterostyly in *Primula veris*.

which determine age- and tissue-specificity of gene action, there may also be a direct influence of the environment. Some effects such as the influence of vitamin deficiency on critical biochemical pathways are reasonably obvious. Some require more careful analysis, such as the subtle triggers to development induced by seasonal changes like increasing daylength. In traits which are influenced by many genes, disentangling environmental influences can become extremely difficult. For example, very tall parents usually have children of above-average height, so human height has a strong genetic basis, and should respond to selection. However, some conspicuous trends such as the increase in the height of Europeans after World War II are largely environmentally based. We can use enforced inbreeding to remove genetic variation in laboratory stocks, but we still see variance in the phenotype. Formally, we can say that the variance we observe in a trait (phenotypic variance, V_P) has both environmental (V_E) and genetical (V_G) components, so that:

$$V_P = V_G + V_E$$

Partitioning of the variance between its environmental and genetic components allows us to distinguish between the role of heredity (critical for selection to operate), and the role of environment. This is the traditional and much-debated dichotomy between nature and nurture. The total genetic variance of a character controlled by more than a single gene can be partitioned into several components. Some of the contributions to genetic variation are determined predominantly by genetic interactions that occur *de novo* in each individual; we can recognise variance components due to dominance of alleles at the same locus (V_D), and epistatic interactions between alleles at different loci (V_I). By contrast, the resemblance of offspring to their parents is determined by the additive effects of alleles within and among loci (additive genetic variance, V_A), so that:

$$V_G = V_A + V_D + V_I$$

It is the additive genetic variance which allows a response to selection. Formally, we say that the heritability of a trait is given by V_A/V_P (Falconer 1981). How can the heritability of a trait be estimated? The easiest way is through an experiment which imposes selection for one generation. We breed from a set of parents whose mean expression of a particular trait differs from the population mean by a given amount (the selection differential, S). We then compare the mean value of the offspring of these parents with the mean value of the population in the previous generation (the response, R). Heritability (usually denoted h^2) is given by R/S. Several estimates are given in Table 2.1. In circumstances where experiments are not possible, as is frequently true for traits of interest to ecologists, it is also possible to estimate heritability through the degree of resemblance between relatives. Such estimates require considerable caution. The resemblance between mothers and

Table 2.1 Estimates of heritabilities for a variety of traits in easily studied species. Note the inverse relation between heritability and importance to reproductive success. After Falconer (1981).

Organism	Trait	Heritability
Homo sapiens (humans)	Stature	0.65
	Serum immunoglobulin	0.45
Bos taurus (cattle)	Body weight	0.65
	Butterfat (%)	0.40
	Milk yield	0.35
Sus scrofa (pigs)	Back-fat thickness	0.70
	Efficiency of food conversion	0.50
	Weight gain per day	0.40
	Litter size	0.05
Gallus domesticus (poultry)	Body weight (at 32 weeks)	0.40
	Egg weight (at 32 weeks)	0.35
	Egg production (to 72 weeks)	0.10
Mus musculus (house mice)	Tail length (at 6 weeks)	0.40
	Body weight (at 6 weeks)	0.35
	Litter size (first litters)	0.20
Drosophila melanogaster (fruitfly)	Abdominal bristle number	0.50
	Body size	0.40
	Ovary size	0.30
	Egg production	0.20

their offspring will often depend greatly on the ability of the mother to transfer nutrients to her offspring (big fat mothers will be able to have big fat young). These maternal effects need not have a genetic basis, and will often depend on whether or not the mother finds herself in favourable environmental conditions. The resemblance between siblings will also be unreliable, as the dominance and epistatic interactions that occur in one sib are also likely in another, so the genetic resemblance need not depend on additive variance. The most reliable correlations are therefore those between half-sibs. If two broods have different mothers but the same father, then any resemblance between them is likely to reflect the influence of genes from the father. Falconer (1981) provides a full account of the calculation of heritability in this way.

A particular complication arises when the effects of genotype and environment are not additive, but instead there is a genotype–environment interaction. Maynard Smith (1989) gives the example of short-sightedness. During this century male myopics have a high fitness because their eyesight can be corrected with spectacles in many societies, yet they are less likely than normal individuals to be put in a uniform and shot at, while in hunter–gatherer societies myopia is likely to be strongly selected against. We will return to the significance of genotype–environment interactions in Chapters 5 and 7.

Our understanding of changes to the form of genes is greatly influenced by the concept of mutation. A point mutation occurs when a single nucleotide is substituted in the original nucleotide sequence, usually by a copying error during DNA replication. The effect of these substitutions can vary dramatically. Each amino acid (the building blocks of proteins) is coded for by three consecutive nucleotides. There are only 20 amino acids but sixty-four possible combinations of the four nucleotides. This 'degeneracy', where different nucleotide combinations can encode the same amino acid, means that changes to some nucleotides will not alter the coding effect of the triplet to which they belong. For example, while methionine is produced by a single combination, so any mutation will alter the amino acid, five distinct sequences can produce leucine. These silent sites are extremely useful is determining the causes and rates of change to the genome because they have no effect on the structure of a protein, and are therefore much more isolated from direct selection. This isolation from the effect of selection is also true of some introns and pseudogenes which are able to accumulate changes which have no effect on either the phenotype of the organism or the function of the DNA.

Even when nucleotide substitution leads to an amino acid substitution within a protein there will be dramatic differences in phenotypic effect associated with the extent to which the amino acid substitution alters the physicochemical properties of the protein.

Because the genetic code is determined by triplets of nucleotides which code for particular amino acids, a far more dramatic phenotypic effect will occur when a nucleotide is lost or inserted during replication (frameshift mutation), as all triplets further along in the nucleotide sequence will be altered simultaneously. Such frameshifts will almost invariably reduce the functionality of the encoded protein, and if the DNA cannot be repaired by the biochemical machinery of the cell, the gene is likely to lose its function. This may be a common source of pseudogenes.

Some DNA sequences are capable of self-replication and may be transposed from the copy site to other places within the genome. These transposable elements often code for a protein which promotes their own replication and transposition. Transposition may lead to an increase in the copy number of copies of a particular gene, and may also interrupt both the regulatory region and structural component of genes. There are several more complicated mechanisms which can change the order and length of nucleotide sequences (for a review see Dover 1986b).

The limited availability of sequence data and the complexity involved in their statistical interpretation undermine attempts to assess the relative importance of the various sources of variation in nucleotide sequences. There is abundant evidence of the accumulation of point mutations through time (Kimura 1983), but also some very limited data which suggest that the majority of visible mutations are caused by transposable elements (Bender *et al.* 1983). The accumulation of tandem repeats may occur with such rapidity that 'mutations' can be detected between members of human families (Jeffreys *et al.* 1988).

The probability of change in any gene will depend strongly both on the external environment (the level of mutagenic agents such as radiation), the physicochemical properties of the DNA, the number of cell cycles per generation (long-lived organisms may experience greater mutation rates; Klekowski & Godfrey 1989) and the size of the molecule (larger stretches of DNA present a greater number of nucleotides to mutagens).

Under unusual circumstances exceptional mutation pressure may occur. The best known case involves a family of transposable elements in *D. melanogaster* known as P elements. These small elements contain only a single gene which codes for their own transposition. Offspring suffer a number of deleterious effects, collectively known as hybrid dysgenesis, when male *D. melanogaster* carrying P elements (P strains) mate with females lacking them (M strains). Any other crosses produce normal fertile progeny. The extent of dysgenesis is temperature-dependent. At high temperatures dysgenic offspring are sterile; at lower temperatures progeny retain some fertility, but later generations show high levels of mutation, transposition and chromosome rearrangements (Snyder & Doolittle 1988). Although fitness effects of these mutations are usually deleterious when homozygous (Mackay 1986), the fitness effects are mild and may contribute greatly to quantitative variation (Mackay 1987).

HOW DO ALLELE FREQUENCIES CHANGE?

The processes described in the preceding section introduce novel DNA sequences into populations. Coupled with the conventional crossing-over of genetic material between homologous chromosomes during recombination, these processes generate novel genotypes. While some genetic changes have no phenotypic effect, it is the continued generation of novel genotypes that allows selection to operate. Selection is not a creative process, but a discriminatory one, which will influence the fate of new genotypes within a population. In the ensuing section I discuss the potency of selection in influencing the frequency of different alleles of genes.

The Hardy–Weinberg Law

In order to assess which are the most important determinants of allele frequencies it has proved useful to design a null model in which none of the factors of interest operate. The most important of these models is known as the Hardy–Weinberg Law and predicts allele frequencies when:

1   The organism is diploid and reproduces sexually.

2   Generations are non-overlapping.
3   Mating is random.
4   Allele frequencies are the same in males and females.
5   Population size is very large.
6   Gene flow and mutation can be ignored.
7   The locus concerned is not subject to selection.

Consider the simplest case where we have two alleles (A and a) at a single locus, and all three genotypes (AA, Aa and aa) can be distinguished phenotypically. The number of individuals of the AA homozygote is n_1, the Aa heterozygote is n_2 and the aa homozygote is n_3. Because the total number of individuals $N = n_1 + n_2 + n_3$, then the frequency of the A allele (p) and a allele (q) can be written as:

$$p = (2n_1 + n_2)/2N; \quad q = (n_2 + 2n_3)/2N$$

Subject to the assumptions given above, the probability that an offspring will receive an A allele from its mother is therefore p, and from its father is also p, so the probability that it will be homozygous AA is $p \times p = p^2$. By the same argument the probability that it will be homozygous aa is q^2. Heterozygotes can arise in two ways. The probability that it will be heterozygous is given by the probability that it will receive an A allele from its father and an a allele from its mother ($p \times q$), plus the probability that it will receive an a allele from its father and an A allele from its mother ($q \times p$). The combined probability will therefore be $2pq$. These proportions form the Hardy–Weinberg ratio, and can be calculated for any number of alleles. It is easy to see that $p^2 + 2pq + q^2 = 1$. This equilibrium frequency is reached in one generation and will remain stable unless acted upon by some other force.

The Hardy–Weinberg equation can be used to calculate genotype frequencies in certain cases. For example, slightly more than one in 1600 Caucasians are affected by a very severe genetic disease called cystic fibrosis, which affects glandular secretions ($q^2 \approx 1/1600$, so $q \approx 0.025$). Therefore $p = 1 - 0.025 = 0.975$, and the proportion of carriers for this rare and very harmful allele is given by $2pq = 0.049$, so about one in every 20 Caucasians is a carrier. These heterozygotes are phenotypically indistinguishable from non-carriers and cannot be purged by selection. A central aim of applied molecular genetics is to devise tests for the cystic fibrosis allele in both adults and embryos.

A far more common application of the Hardy–Weinberg principle is to estimate departures from these frequencies, which indicate that at least one of the assumptions is invalid; this is a useful starting point for further investigation.

Random causes

Sampling errors may arise because parents typically donate only half of their genomes to each offspring. A tossed coin may on average have an equal probability of showing heads or tails, yet in any one run of tosses heads or tails may predominate. Genetic drift describes the effects of these sampling errors on allele frequencies, and will be important when assumptions 2, 3 and 4 of the Hardy—Weinberg model are false. The power of drift is determined by the number of alleles that are sampled in each generation. If a gambler has bet on tails and a coin is tossed three times for three heads the gambler will curse his/her luck. If twenty heads are produced in a row he/she will be very suspicious indeed. In order to estimate the possibility of drift we need to know the number of individuals within a population which actually contribute alleles to succeeding generations, and the evenness of their contribution. Effective population size (N_e) is the size of the ideal population that would undergo the same amount of drift as the actual population (Lande & Barrowclough 1987). The effective population size is always less than the actual population size for several reasons. The most important are:

1 Populations fluctuate, meaning that the population size will occasionally be low, accentuating random drift. An extreme nadir in population size is called a bottleneck. An example of obvious importance is a founder effect, the drift that accompanies the starting of a new population. The frequency of bottlenecks will vary greatly. However, it seems that most populations show some long-term oscillations or trends (see Chapters 9 and 10). Generations with a very small population size have a disproportionate effect on long-term effective population size.

2 The number of males and females contributing to reproduction may not be equal for several reasons. The sex ratio at birth or later during parental investment may be biased, as occurs in many animals and plants (Chapter 7). Males and females may survive differently after the period of parental investment. Further, many surviving individuals may fail to gain access to a mate. For example, farmers often keep a single bull to inseminate a herd of dairy cattle, greatly exaggerating the prospect of significant drift. This effect reached absurd proportions through importation into Australia of sperm from a Friesian bull from Canada. Artificial insemination from this sperm continued until it was realised that the bull was a carrier of a rare recessive allele causing citrullinaemia, a disease which interrupts the urea cycle by halting the production of argininosuccinate synthase. Ten per cent of Australian Friesian stocks are now heterozygous for this allele. Social interactions among members of one sex also limit the numbers contributing to reproduction. For example, fewer male red deer are successful in producing offspring than are females (Chapter 1).

3 In addition to the factors which determined the proportion of males and females that will breed, successful males and females will show variation in reproductive success (Chapter 1), so some parents will donate more alleles than others, reducing the effective population size.

4 The final consideration which might restrict random mating is limited dispersal from the birthplace. For example, most birds and mammals move very little from their birthplace, so their mating opportunities are anything but random (Waser & Jones 1983).

The chief use of the concept of effective population size is to make real populations conform to the predictions of population theory. Formulae which can be used to derive N_e are given by Lande & Barrowclough (1987), though data are usually inadequate for their complete application. There has been an unfortunate tendency to assume that simple formulae that have been developed for non-overlapping populations can be applied to overlapping populations. Often a better general principle is to bear in mind that N_e will usually be much less than the population that can be censused.

Drift will therefore be of great importance in small populations, and may cause their differentiation, because it will operate in a random direction in each population, and lead to the fixation and loss of alleles at a single locus. The effect on quantitative genetic variation is a little more complex. Much attention has been concentrated on the effect of an extreme bottleneck where the population is reduced to only a few individuals. These events are ecologically important because they occur whenever a new population is founded in unoccupied habitat such as islands, of applied importance in the management of endangered species (Chapter 10), and of great evolutionary importance because they form the basis of many models of speciation (Chapter 8).

It seems obvious that bottlenecks should depress additive genetic variance through the loss of rare alleles (Lande 1980). However, a number of empirical and experimental studies have recently reported increased genetic variance after a bottleneck (Bryant *et al.* 1986; Carson 1990). Goodnight (1987, 1988) suggests that this is because some of the epistatic variance can be converted into additive genetic variance by drift and inbreeding in the resultant small population. Provided epistatic variance is significant, the total additive variance available for selection may actually increase, so the results of bottlenecks may vary according to the relative importance of additive and epistatic variance.

The Neutral Theory of Molecular Evolution

Drift has been implicated to be of extreme importance in the fixation and presence of different alleles at a single locus. Electrophoretic data on allele frequencies accumulated rapidly towards the end of the 1960s and presented a major puzzle to geneticists, and the theoretical edifice they had hoped to test. Rather than detect rare cases of polymorphism which might be the target of selection, more than one allele at a given locus was repeatedly detected both within individuals and within populations. In hindsight, it is also clear that primitive electrophoretic procedures greatly underestimated the number of alleles at each locus. For example the number of alleles originally detected for esterase-5 in *Drosophila pseudoobscura* was a troublesome twelve. More recent

studies using sequential electrophoresis recognise 41 (Keith 1983)! The
Neutral Theory of Molecular Evolution is one attempt to account
for this variation. Motoo Kimura was the original proponent and chief
advocate of this view, which claims that the great majority of evolutionary changes at the molecular level are caused not by selection
acting on advantageous mutants, but by random fixation of selectively
neutral or nearly neutral mutants by random genetic drift (Kimura
1983). This hypothesis does *not* deny that natural selection plays a
significant role in shaping adaptive evolution. However, it is very
critical of the view that selection is of widespread importance in the
maintenance of polymorphism, and instead contends that heterozygosity is usually a transient phase in molecular evolution, caused by
the balance between mutational input and random extinction by drift.
Selection is viewed as a conservative force maintaining function, with
drift contributing much of the divergence between populations, and
ultimately species.

Major- versus minor-axis polymorphism

Early analyses and reviews failed to resolve this controversy (Lewontin
1974). Considerable additional sophistication has emerged in studies of
genetic variation with the application of techniques which enable the
sequence of DNA to be measured directly. These data seem eventually likely to resolve the debates that have preoccupied geneticists
throughout this century. Loci which have been examined using
electrophoresis fall into four classes (Table 2.2), depending on whether
they exhibit major polymorphisms, with two or more common alleles,
or minor polymorphisms, with a number of quite rare minor variants
(Lewontin 1985a, 1985b). The majority of loci (about 70 per cent) are
monomorphic. Lewontin argues that minor axis variation is most

Table 2.2 Classification of genetic loci by the form of polymorphism that they
exhibit, with examples. After Lewontin (1985a).

		Major polymorphism	
		Monomorphic	Polymorphic
Minor polymorphism	Monomorphic	Most loci	Alcohol dehydrogenase
			Molecular-structure dependent
	Polymorphic	Xanthine dehydrogenase	Esterase-6
		Environment-dependent	

common in very large proteins with high molecular weights, and is likely to reflect the tolerance of the protein to the amino acid substitutions (see also Koehn & Eanes 1978). In contrast, major axis variation is likely to be influenced by selection, and should show predictable variation with the environment.

Two of the genetic loci about which we know most are the alcohol dehydrogenase and esterase-6 genes of *Drosophila melanogaster*, which can be used to illustrate both major and minor axis variation. *D. melanogaster* feed on rotting fruit. Both larvae and adults use ethanol vapour as an important source of energy, and alcohol dehydrogenase is crucial in the utilisation of ethanol. In *D. melanogaster* the alcohol dehydrogenase locus codes for a small polypeptide of 255 amino acids. There are two introns in the coding regions, which contain 65 and 70 bases respectively. In *Drosophila melanogaster*, esterase-6 is produced in the sperm ejaculatory duct of the adult male, from where it is transferred to the female during mating (Richmond & Senior 1981). The protein is larger, and there is a single intron. Transfer of esterase-6 triggers a complicated series of changes to the behaviour and physiology of the female, of which the most important are increased egg-laying and decreased receptivity to remating, at least in the short term (Scott 1986).

In both genes there are two main alleles, designated fast and slow according to their electrophoretic mobility. The frequency of the two major axis alleles within a population is highly correlated with latitude in a pattern that is repeatable between continents, suggesting that there is strong selection acting on the alleles (Oakeshott *et al.* 1982; Anderson & Oakeshott 1984). There are also biochemical differences in the activity of the alleles. For example, changes to substrate concentration reverse the relative catalytic efficiencies of the two major variants of esterase-6, suggesting a phenotypic effect which could be distinguished by selection (White *et al.* 1988). In variants of alcohol dehydrogenase from six populations on four continents there is a six per cent polymorphism of bases for the introns and a seven per cent polymorphism of bases in exons when silent substitutions are considered (Kreitman 1983; Table 2.3). However, nucleotide substitutions only very rarely cause amino acid substitutions. There is almost no neutral amino acid substitution possible in this system. However, unlike alcohol dehydrogenase, sequence data, high resolution electrophoresis and thermostability studies reveal the existence of large numbers of minor alleles in esterase-6 (Cooke *et al.* 1987; Cooke & Oakeshott 1989). Although the silent substitution rates are similar between the two loci, esterase-6 tolerates four times as many amino acid substitutions, even though the difference between replacement and amino acid substitutions suggests there is strong selection on esterase-6 for conservation of function (Table 2.3).

If we consider the evolution of proteins between species and higher taxa, there also appears to be a negative correlation between functional importance and rate of divergence of those sequences which code for

Table 2.3 Silent nucleotide substitutions and amino acid substitution in exons of the alcohol dehydrogenase (Adh; 12) and esterase-6 (Est6; 13) genes of *Drosophila melanogaster*. Data from Kreitman (1983), Collet (1988) and Cooke & Oakeshott (1989).

	Adh	Est6
Number of isolates compared		
Silent substitutions		
Number of sites compared	192	361
Number of polymorphisms	13	29
Percentage of polymorphism	6.8	8.0
Replacement substitutions		
Number of sites compared	573	1274
Number of polymorphisms	2	13
Percentage of polymorphism	0.3	1.3

function. Small messenger molecules often show great homology in organisms as diverse as yeast and mammals, suggesting that functional constraints may be so great that no neutral evolution is possible at all (e.g. Roth *et al.* 1982; Kolata 1984; Thorpe & Duve 1984). By contrast, fibrinopeptides which are spliced from the functional protein during the clotting of blood diverge very rapidly.

Active sites in serine protease inhibitors

There are often great difficulties in distinguishing between selection and drift, because the choice of an appropriate null model is not always obvious. Rapid change in sequences need not always reflect relaxation of selection. One area where conservation of sequence would be expected is the site which codes for the part of the protein where any interaction with target molecules takes place (usually called the active site or reactive region, depending on its mode of action). In contrast to this prediction, in one very diverse class of molecules called serine protease inhibitors the sequence of the reactive centre region apparently evolves much more rapidly than the rest of the protein (A.L. Brown 1987; Hill & Hastie 1987). These inhibitors interact with the active site of their target enzyme to inhibit its action in breaking up proteins. Hill & Hastie (1987) suggest that among the many targets of the inhibitors are proteinases synthesised by parasites or by infectious agents entering wounds. Both protozoan and bacterial parasites synthesise proteases that increase virulence and the efficiency of infection. By contrast, Graur & Li (1988) have suggested that the amino acid constitution of the active sites is such that mutations are unlikely to alter the physicochemical properties of the amino acid or protein. This allows the active site to evolve free of any compositional constraint. Studying the dynamics of intrapopulation polymorphisms would prove illuminating, but no data are currently available.

Molecular convergence

A second case of very rapid sequence evolution differs because instead of causing rapid divergence, it suggests that adaptive convergence can influence molecular evolution in a similar way to its influence on morphological, physiological and behavioural evolution. The neutral model predicts that this is unlikely. Sequence similarity in unrelated organisms could arise as a result of several processes, including strong conservation of sequences, or molecular stasis. For example, observations of the role of viruses in infiltrating host DNA and the development of the techniques for production of recombinant organisms by fusing eukaryote genomes with bacterial and viral genetic material has led to the suggestion that horizontal transfer may operate in nature, allowing, for example, animal genes to be inserted into plants. However, increased knowledge of the genetic structure of plants has dramatically weakened the only suspected example of this phenomenon in eukaryotes, the acquisition of haemoglobin molecules by plants which have a symbiotic association with the nitrogen-fixing bacteria *Rhizobium* and *Frankia* (Bogusz *et al.* 1988).

Good evidence for true molecular convergence comes from the stomach enzymes of herbivorous mammals which ferment food in their forestomachs. Foregut-fermenters use bacterial brews in their complex stomachs to break down the otherwise indigestible structural carbohydrates in plant tissues. Two groups of these mammals, the ruminants such as the cow, and the leaf-eating colobine monkeys, can apparently 'have their cake and eat it too'. Not only is there a close symbiosis between the mammals and the bacteria, but the mammals 'farm' the bacteria and digest their cell contents with a lysozyme produced in the foregut which breaks down the bacterial cell walls. The amino acid sequence of the lysozymes of the langur *Presbytis entellus*, a colobine, are more similar to the lysozymes of the cow, a ruminant, than to other primates, and these two unrelated species share a number of unique residues not found in other mammals, and convergence appears highly likely (Fig. 2.3; Stewart *et al.* 1987). Horizontal transfer seems very improbable. Foregut fermentation is of more recent origin within the monkey lineage (15 to 20 m.y.a.) than within the ruminant lineage (about 55 m.y.a.), so horizontal transfer should have been from ruminant to monkey rather than vice versa. Yet phylogenetic analysis places the cow lysozymes in the primate part of the tree, and this only occurs when the langur sequence is included in the comparison. Neither pattern is predicted by a horizontal transfer model. Gene duplication is also unlikely because there is no electrophoretic or immunological evidence for diversity of lysozymes within primates. However, five similar amino acid replacements have occurred within both the colobine and ruminant sequences. Not only does this suggest that convergence may influence sequence similarity, but it suggests that a doubling of the rate of divergence of the colobine sequence from that found in its

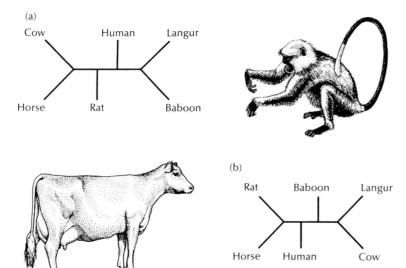

(a)

Cow   Human   Langur

Horse   Rat   Baboon

(b)

Rat   Baboon   Langur

Horse   Human   Cow

Fig. 2.3 Convergent molecular evolution in herbivorous mammals using forestomach digestion. (a) The biological tree for six species of mammals ('true' phylogeny); (b) the most parsimonious tree based on lysozyme sequence data (lysozyme phylogeny). Modified from Stewart *et al.* (1987).

primate relatives has been driven by strong positive selection, and that rapid divergence of proteins need not always indicate the impotence of selection in restraining random genetic drift.

Gene flow and selection

Any local adaptation or population differentiation will be restrained by gene flow from one site to another (gene flow is sometimes called migration, but I shall avoid this very confusing usage). Indeed, several authors have argued that flow of alleles from one site to another will be one of the strongest constraining forces on evolutionary change. The strength of gene flow in breaking down selection can be assessed theoretically by considering the balance between genetic drift and gene flow. Both drift and gene flow should have the same average effect on all nuclear genes. Drift will tend to eliminate alleles from isolated populations, but gene flow can restore them. The amount of gene exchange between populations necessary to prevent differentiation as a result of drift is not particularly great (about one individual per population per generation). Natural selection is much more effective at overcoming drift in both preventing or establishing local differences. Geographic uniformity can be maintained by selection regardless of gene flow, and local adaptation of subpopulations can be maintained provided that the fitness differences exceed the fraction of immigrants in the population (Slatkin 1987).

One of the greatest puzzles created by the abundance of nucleotide sequence data which have become available is the high degree of similarity within species between genes in multigene families. Because sequence similarity within a multigene family is greater within than between species, concerted evolution of the entire family must take place after speciation. This homogeneity is puzzling because in many cases the number of gene copies within a family appears to exceed any requirement for the gene product.

We now understand a variety of molecular mechanisms which promote concerted evolution. For example, if transposition lead to the excision of one element when the new element was inserted, homogeneity will increase. There may also be 'communication' between nearby members of a gene family, which can slowly enhance the frequency of one type of gene within the family relative to another (Dover 1982). Perhaps most importantly, there may be systematic biases in the direction of conversion of one gene form to another. Although one of the beneficial functions of gene conversion may be to repair damaged portions of DNA by using an existing sequence as a model for the reconstruction of a damaged strand (Bengtsson 1985), conversion can also lead to the concerted evolution of novel sequences. It is worth noting that the latter effect is likely to be more conspicuous regardless of its frequency. Concerted evolution is clearly important in introducing genetic differences between isolated populations, but is likely to work more rapidly between genes that are spatially contiguous on a chromosome than those that are scattered throughout the genome (Sharp & Li 1987).

Dover (1988) use the collective term molecular drive to describe the molecular mechanisms which contribute to concerted evolution, and champions the view that they must be assigned equal place with genetic drift and selection in analysis of evolutionary phenomena. This view relies on their potency in changing allele frequencies in addition to their role in generating variation within genomes. For example, despite the deleterious mutagenic effects of P-elements in *D. melanogaster* they have spread rapidly between and among populations (Fig. 2.4). Among most population geneticists, there is considerable enthusiasm that these changes may be encompassed in the old models. In particular, it seems that most of these changes can be treated mathematically in the same way as point mutations (e.g. Charlesworth 1987a), though strongly biased conversion in one gene family might have sufficiently dramatic effects that they impair the function of other genes, necessitating evolution to accommodate those changes (Dover 1988). Alternatively, others regard gene conversion as a process which restricts the kinds of variation that are possible, rather than a process functionally equivalent to drift and selection (Endler & McLellan 1988). This debate awaits resolution in the light of the failure or success of theoretical population genetics in dealing with the explosion of sequence data becoming available (Chapter 3).

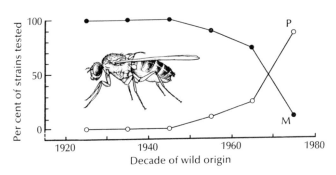

Fig. 2.4 Increase in the occurrence of P-strains of *Drosophila melanogaster* in collections made worldwide. After Kidwell (1983).

THE MAINTENANCE OF VARIATION

Having seen how variation can arise, and how populations can change, it is important to address the question of how variation can be maintained in natural populations. Selection depends on variation, but will often tend to eliminate that variation by fixing favoured variants. Similarly, drift has the potential to eliminate alleles even if they are selectively advantageous. We have also seen that the Neutral Theory predicts that variation will be maintained in natural populations because many alleles will be in a transient stage between elimination or fixation by genetic drift, and that selection will be irrelevant in determining the fate of many alleles. However, I have also shown variation in the frequency of major alleles in alcohol dehydrogenase and esterase-6 appears to be influenced by selection, a deterministic process. Selection depends on variation, and it is important to understand these deterministic influences. The remainder of this Chapter is devoted towards understanding the balance between these deterministic and stochastic processes, with special consideration of which traits and environments can be expected to promote high levels of additive genetic variation.

The mutation—selection balance

The mutation rates in eukaryotes at individual loci are usually estimated to be in the vicinity of only 10^{-6} to 10^{-5} per gamete per generation, so that mutation pressure will be ineffective in resisting selection unless the mutants are approximately neutral in their phenotypic effect (see p. 37). In the case of polygenic traits, mutation is potentially a much more potent force, as all the loci which contribute to the phenotype are available as targets for mutation. There are two ways the effect of mutation rate per character per generation can be evaluated. The first is to measure the amount of new variance attributable to mutation (Lynch 1988), for which a range of estimates are summarised in Table 2.4. This tells us the scale of mutation pressure, but does not give us reliable estimates of the number of loci which typically affect most quantitative characters. Many of the models of polygenic evolution that I will discuss throughout this text are based on the assumption

Table 2.4 Estimates of the extent of heritable variation which can be attributed to mutational input per generation (V_M). This estimate is scaled against the environmental variation (V_E) in order to give an estimate of the expected heritability after one generation of mutation of a trait in a population where heritability is zero before mutation. Average values from Lynch (1988).

Organism	Trait	V_M/V_E
Daphnia pulex	Life history traits	0.0017
Drosophila melanogaster	Abdominal bristle number	0.0019
	Sternopleural bristle number	0.0025
	Viability	0.00003
	ADH activity	0.0006
Tribolium castaneum	Pupal weight	0.0091
Mus musculus	Skull metrics	0.0046
	Ilium length	0.0158
	Ulna length	0.0311
Barley (inbred lines)	Biomass	0.0003
	Grain yield	0.0003
Rice	Head formation	0.0038
	Plant height	0.0015
Corn	Nine traits	0.0078

that very many loci are involved, all of which make small additive contributions to the character of interest. The very limited experiments which estimate the second measure, the number of new mutations per character per generation, give estimates for eukaryotes in the vicinity of 10^{-2}, implying that hundreds of loci may contribute to each character (Lande 1976; Barton & Turelli 1989). While these data support the assumption that many loci are involved, we still lack enough reliable general empirical measures of the parameters of interest to be certain that we are not dealing with artefacts. Some authors retain considerable scepticism that the assumptions are valid, and that mutation is a significant contributor to quantitative variation (Turelli 1984; Barton & Turelli 1989; Bulmer 1989). It seems likely that high heritabilities of the magnitude revealed in Table 2.1 could only be maintained if selection is very weak, the mutation rate is high, and a very large number of loci affect the character.

Indeed, it now appears that after some decades when the development of quantitative genetics was dominated by the obvious applications in improving agricultural stocks (Falconer 1981), we are in the midst of an explosive development of theory relating to phenotypic evolution (Lande 1976, 1988a; Barton & Turelli 1989; Bulmer 1989; Bürger et al. 1989; Houle 1989), so that theory is clearly outstripping our empirical base. Excessive reliance on studies of agricultural species with a long history of inbreeding raises doubts about how representative are the scant data that are available.

In cases of overdominance, or heterozygote superiority, a combination of alleles is of greater fitness than any homozygous alternative. The best known examples are intricately linked to the presence of disease as a selective agent. A rare allele (sickle-cell) of haemoglobin in humans causes anaemia by affecting the function of erythrocytes, and would ordinarily be strongly selected against. However, the sickle-cell allele occasionally reaches high frequency, inevitably in association with a high incidence of malaria, as heterozygotes are resistant to the disease. The allele is said to be maintained at a high equilibrium frequency by balancing selection.

Overdominance does not explain all examples where homozygosity causes reduced fitness. Despite claims by some authors to the contrary, behavioural, morphological or physiological barriers to incest occur in many species of plants and animals (Fig. 2.2; Table 2.5). The existence of these barriers is evidence that inbreeding and its consequence of increased homozygosity can be costly, a view supported by breeding experiments which repeatedly demonstrate inbreeding depression. Two hypotheses attempt to explain the incapacity of inbred individuals (Charlesworth & Charlesworth 1987). The first suggests that outbred lines are superior because they are more likely to express heterozygosity at loci showing overdominance. The second argues that inbreeding will expose deleterious recessive or partially recessive alleles. Data overwhelmingly favour the latter hypothesis, creating the result that inbreeding avoidance prevents selection from reducing the frequency of deleterious alleles.

Antagonistic pleiotropy

Provided that there is a negative genetic correlation between two traits contributing to fitness, significant heritability could persist for those traits even though additive genetic variation for fitness is zero. The cause of this negative correlation would be antagonistic pleiotropy. Selection experiments provide substantial evidence for this hypothesis, provided that appropriate experimental designs are used (Charlesworth 1984a; Service & Rose 1985). Berven & Gill (1983) examined heritabilities of larval development time and larval body size of *Rana sylvatica*, a frog with a broad distribution in eastern North America (Fig. 2.5). Frogs from three sites which differed in latitude and altitude were studied (Table 2.6). In tundra sites shortened developmental time is strongly selected, presumably because of the short time available for larval development and metamorphosis; the length of the larval period is invariant relative to larval body size, where some heritability is retained. In more equable lowland sites in Maryland, large animals compete more effectively. Body size is strongly selected and exhibits very low levels of heritability. In high altitude populations in Virginia, both traits are apparently important and have been canalised to an

Table 2.5 Some mechanisms promoting incest avoidance in plants and animals. In hermaphrodites, particularly common among plants, incest avoidance mechanisms also prevent selfing.

Spatial separation of relatives	In birds, females are more likely to disperse than males. In mammals, the converse is true (Greenwood 1980). Although several explanations of this trend have been proposed, in some species the only explanation for sex-biased dispersal appear to be incest avoidance (e.g. *Antechinus*, Cockburn *et al.* 1985)
Temporal separation of relatives	Where plants and animals are capable of expressing male and famaleness, maleness is often expressed before femaleness (protandry), or vice versa (protogyny)
Genetic incompatibility	Many angiosperms have genetically determined systems which ensure that seeds cannot be produced from their own pollen (e.g. *Nicotiana*). In gametophytic systems, the pollen type is controlled by its own genotype. In sporophytic systems, the pollen type is determined by the genotype of the pollen-producing plant
Morphological incompatibility	Some plants with sporophytic incompatibility systems have heteromorphic flowers, in contrast to the homomorphic flowers typical of gametophytic systems. Heterostyly (Fig. 2.2) maximises the probability of pollen transfer from anther to stigma of a plant of different incompatibility type. Because sporophytic incompatibility does not occur in any of the homostylous families which contain heterostylous members, it seems likely that heterostyly evolved before the sporophytic incompatibility (Darwin 1877; Charlesworth 1985; Barrett 1990)
Behavioural incompatibility (mate choice)	In some species mating with close relatives appear to be actively avoided. For example, female large cactus finches (*Geospiza conirostris*) prefer to mate with a male which sings a song other than that of their father (Grant 1984). Female mongolian jirds (*Meriones unguiculatus*) often leave their home nest in order to mate with another male, even though they then return and rear young with their relatives, often brothers (Ågren 1984)

equal extent. Gene effects on the two traits are negatively correlated, presumably as a consequence of antagonistic pleiotropy. This negative correlation sustains high levels of genetic variability, and therefore there is potential for rapid response to selection.

Fluctuating environments — temporal change

Rapid fluctuations in the environment could maintain heritability because reversals in the direction of selection exceed the capacity of

Fig. 2.5 The wood frog (*Rana sylvaticus*). After Stebbins (1966).

Table 2.6 Geographic variation in larval development traits of the wood frog (*Rana sylvatica*), and the heritability and genetic correlations between those traits. From Berven & Gill (1983).

Site	Lowlands	Mountains	Tundra
Environmental sensitivity			
Length of larval period	High	Medium	Low
Larval body size	Low	Medium	High
Heritability			
Length of larval period	0.27	0.34	0.07
Larval body size	0.08	0.58	0.27
Genetic correlations			
Development rate versus body size	0.65	−0.86	+0.09

selection to depress additive genetic variation. A classic example of selectionist analysis is provided by the melanic and non-melanic forms of the moth *Biston betularia*, which is depicted in this chapter's vignette (Kettlewell 1973). One of the fruits of the industrial revolution was widespread pollution, including reduction and darkening of the lichens on tree trunks, a favourite roosting point of the moths. Once cryptic moths became obvious to predators, and there was a rapid and sustained rise in the frequency of a hitherto rare dark form. In areas where pollution has been reduced, the fortunes of the white form have been revived, and it has started to increase in frequency again.

Some of the best evidence for the role of environmental capriciousness comes from those constant sources of inspiration to evolutionary biologists, Darwin's finches. Rainfall on the Galápagos islands is usually fairly unpredictable. During dry years, one species, *Geospiza fortis*, is subject to selection for increased bill size and depth (Fig. 2.6). This is because small soft seeds are eaten preferentially by both small and large birds. When the food supply shrinks these seeds are depleted rapidly and the food supply comes to be dominated by large hard seeds that are more easily cracked by large birds. Between November 1982

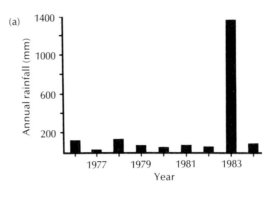

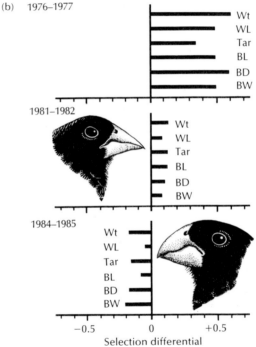

Fig. 2.6 Temporal variation in the direction of selection on bill size in *Geospiza fortis*. (a) Rainfall on Isla Daphne in the Galápagos from 1976 to 1984. After Gibbs & Grant (1987a). (b) Selection differentials (the difference in the trait before and after selection, standardised in standard deviation units) for morphological traits of adults during two years of drought and low food supply (1976–1977 and 1981–1982), and during the years following the torrential rain accompanying the El Niño event of 1982–83. After Gibbs & Grant (1987b). Wt, weight; WL, wing length; Tar, tarsus; BL, bill length; BD, bill depth; BW, bill width. Large-billed and small-billed finches from a photograph in Grant (1986).

and July 1983 the Galápagos were subjected to drenching record-breaking rains because of an unusually intense occurrence of the El Niño Southern Oscillation, a disruptive alteration to ordinary oceanic patterns. The heavy rainfall led to unusually vigorous plant growth, abundant seeding, and reversal in the direction of selection.

In contrast to balancing selection, diversifying selection could maintain polymorphism if different alleles or combinations were favoured in different microhabitats. Heterozygosity would be sustained in part by gene flow and would be anticipated to produce clinal variation in gene frequency.

The selection pressures faced by organisms are unlikely to be equivalent across their geographic ranges. Indeed, one of the many good pieces of evidence for selection comes from strong and repeatable trends in a trait across the geographic range of an organism (Endler 1986a). For example, the latitudinal variation in allele frequencies of esterase-6 and alcohol dehydrogenase indicates the presence of clinal selection.

The scale at which habitat selection contributes to clinal variation in allele frequencies varies greatly. White clover, *Trifolium repens*, has been planted in pastures throughout the world, and probably occurs in the lawns or playing fields outside your window. Virtually anywhere through its range there is local clinal variation in the response of plants to herbivory. Some plants release cyanide when damaged, and others do not. The frequency of cyanogenic and acyanogenic morphs varies over a large geographic scale, but also over a distance of a few metres in a typical suburban garden. Plants which grow in locations which are protected from frosts and which harbour slugs and snails are often uniformly cyanogenic, but where predation by molluscs is reduced the plants may be completely acyanogenic. Cyanogenesis depends on alleles at two loci, and clearly confers protection against molluscs. However, the cyanogenic morph grows less vigorously, and is more susceptible to frost and attack by rusts (Fig. 2.7; Dirzo & Harper 1982a). Intermediate allele frequencies are most common, and fixation is prevented by gene flow.

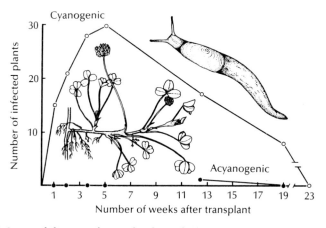

Fig. 2.7 Susceptibility to infection by the pathogenic rust *Uromyces trifolii* in cyanogenic and acyanogenic forms of white clover, *Trifolium repens*. After Dirzo & Harper (1982a).

Frequency-dependent selection has been introduced in Chapter 1. While positive frequency-dependence will depress genetic variation by favouring common genotypes, negative frequency-dependence will increase the frequency of rare phenotypes, and is potentially of profound importance in maintaining genetic variation whenever it occurs. As we shall see repeatedly throughout this text, interaction with natural enemies (particularly disease) is thought to be a potent source of negative frequency-dependence. Selection on the hostile species for increased exploitation will be influenced by the frequency of phenotypes among the species it harms. Common phenotypes will exert strong pressure because they are encountered frequently, allowing a selective advantage for rare phenotypes.

Multiple causation, and partitioning variation

My principal aim in the preceding discussion is to show that several factors can lead to intrapopulation variation in the frequency of traits controlled by single alleles and those under polygenic control. In practice, it is rarely possible to assign the maintenance of heritable variation to a single cause. In two extensive studies it has been demonstrated that several factors conjointly influence allele frequencies.

Banding in snails

One of the best demonstrations of adaptationist analysis comes from the European land snails *Cepaea nemoralis* and *C. hortensis*, which exhibit conspicuous polymorphism in shell colour (yellow, pink or brown) and number of bands (one to five; Fig. 2.8). The following account is derived from Jones *et al.* (1977). Both species exhibit the same pattern of inheritance of colour, suggesting that the polymorphism arose before speciation. Many populations exhibit variability at up to six loci influencing shell colour. This great within- and between-population variability was used by the early architects of the Modern Synthesis as an example of variation unlikely to be influenced by selection. For example, Dobzhansky (1941) comments that the frequencies of genes affecting the polymorphism were 'haphazard' differences attributable to drift; Huxley (1942) claimed that 'the distribution of types appeared to be wholly random', and Mayr (1942) felt that variation was 'obviously of very insignificant selective value'. These comments stand in sharp contrast to the claim that these authors were excessively pan-selectionist (e.g. Antonovics 1987).

The genes controlling the major polymorphisms are closely linked, though recombination between the linked genes are occasionally recorded. Other unlinked loci modify the basic patterns, and there are epistatic interactions between the polymorphic loci. Populations only a few hundred metres apart may be fixed for different alleles.

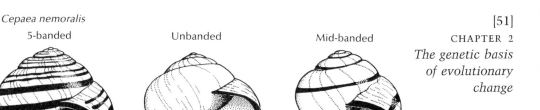

Cepaea nemoralis
5-banded · Unbanded · Mid-banded

Cepaea nemoralis 3-banded · Punctate · Cepaea hortensis

Fig. 2.8 Polymorphism in banding patterns in *Cepaea nemoralis* and *C. hortensis*. After Jones *et al.* (1977).

Successive analyses have demonstrated a number of selective influences preserving or reducing intrapopulation variability in shell colour. Predators including birds, mammals and insects select conspicuous shells, so cryptic coloration is strongly selected, but the least conspicuous colour varies with both habitat (spatial heterogeneity) and season (temporal heterogeneity). Spatial and temporal heterogeneity also arise in the climate to which the snails are exposed. Dark (banded) shells heat more quickly and attain a higher equilibrium temperature than light shells, so the amount of sunshine in a habitat will influence the thermal fitness of snails, and will vary over large and small geographic distances, as well as seasonally. This generates temporal oscillations in selective pressure and clinal variation along complex climatic and microhabitat gradients. Both positive and negative frequency-dependent selection may be involved. Some authors have argued that thrushes form a search image for *Cepaea* morphs, so rare forms are more likely to remain undetected than the common ones most likely to be used a search model (Harvey *et al.* 1975). Alternatively, some birds prefer rare forms, particularly when prey density is high (Allen & Anderson 1984). Jones *et al.* (1977) suggest that detailed analysis has led inexorably to an increase in the complexity of explanations of genetic variation, and that in systems for which less data are available we should beware assuming that a single unifying explanation will be satisfactory.

The major histocompatibility system

One class of genes where very large numbers of alleles appear to be maintained by selection occurs in the major histocompatibility system

(MHC) of mammals. These genes have diverse functions, but among the most important are the generation of immune response differences. The human leucocyte antigen (HLA) system of humans and the H2 system of mice are understandably better studied than the MHC in other species. Several features of these systems are of special interest. First, recombination within the MHC appears to be heavily concentrated in several hotspots (Bodmer *et al.* 1986). Next, some of the genes within the region are highly polymorphic, though the polymorphism is localised to parts of the gene, but other genes show very low levels of polymorphism. This maintenance of different levels of polymorphism is strong evidence for selection, since drift or migration should affect all genes equally.

Several selective forces that might drive the extreme polymorphism have been suggested. First, there might be different susceptibility to disease associated with immune response differences, caused by frequency-dependent selection for new alleles that give protection against new or varied pathogens (Haldane 1949). Compelling support for this hypothesis comes from the observation that the codons which determine the regions that bind antigens are much more polymorphic and accumulate many more substitutions than the codons elsewhere in the MHC (Hughes & Nei 1988, 1989; Potts & Wakeland 1990). Second, particular combinations of alleles might be sustained through balancing selection. Polymorphisms are typically much more evenly distributed across the geographic range of the species than are allozyme polymorphism, and polymorphism appears to antedate the formation of subspecies or even species (Hedrick & Thompson 1983; Figueroa *et al.* 1988; Lawlor *et al.* 1988; McConnell *et al.* 1988; Nadeau *et al.* 1988). Because these polymorphisms are evidently of great duration and very stable, strong balancing selection is implicated. Third, maternal–foetal interactions may be important (Hedrick *et al.* 1987). Human couples with a history of recurrent spontaneous abortion are much more likely to share antigens at different HLA loci than are normal couples. This could either be because the presence of an immune response between mother and foetus is necessary for proper implantation and growth, and the immune response depends on maternal/foetal differences at HLA loci; or because of association of recessive detrimental alleles with HLA antigens. Occasionally humans are susceptible to the antigens produced from the immune response to an infectious disease. For example, ankylosing spondylitis, is almost always associated with the presence of an antigen called B27.

Under these latter conditions some selection for non-random mating might be anticipated. Although unknown from humans, male and female mice prefer to mate and nest with individuals with different H2 haplotypes (Beauchamp *et al.* 1985; Egid & Lenington 1985). Yamazaki *et al.* (1983) have demonstrated that females are also more likely to abort in the presence of a strange male (the Bruce effect) if the original stud and the strange male differed in H2 haplotype. Such preferences may be both a cause and a consequence of allele diversity.

Just as some aspects of molecular structure tolerate variability more than others, it seems likely that quantitative variation will be tolerated in some traits more than others. In both Table 2.1 and Table 2.6, the most important life history traits showed the least heritable variation. Understanding the causes of this reduction of variance is crucial to much of behavioural and ecological theory.

Fisher's fundamental theorem

Fisher (1930) showed in a famous observation that increase of fitness in one generation is equal to the additive genetic variance in fitness at that time. A crucial implication of this conclusion is that traits with a close proximate connection to fitness should exhibit low heritabilities, as selection will act to maximise fitness and at equilibrium fitness differences will approach zero. Several lines of evidence support this claim. Gustafsson (1986) has used long-term data from an island population of collared flycatchers (*Ficedula albicollis*) to compare the heritability of lifetime reproductive success (fitness) with heritabilities of morphological traits and components of reproductive success such as life span and production of young. In general, heritability increased rapidly as the correlation of the trait with lifetime reproductive success declined (Fig. 2.9). It is important to remember that this conclusion only applied to the sum of all fitness components, and antagonist pleiotropy between fitness components may permit the retention of heritability for individual components, so that they respond to both artificial and natural selection (Table 2.6).

Larger scale reviews of heritability data generally support the conclusions from flycatchers and wood frogs. Life history traits, which have a close proximate relation to fitness, show much lower heritabilities than morphological traits in both *Drosophila*, the genus of animals whose genetics we understand best (Roff & Mousseau 1987), and in a compilation of measures of narrow sense heritabilities for wild, outbred animals (Mousseau & Roff 1987).

Nonetheless, the compilation of Roff & Mousseau (1987) shows that significant heritabilities are repeatedly reported for traits with a very close proximate connection to fitness. Similar results are available for plants (Mitchell-Olds 1986), though a full compilation is unavailable (Mitchell-Olds & Rutledge 1986). In principle, any of the effects discussed above could be maintaining this variance, and its seems unlikely that all can be attributed to antagonistic pleiotropy between fitness components (Bell 1984a, 1984b). Because the heritabilities are usually low it is uncertain whether it is ever necessary to invoke more than random mutational input to genetic variance (Turelli 1984; Lynch 1985, 1988; Charlesworth 1987b). One of the best pieces of evidence that selectionist explanations are occasionally necessary to explain even very low heritabilities comes from the observation that herita-

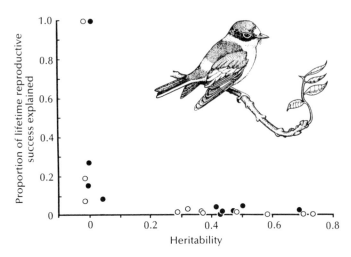

Fig. 2.9 Relation between heritability of a trait and its correlation with fitness in the collared flycatcher (*Ficedula albicollis*). ●, Males; ○, females. Modified from Gustafsson (1986).

bilities for egg-to-adult viability decrease with latitude in *Drosophila melanogaster*, suggesting that ecological factors influence heritability directly (Mukai & Nagano 1983; Mukai 1985). Charlesworth (1987b) argues convincingly that the greater variance in equatorial populations suggests a role for biotic heterogeneity such as host−pathogen interactions (see Chapters 5 and 7).

SUMMARY

Natural selection depends on the existence of heritable variation in natural populations, which is the concern of the science of genetics. Although the basis of variation can be studied directly by the examination of DNA sequences, almost all traits of ecological interest are subject to polygenic control, and the molecular basis of variation is unlikely to be known in any detail. The retention of variation at both the molecular level and polygenic level is usually indicative of a loose proximate connection to fitness, as fitness differences should approach zero in populations at equilibrium. This creates the paradox that selection will reduce the potential for further selection. Selection has a predictable influence on genetic variation, and operates in concert with two other influences where the direction of change is less predictable. The first, mutation, introduces variation upon which selection can act. Although best understood at the level of single gene, it is clear that mutation can contribute to the level of quantitative variation. The second factor, random genetic drift, is of great importance in small populations, and can generate differences between subpopulations. Several types of selection can lead to the retention of heritable variation in natural populations. These include antagonistic pleiotropic effects of genes, high fitness for heterozygotes, and fluctuations in the environment which prevent equilibrium being maintained. In most populations and traits several influences will operate concurrently.

FURTHER READING

This Chapter abstracts a vast literature rich in both formal mathematical theory and biochemical detail. Introductions to the methods of population genetics are provided by Hartl (1981), Wilson & Bossert (1971) and Roughgarden (1979). A very readable account of evolutionary genetics is given by Maynard Smith (1989). Lande & Barrowclough (1987) is an excellent discussion of the concept and calculation of effective population size, and Falconer (1981) is the best introduction to quantitative genetics. Nei (1987) is a good discussion of the population significance of molecular genetics, and Watson *et al.* (1987) an excellent review of the recent data on the molecular basis of genetic change. Lewontin (1974) and Kimura (1983) are lucid descriptions of the controversy over variation at the molecular level, which remain worth reading despite the enormous accumulation of data since their publication.

TOPICS FOR DISCUSSION

1 Progress in genetics has been influenced by a few species which have been chosen as general models. Among eukaryotes, the fruitfly *Drosophila* and the nematode *Caenorhabditis elegans* are obvious examples. Research the reasons why these animals were chosen, and discuss any problems they present as general models.

2 A central premise of much of the argument presented in this Chapter is that mutations are a randomising agent, and do not occur according to the 'needs' of the organism. Debate the recent claims by Cairns *et al.* (1988) that the rate of mutation is directed by the exigencies of the environment (see Lenski 1989 for a critique).

3 There is now broad enthusiasm for a massive project to sequence the human genome. Discuss the types of data that are likely to arise in relation to some of the unresolved dilemmas described in this Chapter.

Chapter 3
The Power and the Units
of Selection

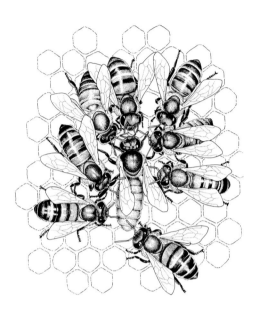

Although changes to the genome provide the mechanistic basis for evolution, genetics and evolutionary biology are not synonyms. First, it has not yet proved possible to describe the phenotype completely from knowledge of the genotype. Even in organisms whose genome is excellently understood like the nematode *Caenorhabditis elegans* we only have a superficial understanding of the 'why' of many aspects of its ecology and reproductive biology (e.g. Bell 1982), and negligible ability to generalise the understanding to the extraordinarily diverse taxon to which it belongs. Second, this expectation seems unreasonable given the large number of non-genetic factors which impinge on the phenotype. These may be extrinsic environmental pressures, or physical limitations to the possible pathways of evolution. Third, it is not clear that the gene is the appropriate focus for ecological investigations. Organisms interact with their environments, genes do not. Last, much of the best data on evolutionary change come from the fossil record, and are necessarily collected in a genetical vacuum.

In this Chapter, I describe some of the limitations to the power of selection, and features which uncouple the phenotype and genotype. This discussion is used to introduce one of the most troublesome practical and philosophical questions in evolutionary biology — what are the units which can be discriminated by selection.

CONSTRAINTS ON SELECTION

Some biologists have implicitly assumed that given sufficient time mutation and other alterations to the genome will generate enough variety for any end to be achieved, leading, ultimately, to precise adaptedness. According to this view (sometimes called 'the adaptationist programme'; Gould & Lewontin 1979), it is only the rapidity with which environments change that makes adaptedness unlikely. Failure to achieve precision in stable environments is thought to be a consequence of trade-offs, with improvement in one trait leading to harmful changes in another, because of genetic or energetic interactions. For example, a feeding bird alternates between searching for food and scanning for predators. Both activities are necessary for survival, and both need to be set aside for some of the activities needed for reproduction (Chapter 4). If natural selection is optimising the result of some trade-off, strong directional selection is anticipated if one of the opposing forces was removed. For example, scanning by birds might decline in a population inhabiting an island devoid of predators. Trade-offs are quite different from constraints, which suggest that many phenotypic outcomes are unlikely ever to be realised as a result of evolution, because of the inherent properties of living systems.

Historical constraints

First, and most important, evolution does not generate organisms from scratch to fit each of the environments that they inhabit. Rather, each step along the tortuous path of evolution is contingent on each preceding step. The importance of history provides a context which is all important in interpretation of the present. For example, it is pointless to ask why mythical dragons with four legs and two wings are not one of the real experiments of vertebrate evolution, for all flying vertebrates are descended from a tetrapod ancestor, and the acquisition of wings has always occurred at the expense of the forelimbs. The constraint is four limbs, the trade-off is between legs and wings. Alternatively, it is worth noting that birds and bats which have returned to a predominantly terrestrial habit retain their wings, even when they must be a substantial encumbrance (e.g. in the bat *Mystacina*, which lives and forages in the leaf litter; Pierson *et al.* 1986). The contingencies of phylogeny have been called 'historical constraints' by Gould (1989a) and 'local constraints' by Maynard Smith *et al.* (1985).

Formal constraints

The second set of constraints have been called 'formal constraints' by Gould (1989a) and 'universal constraints' by Maynard Smith *et al.* (1985). These describe the universal physical and chemical laws which influence all structures, and dictate what is and what is not possible. Pigs do not fly. Indeed, some developmental biologists have argued strongly against the Darwinist view that evolutionary models should be based on the fitnesses of genetically determined alternatives (Goodwin 1988, 1989; Ho *et al.* 1986). They argue that the laws of physics so tightly govern the expression of form that it is the dynamics of cellular interactions rather than the effects of gene products that determine the phenotype, and that the principal role of genes is the evocation of developmental responses. According to this view, evolution therefore has an epigenetic rather than a genetic basis (see Alberch 1980, 1982 for a balanced discussion of the distinction between epigenetic and genetic models of form). They portray Darwinist theory as demanding that all phenotypes are equally probable, which of course is untrue, and a view which has never been part of the Modern Synthesis (Dawkins 1986; Futuyma 1988).

Quite apart from this unjust caricature, there are problems with this view of immutable development. There are many examples in which similar phenotypes and stages of development are attained by different developmental pathways (Wagner 1986; Levinton *et al.* 1986; Buss 1987). For example, all winged adult insects have a reasonably similar form but may be divided into exopterygotes and endopterygotes according to the radically different means by which they acquire that form (Fig. 1.1; Sander 1983). The coelomic cavity in the tentaculate phoronids can arise from either folding of cells with an epithelial character, or by aggregation of mesenchymal cells (Salvini-Plawen & Splechtna 1979).

Lewontin (1983) accepts that there is absence of gene/development correlation, but also savages the idea of physical determinism, suggesting

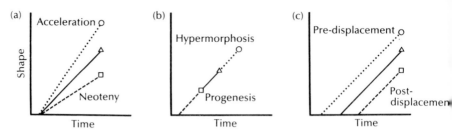

Fig. 3.1 The different modes of heterochrony, resulting from changes in (a) the rate of development, (b) the termination of development and (c) the onset of development. △, ancestral shape; ○, peramorphic descendant shape; □, paedomorphic descendant shape. After Alberch *et al.* (1979).

that the gene/development/environment/organism interaction is best viewed as a branching set of pathways allowing movement from many states to many other states. Alternatively, Gould (1989a) portrays organisms as being influenced simultaneously by historical and formal constraints and adaptation.

Shape and developmental constraints

The study of constraint has focused heavily on development, and indeed developmental pathways or ontogeny are the usual way that these constraints express themselves (Gould 1989a). Vertebrates have four limb buds which generate structures ranging from the human hand to the wing of an albatross.

Gould (1989a) strongly advocates the view that we should view developmental constraints not just as a limit but also as an important force compelling evolution in particular directions. Because developmental trajectories are highly integrated, major phenotypic changes may often be accomplished more easily by changes in the timing of existing developmental processes (heterochrony), than by the generation of completely new pathways. Alberch *et al.* (1979) have presented a simplified classification of heterochronic changes which occur through changes in the onset, offset and rate of development (Fig. 3.1). Although the mechanisms are very different, there are two outcomes of importance. The first, paedomorphosis, occurs when the descendant has juvenile features relative to its ancestor, and is achieved through neoteny (relative to the ancestor, a decrease in the rate of development without a change in the time at which the adult shape is acquired), progenesis (an early termination of development, without change in rate) and postdisplacement (a delay in the onset of development, without a change to the rate or time at which the adult shape is achieved). The second, peramorphosis, occurs when the descendant has more fully developed features than its ancestor, and is achieved through acceleration (an increase in the rate of development without a change in the time at which the adult shape is acquired), hypermorphosis (a prolongation of development, without change in rate) and postdisplacement (an earlier onset of development, without a change to the rate or time at which the adult shape is achieved).

Plants of the genus *Delphinium* (Ranunculaceae) illustrate how heterochronic changes may facilitate evolutionary innovation (Guerrant 1982). Most species are pollinated by bumblebees, and possess an open flower well-suited to ensure transfer of pollen to their visitors (Fig. 3.2). Two species are pollinated by hummingbirds, which prefer to visit red flowers with long floral tubes. *Delphinium nudicaule* has just such a flower, which bears an uncanny resemblance to the buds of its bumblebee-pollinated ancestors. Guerrant has shown that although floral development takes a similar time in both types of *Delphinium*, the rate of differentiation is much slower in most parts of the hummingbird, pollinated flowers, so changes in shape are not nearly as great. Some developmental dissociation has been possible, as the petal which forms the nectar cup does develop rapidly, allowing storage of the dilute and copious nectar favoured by hummingbirds. However, in general, evolution will follow 'the line of least resistance' (Stebbins 1970), and involve the modification of existing pathways rather than the creation of totally new developmental trajectories.

Development remains one of the most weakly understood areas of biology, and there has consequently been an unfortunate tendency to ignore it in the hope that it might go away. However, there is little evidence from recent advances in the study of development that the Darwinist model is under threat (Futuyma 1988). Instead, the study of

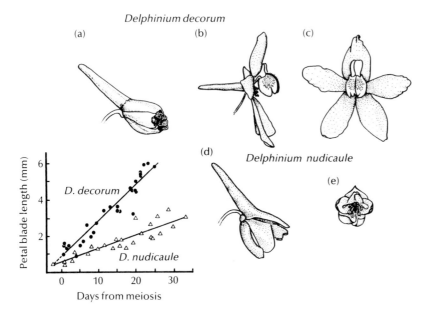

Fig. 3.2 Petal blade length versus time from meosis in buds of the bee-pollinated *Delphinium decorum* (●) and the neotenic hummingbird-pollinated *D. nudicaule* (△). *D. decorum*, (a) bud, (b) flower, side view, (c) flower, front view. *D. nudicaule* (d) flower, side view, (e) flower, front view. After Guerrant (1982).

[62]

CHAPTER 3
*The power and
the units of
selection*

development should be viewed as representing a fascinating and increasingly important part of mainstream theory.

Body size

Just as changes in shape are constrained by developmental integration,
changes in size exert profound and highly correlated influences on the
phenotype. The study of size is called allometry. Whole books have
been devoted to documenting the sets of characters which are highly
correlated with body size (Peters 1983; Calder 1984). Many of these
associations conform to the power law, which has the form:

$$Y = aX^b$$

where Y is the trait of interest, a and b are constants, and X is body
mass.

An excellent example illustrating how formal (allometric) and historical constraints operate in concert comes from birth weight in
primates (Fig. 3.3). Within the two chief lineages among the primates
(strepsirhines and haplorhines) there appears to be little opportunity to
vary birth weight of young, even in the case of twinning. Nonetheless,
the rule constraining the haplorhines is very different from the rule
affecting strepsirhines, and allows larger babies at all birth weights.
One of the reasons that humans appear to be very distinct from their
close relatives among the great apes is difference in the rate of development (Gould 1977). Humans show the greatest departure of any haplorhine from the allometric curve, producing very large babies.

Departures from allometric rules of this sort may often be interpreted
as a reflection of adaptedness, a theme which will be developed further
in Chapter 4.

Adaptation is capable of producing complex design

One further attack on the adaptationist or neo-Darwinist model requires
brief treatment because of the frequency with which it is invoked.

Paley's watchmaker

This argument is much beloved of those who believe we must invoke
divine or some other extrinsic intervention in order to explain the
pattern obvious in nature, and appeals incorrectly to the laws of probability (e.g. Hoyle & Wickramasinghe 1981; Hitching 1982). It runs as
follows. As complexity increases, the probability of an event being the
outcome of chance diminishes to the point when some conscious and
deliberate designer must be invoked in explanation. Dawkins (1986)
illustrates this argument with the influential theologian Paley's (1828)
discussion of the probability of a watch and its intricate mechanisms
arising by chance. The misapplication of probability arises because of
two incorrect assumptions. First, the naive belief that the 'random'

[63]

CHAPTER 3
*The power and
the units of
selection*

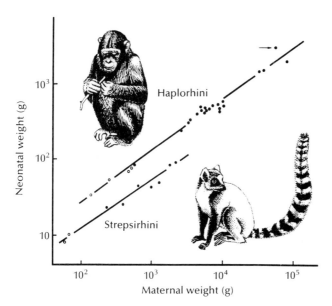

Fig. 3.3 Allometric relation between weight of primate litters at birth and maternal weight in haplorhine (monkeys, baboons and apes) and strepsirhine (galagos, pottos and lemurs) primates. ○, species where multiple births are common; ●, species producing singletons. The arrow indicates *Homo sapiens*. Modified from Leutenegger (1979).

nature of mutation requires that the 'end' phenotype needs to be assembled from scratch, by the random shuffling of simple components. Second, as a corollary, that intermediate stages towards a beautiful Swiss timepiece have no intrinsic value, and that only the finished product can function. The most commonly invoked biological analogy is the development of the human eye. Dawkins catalogues a full set of intermediate photoreceptors, and might well have gone on to include the superior (in their context) eyes of birds of prey. Instead, it is easy to show that the evolution proceeds by cumulative change rather than by random shuffling.

Evolution towards adaptedness does not always depend on adaptation

Because evolution follows the line of least resistance, features which have evolved in response to one set of selection pressures are frequently co-opted for use in response to another set of selection pressures. This raises the curious possibility that much of the development of an adaptive trait may be independent of selection for its adaptive function. Some authors have argued that this problem is sufficiently great to that a separate term, exaptation, is necessary for evolution of this sort (Gould & Vrba 1982), but I believe that their argument results from the repeated tendency to use adaptation to describe both a state (adaptedness) with a process (adaptation) (see also Endler 1986b; Endler

[64]

CHAPTER 3
*The power and
the units of
selection*

& McLellan 1988). Indeed, the accumulation of all evolutionary change from substitution of single amino acids to accumulation of complex morphological traits will change the future response to selection in novel environments. The rapid co-option of some features is a reflection of latent selection potential, the potential to respond should selection pressures suddenly emerge (Stebbins & Hartl 1988).

Those quintessentially Australian animals, the kangaroos, illustrate the problem well. Two characteristics distinguish the large kangaroos from their marsupial relatives. The first, common to all kwarks (kangaroos, wallabies and rat kangaroos), is bipedal hopping. In most mammals, locomotion becomes more costly as speed increases. Bipedal hopping is initially more costly than quadrupedal locomotion, but in kwarks above a certain size (about 1.5 kg), the costs of locomotion actually decline as speed increases (Baudinette 1989). This is because of the use of elasticity of tendons and the pumping of the lungs of the viscera to generate a sort of perpetual motion machine. Such extraordinary economy of movement could not be responsible for the evolution of bipedality, which originated once, and in species well below the critical size at which economy is achieved. The evolution of bipedality remains unexplained (Lee & Cockburn 1985).

A second trait of obvious adaptive value found in large kangaroos is molar progression, which brings a sequence of molars into close occlusion, and overcomes the problems of molar wear which arise when an abrasive diet of grasses is consumed. In a series of ingenious experiments, Sanson (1982, 1989) has demonstrated that molar progression can only arise when the size of the premolar is sufficiently small to allow its displacement by a row of migrating molars (Fig. 3.4). The premolar of the smaller kwarks is large and blocks molar migration, but in large kangaroos it is small and soon lost. So an obvious contributor to adaptedness, molar progression, has originated as a result of an unexplained long-term evolutionary trend towards reduction in the size of the premolar.

Large kangaroos are the most common and conspicuous members of the kwark radiation, and have attracted most attention from ecologists, physiologists and morphologists. However, two of the outstanding examples of their adaptedness could not have arisen in response to selection for their current function, nor could this have been realised without analysing the context in which the trait originated.

Both molar progression and energetically efficient locomotion are clearly adaptive, yet adaptation was not the process generating adaptedness, at least initially. Unfortunately, it is not clear what process did induce these complicated morphological changes. The eggshells of silkworms provide an example where we do understand the development and genetical control of a trait sufficiently well to guess at a process other than adaptation, which nevertheless underlies the evolution of a complex adaptive trait. The eggshells of silkmoths comprise a series of proteins which are produced at precise developmental stages by the cells surrounding the developing oocyte. The shell protects against

[65]

CHAPTER 3
*The power and
the units of
selection*

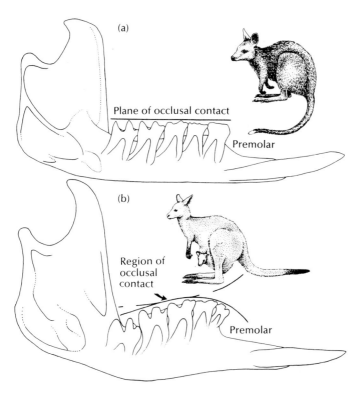

Fig. 3.4 Morphology of the lower jaw and molars in the browser (a) *Wallabia bicolor*, and the grazer, (b) *Macropus giganteus*. The grazer shows molar progression. Note the reduction of the premolar. Jaws after Sanson (1982).

predators, parasites and desiccation, and allows gas exchange. Comparison of the cultivated silkmoth *Bombyx mori*, and a wild silkmoth, *Antheraea polyphemus*, reveals great differences in the outer surface of the shell. In *Bombyx*, the surface is covered by an impermeable layer, as would be anticipated in eggs which must survive diapause. By contrast, the surface of the eggs in *Antheraea* includes chimney-like projections which facilitate gas exchange, but restrict the potential for diapause. This is a classic example of a relation between the life history of the species and the structure of an adaptive trait. Detailed sequence data of the proteins which comprise the impermeable layer in *Bombyx* show that the proteins are coded by two subfamilies of a multigene family which produces all of the eggshell proteins (Goldsmith & Kafatos 1984). Each subfamily comprises fifteen near-identical genes arranged in pairs (one from each family). Each of the genes is characterised by a repetitive motif rich in cysteine, and the subfamilies appear to be derived from an ancestral subfamily present in *Antheraea*. The initial cysteine enrichment is unlikely to have arisen as a consequence of selection, but probably represents continued accumulation of copy number through slippage replication (Dover 1986a). Further, the development of cysteine enrichment in all the members of a subfamily is likely to have come about by gene conversion. The major shift in

[66]

CHAPTER 3
*The power and
the units of
selection*

protein function appears to be dependent on many changes, and the initial changes are unlikely to have had any phenotypic effect, so it appears that the acquisition of the many steps required to allow egg diapause are independent of selection.

Dover (1986a, 1986b, 1988) uses this argument to support his claim that molecular drive may be sufficiently potent to demand a response from organisms. Instead of organisms changing as a result of selection imposed by their environment, the changes imposed by molecular drive demand a response on the part of the organism which must seek a new habitat — in this case, one where diapausal eggs are useful. However, it seems unnecessary to treat these changes in a different way to any other mechanism which generates latent selection potential.

THE UNITS OF SELECTION

The definitions of natural selection, adaptation and fitness suggest that selection should occur in any system which has the properties of variability, heritability and fitness differences. Debate over to the unit of selection has preoccupied philosophers and biologists for many decades, and there is little evidence of respite from this debate. On the one hand there has been strong reaction against the forceful claim by Oxford zoologist Dawkins (1976, 1981) that the gene should be viewed as an important unit of selection, and on the other hand a plea that we should not restrict our view of evolution to a single level in the hierarchy of life, but instead need to accept the possibility that selection could operate simultaneously on several levels (Vrba & Eldredge 1984; Gould 1985a). In the following discussion I attempt to establish the plausibility of selection operating at many levels.

The hierarchy of life

Cursory examination of any introductory text of biology leaves little doubt that biological questions can be examined from several perspectives which differ predominantly in the scale at which questions are framed. From molecular geneticists who are interested in fractions of molecules, to some systematists who focus on the relationships of the evolutionary radiations represented by higher taxa (clades), we can array levels of investigation in an ascending hierarchy:

Genes (Molecules) → Cells → Organisms → Populations
→ Species → Clades.

There have been attempts to ascribe the properties of variability, heritability and fitness differences to each of these groups. Given these properties, selection should take place. The following discussion addresses two questions of fundamental importance.
1 What happens if the direction of selection differs between the levels in the hierarchy?
2 If selection at various levels is commonly in the same direction, is

it heuristically useful to assume that one level is pre-eminent? In order to understand the development of the argument it is useful to first discuss the historical reasons for much of the recent controversy.

An historical overview: the demise of group selection

Until the middle of the 1960s it was widely accepted that selection acted for the good of the group, or perhaps the good of the species. This view is summarised superbly by Williams (1966a), its most eloquent critic at that time, and I follow Wilson (1983) in being happy to quote him at length:

> Always when biotic adaptation is postulated, its immediate or ultimate effect is the improvement of the situation from a traditional aesthetic point of view. It is assumed that: a population of vigorous animals under heavy predation pressure is better adapted than one that is sickly and chronically starved; a population that divides its resources into stable individual territories is better adapted than one in which there is a chaotic scramble for resources; a population in which territory or social position is held by threat-display and recognition by neighbours is better adapted than one that maintains the social structure by frequent combat with effective weapons; a population with stable density, stable age distribution, etc. is better adapted than one in which such factors vary widely; a population in which older individuals regularly yield to promising youths is better adapted than one dominated by fecund but slowly displaced oligarchs; populations in which individuals, such as worker bees, often jeopardize their own well being for a larger cause are better adapted than those whose members consistently act only in their own immediate interests; those in which individuals normally live in peace or active cooperation and mutual aid are better adapted than populations in which open conflict is more in evidence; on the other hand, when active mutual destruction must take place, infanticide is preferable to the killing of peers.

The essence of this view is that selection will favour properties that enhance the stability and long-term survival of groups, even if individuals suffer as a consequence of that selection. Williams destroyed this view of the world in a monograph which has lost none of its cogency in the ensuing decades. His argument runs as follows (1966a):

> Suppose that in a certain park, of a sort suitable for the nesting of robins, the population is being regulated in the manner envisioned ... Because it is effectively regulated in relation to the food resources provided by the park, this population will never overexploit its resources and will therefore last a long time ... Therefore things are going well for the robins ... but suppose an individual appears in which the mechanisms of reproductive constraint are absent or less well developed than is normal. In times of increased numbers, when many of the

[68]

CHAPTER 3
*The power and
the units of
selection*

robins are dutifully neglecting to raise a family, the abnormal
bird will be among the active breeders. It will produce more
than its fair share of offspring, compared to normal birds, and
the behavioural abnormality will be more abundant in the next
generation. When that generation reproduces, the same thing
will happen again. Selection at the individual level will cause
the population to evolve a decreased level of control on its
numbers. This will be true not only in one park but in all of
them, and even the best controlled populations would be
evolving a loss of control. Selection among populations cannot
cause evolution to go in one direction, when each of the
populations is evolving in the opposite direction.

There is now consensus that group selection of this sort is unlikely
to restrain selection at the level of the individual (Wilson 1983).

An historical overview: the problem of altruism

One of the problems caused by this consensus was the need for an
explanation of the occasional occurrence of altruism, behaviour which
benefits a recipient but causes some reduction in the fitness of the
donor. Some altruistic acts may be very subtle, but others are striking
indeed, of which the most impressive is eusociality, where there is
reproductive division of labour, and sterile castes assist adults (often
their parents) to rear young. For example, one of the phenotypes which
develops in a honeypot ant (the replete caste) functions as a food
repository for its nestmates; termite colonies contain a single queen
which is assisted by both workers and soldiers, and a colony of naked
mole-rats usually contains a single queen which is fed by colony
members through the charming habit of communal coprophagy
(Fig. 3.5). There are other examples of altruistic behaviour almost as
impressive as eusociality. For example, some hymenopteran larvae
never grow into adults but have specialised mouthparts which enable
them to attack anything which might harm their sisters, which would
be incompetent at self-defence (Fig. 3.5). All these behaviours appear
directly contrary to the interests of the donor of the altruism.

The solution to this problem comes from the observation that
sexually reproducing organisms do not reproduce themselves exactly;
instead, they leave copies of parts of their genome in their offspring.
These gene copies are not just shared by parents and offspring, but by
all relatives which have similar copies of a gene through descent. The
probability of sharing a gene copy is a function of the degree of relatedness; sisters share more genes than cousins. Hamilton (1964) developed
some early ideas of Fisher (1930) and Haldane (1949) to show that
selection favours genes for altruism between relatives provided that:

$$rB - C > 0$$

where r is a coefficient of relatedness (Box 3.1), B is the benefit received
by the recipient of an altruistic act, and C is the cost suffered by the

[69]
CHAPTER 3
*The power and
the units of
selection*

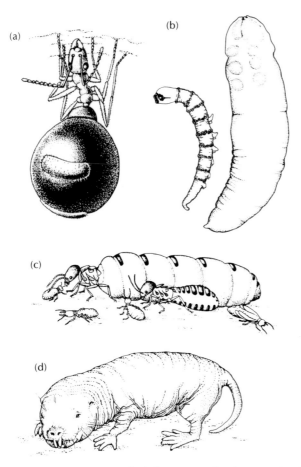

Fig. 3.5 Examples of worker castes which forego reproduction to assist their relatives; (a) the replete caste of the honeypot ant acts as a food storage organ; (b) a normal larva and a sterile defender larva of the wasp *Pentalitomastix* sp. (after Cruz 1981); (c) termite workers and soldiers defend their queen (after Skaife 1955); (d) a worker in a colony of naked mole-rats (*Heterocephalus glaber*).

donor. This has led to the concept of inclusive fitness, which may be defined as the lifetime reproductive success of an organism, with the costs and benefits it receives from its social environment removed, plus the benefits in fitness it provides to its relatives, devalued by the coefficient of relatedness. Although heuristically invaluable, in practice the three components of inclusive fitness are usually impossible to measure (Grafen 1982), and the concept has generated profound confusion. Dawkins (1979) and Grafen (1984) catalogue the more common misunderstandings, some of which are sufficiently ingrained to warrant discussion. First, altruism is not simply proportional to the degree of relatedness. Although a female mammal is as closely related to her sister as she is to her daughter, it is unlikely that altruism directed towards an adult sister will produce the same increments in fitness for the sister as parental care directed towards a neonate, because of the

Box 3.1 Calculating the coefficient of relatedness (r). From a diagram showing the individuals of interest and their common ancestors, the number og generation links can be counted. Assuming no segregation distortion, at each generation link there is a $50:50$ chance that a copy of a particular gene will get passed on. For L generation links the probability is $(0.5)^L$. To calculate r, sum this value for all possible pathways between the two individuals, so that $r = \Sigma(0.5)^L$. Solid lines are generation links used in the calculations. The dotted lines are the sources of genes which dilute relatedness. After Krebs & Davies (1987).

Parent and offspring   Full sibs

$r = 1 \ (0.5)^1 = 0.5$ $r = 2 \ (0.5)^2 = 0.5$.
(genes identical by descent come from both father and mother)

Grandparent and grandchild   Half sibs

$r = 1 \ (0.5)^2 = 0.25$ $r = 1 \ (0.5)^2 = 0.25$
(genes identical by descent come from only one parent)

greater dependence of the neonate on some extrinsic assistance, and because of the greater reproductive span remaining to the daughter. Second, altruism is not expected just because relatedness is high, as occurs for example in clonal organisms. The degree of relatedness is not the only element of Hamilton's equation.

This second source of confusion was probably accentuated because of the satisfying result that eusociality has evolved repeatedly in the Hymenoptera, which are haplodiploid, where males are produced from unfertilised haploid eggs and females are diploid, produced from fertilised eggs. This produces the paradoxical effect that sisters are more closely related to each other than mothers are to their daughters (Table 3.1), so the conditions for the spread of altruism between sisters are relaxed. Although they are not haplodiploid, it also appears that

Table 3.1 Degree of relatedness in a haplodiploid species. The extent of relationship between sisters drops sharply if the mother mates with more than one male, as occurs frequently in the social Hymenoptera. After Wilson (1975) and Brockmann (1984).

	Mother	Father	Sister	Brother	Son	Daughter	Niece or nephew (via sister)
Female	0.5	0.5	0.75	0.25	0.5	0.5	0.375
Male	1	0	0.5	0.5	1	1	0.25

other eusocial species such as termites (Rowell 1987) and naked mole-rats (Reeve *et al.* 1990) live in colonies where relatedness between some individuals is extremely high, though in both cases the high coefficient of relatedness can be interpreted equally well as a consequence of eusociality rather than a major contributor to its evolution.

Collectively, models of selection based on individual fitness adjusted for the influence of inclusive fitness have been called models of kin selection, and have led to the argument that the target of most selection is the gene, and not the organism in which the gene is housed (Dawkins 1981). This is because the gene acts as the replicator which is duplicated and preserved through time, while the organism is only a temporary vehicle for that gene.

A straightforward way to confirm that evolution can be driven by selection at the level of the gene is to show that spread of a particular genetic element is inimical to the interests of the organism. Such genetic material has been called 'selfish' (Doolittle & Sapienza 1980; Orgel & Crick 1980).

Transposable elements

In Chapter 2 the accumulation of P-elements through time in *Drosophila* populations is described. There are three hypotheses for the accumulation and maintenance of transposons in host populations (Charlesworth 1987a). These ideas vary to the extent that they implicate positive or negative effects on the host:

1 Transposable elements play a positive functional role for the host, such as regulation of genetic activity. This hypothesis was suggested by the discoverer of mobile genetic elements, McClintock (1956). Although benefits to bearers of transposable elements have occasionally been reported in cultures of bacteria (Hart *et al.* 1986), the evidence for this hypothesis is very weak, particularly for eukaryotes, and most authors conceded that fitness effects are usually deleterious.

2 Transposable elements cause mutations and chromosome rearrangements which can be utilised by natural selection, so the presence of elements facilitates and accelerates adaptive evolution. Many bacterial transposons were first isolated from plasmids carrying genes conferring drug- or metal-resistance. However, abundant evidence sug-

[72]

CHAPTER 3
*The power and
the units of
selection*

gests that the effect of mutations with noticeable phenotypic effects is deleterious, and that bacteria and *Drosophila* with an elevated mutation rate because of large numbers of transposable elements have low fitness (Mackay 1986).

3   Transposable elements are intragenomic parasites, maintained by intragenomic replication which occurs at least at the rate at which they are excised or eliminated by selection. This is clearly a model of gene selection (Doolittle & Sapienza 1980; Orgel & Crick 1980). P-elements, which code only for their own transposition, appear to fit this model well. Sessions and Larson (1987) have sought correlations between developmental rate and volume of DNA per nucleus in plethodontid salamanders, in which DNA varies greatly as a consequence of the accumulation of large amounts of repetitive DNA. There is a strong negative correlation between tissue differentiation in limb regeneration and the volume of DNA per cell. On the basis of careful phylogenetic comparisons they conclude that genome size sets a limit on differentiation rates, and suggest that there may be tension between selection for rapid differentiation (something easily imagined as favourable to the organism) and selection favouring the accumulation of junk DNA (something which appears only to favour the genes themselves).

t-*Alleles in mice*

One genetic system indubitably harmful to the individual occurs in mice as a complex of genes known as *t*-alleles. Most populations of mice contain individuals which are either homozygous for a wild-type allele, or heterozygous for a variable recessive allele. Most of these recessive *t*-alleles are homozygous lethals, but even dissimilar recessives may cause sterility. The maintenance of these alleles, which may approach frequencies of 0.5, is a considerable problem (Lacy 1978). One factor promoting the high allele frequencies is segregation distortion, the non-random partitioning of the chromosomal complement at meiosis. Males often transmit their *t*-allele to 90 per cent or more of their progeny. This ensures the transmission of the allele, but is problematic for heterozygous females and the males which mate with them because many of their offspring will be sterile and inviable.

Selfish chromosomes in wasps

An even more extreme case of efficient gene replication inimical to the interests of individuals has recently been documented in the parasitic wasp, *Nasonia vitripennis* (Nur *et al.* 1988). Males are haploid and usually develop from unfertilised eggs, and females are diploid and develop from fertilised eggs. Some males carry a genetic element borne on a supernumerary (B) chromosome which causes condensation and loss of paternal chromosomes from fertilised eggs, converting females into males. This has the extraordinary effect of ensuring transmission of the male supernumerary chromosome at the expense of the rest of the male genome.

In a recent review, Werren *et al.* (1988) conclude that complex genetic systems are probably generally vulnerable to subversion by such selfish genetic elements.

[73]

CHAPTER 3
*The power and
the units of
selection*

Replicators and interactors

This argument should not be construed to mean that the interests of genes and organisms are always inimical, but does suggest that genic selection is a viable evolutionary force. Whether genic selection is all-embracing, and adequate to resolve the debate over the target of selection is less clear. Selfish DNA may be a special case of uncertain phenotypic importance. Two chief objections are advanced to support the view that the individual organism is usually the special target of selection (Sober 1984):

1   The directness objection, which suggests that selection works on phenotypes, and genes or replicators are usually invisible to selection. In addition to the distinction between replicators and vehicles, they invoke the concept of interactors, the unit which is distinguished by the selective agent. If similar organisms have different genotypes, their genetic content cannot be discriminated by selection, because only the phenotype is exposed to selective elements;

2   The context-dependence objection, which points out that the fitness of an individual gene is entirely dependent on context, or genetic background. The same gene may have high fitness in one context, and be lethal or deleterious in another (Lewontin 1983). Once again the sickle-cell allele of haemoglobin is an excellent example. This allele is vanishingly rare in many human populations but reaches high frequencies wherever humans are subject to a high incidence of malaria. Any human with two sickle-cell alleles suffers debilitation and even mortality, yet individuals with one sickle-cell allele and one normal allele have some resistance to malaria. It is not the sickle-cell allele which is enhanced by selection, but the combination of alleles.

The cell as the unit of selection: modular organisms

One of the most potent forces supporting restriction of selection to individuals and genes has been Weismann's (1893) doctrine and his observation that there is a barrier functionally separating the somatic tissues from the germ line. He argued that gametes were sequestered during early development and protected from any influence of the environment. This conclusion was highly influential, and helped to destroy the hypothesis of evolution through the inheritance of acquired characters, which had been advanced by Jean-Baptiste de Lamarck, and also toyed with by Darwin. The absolute nature of Weismann's doctrine does not apply to one important form of life, which ecologists call modular organisms. Under modular organisation, the zygote does not grow to an obvious and determinate form but instead develops into a unit of construction (a module) which then produces further modules like the first. Individual organisms usually have a branching structure,

[74]

CHAPTER 3
*The power and
the units of
selection*

their form is indeterminate and strongly influenced by the environment, and it is often, but by no means always, the terminal modules along a branch which differentiate into tissues capable of gametogenesis. In these cases gametogenesis is usually from previously unsequestered somatic tissue, and reproductive modules on one individual may be genetically heterogeneous because of somatic mutations within a branch of development. Selection might conceivably act within modular organisms as well as between modular organisms (Whitham & Slobodchikoff 1981; Gill & Halverson 1984), though the conditions are rather restrictive (Slatkin 1985a). The best-known modular organisms are plants, but this form of development is found in fungi, protists and at least nineteen phyla of animals, including such important groups as poriferans, hydrozoans, anthozoans, bryozoans, and colonial ascidians (Fig. 3.6; Buss 1983; 1987; Begon *et al.* 1986).

Modular organisms challenge our confidence in claiming that selection operates at the level of the individual in other important ways. If the branches of a genetic individual are severed, the two daughter individuals can then exist independently, although until that time transport of nutrients between modules and branches may have been important in physiological integration of the individual.

Adaptive landscapes and the shifting balance

Wright (1931, 1977) argued that one consequence of the context-dependent selective value of genes, particularly epistatic interactions, is that some evolutionary transitions are unlikely to occur as a consequence of selection on genes and individuals, even if fitness is raised. Consider a single locus with two alleles, where heterozygotes are inferior to either homozygote. Even if one allele is fitter than the other when homozygous, transitions from one state to another are hindered by heterozygote inferiority. As the number of alleles and dominance and epistatic interactions increase, the complexity of these transitions will increase greatly. Wright proposed the useful metaphor of an adaptive

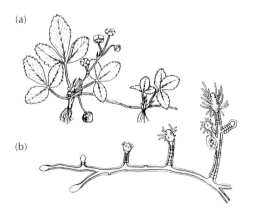

(a)

(b)

Fig. 3.6 Modular growth (a) in a plant (a strawberry) and (b) in an animal (a hydra).

[75]
CHAPTER 3
*The power and
the units of
selection*

landscape with a selective surface of peaks and valleys, the peaks corresponding to gene combinations of high fitness. Selection should act to push populations to the top of the peaks and frustrate the movement across valleys from one peak to another. Wright suggests that when the population is divided into small groups, drift and inbreeding can fix unfavourable gene combinations, which are not selected against because there is no variation within subpopulations on which selection can operate. The population can then move across the adaptive landscape towards another adaptive peak, possibly of greater fitness. In time, evolution proceeds through the shifting balance between one adaptive peak to another. Drift permits exploration of adaptive space in a way that selection cannot. There is then a third phase where the rapid population growth and concomitant high emigration rate from the population lead to the spread of the new genotype into other subpopulations. Laboratory data tend to confirm the feasibility of this model (Wade 1977, 1982; McCauley & Wade 1980; Wade & McCauley 1980).

Whether this is a type of selection on group properties has proved controversial, as drift rather than selection is the dominant cause of evolutionary change, and gene flow and conventional selection the cause of the final spread of the genotype (Crow *et al.* 1990). However, in a paper written at the extraordinary age of 99, Wright affirms his own view that the shifting balance hypothesis is a model of group selection, as the new group should make a differential contribution to the global population if its fitness is enhanced by shift between peaks (Wright 1988).

Intrademic group selection

There has recently been a cautious rehabilitation of group selection, based on a class of models which have attracted great attention since their independent development by several workers in the 1970s. They have a number of characteristics (after Wilson 1983; Harvey 1985). First, a global population, or deme, containing genetically distinct individuals, divides into trait groups for a period which may be a fraction of a generation, or for many generations. These trait groups may contain relatives or may be drawn randomly from the population. The trait groups must occasionally disperse. Consider an allele that reduces individual fitness within the group, but enhances productivity of the group by helping group members. When the global population coalesces the frequency of the allele can rise. If an allele of this sort is selected against within mixed groups but persists in the population, it is said to be maintained by intrademic group selection.

One of the difficulties of these arguments arises because empirical applications remain speculative. First, there are two problems with concluding that the appropriate trait group variation occurs in nature. Temporary associations in populations exhibiting 'altruistic' behaviour are often found in groups of relatives, where altruistic behaviour may

[76]

CHAPTER 3
*The power and
the units of
selection*

depend on kin selection (Grafen 1984). Second, different productivity of local groups may be an artefact of selection due to differences in microhabitat (Wilson 1977). In addition, the predictions of alternative models often differ to only a slight extent (Harvey 1985). So despite the existence of catalogues of possible examples (Wilson 1980), the chief values of the models currently lies in their ability to provide an alternative method of theoretical analysis of a wide variety of phenomena.

Selection on species and clades

Clearly one of the most important processes in evolution is speciation, the formation of new species. The species descended from a single ancestral species are said to belong to a clade. Variability between clades is a property of higher taxa, almost by definition. It has been suggested by many authors that aspects of this variation are often linked to survival and fecundity of those clades because they influence the propensity to speciate and resilience to extinction. Heritability is often more troublesome (Slatkin 1985b; Jablonski 1987), but there has nonetheless recently been great enthusiasm for the possibility that selection between clades (usually but confusingly called species selection), may be of great importance in evolution.

There is no real philosophical problem with this argument. Species can be thought of as both replicators and interactors. However, models of species selection share some of the same problems that plague models of group selection. The force of natural selection on a given level depends on three things (Stearns 1986): the amount of variability for the trait of interest among units within that level relative to the amount of variability at all levels, the cycle time of those units, and the degree to which reproductive success of those units is related to variation in the trait among the units. The weakness of selection at higher levels of the hierarchy stems from the extremely slow cycle time. The cycle time is a single generation for genes and individuals. For species, the average cycle time may be many orders of magnitude greater. Thus the power of selection on individuals will usually be immeasurably greater than the power of selection on higher elements in the hierarchy. However, Stearns (1986) goes on to point out that selection can only operate on variability, and in some cases variation will be confined to higher levels of organisation. This would occur if selection at the level of the individual has tended to fix traits within individual species, but differences remain between species.

Some palaeontologists have gone so far as to suggest that the overwhelming determinants of patterns in nature are not the tinkering which results from natural selection on individuals (microevolution), but those aspects of the history of life which determine the diversification and resilience to extinction of clades (macroevolution). Broad scale events may reset the competitive environment to such an extent that the emergence from an extinction proceeds in a near random fashion, inevitably favouring those taxa with a propensity to survive massive perturbation, and not those populations which have gradually accumulated special adaptations to their local environment (Benton 1987). There is little doubt that the history of animal life has been profoundly affected by a number of mass extinctions, characterised by startling declines in diversity, or at least the diversity of marine invertebrates. Estimates of the number of mass extinctions vary from five events of unquestioned significance (Fig. 3.7), to twenty-nine (Sepkoski 1986). The most famous of these extinctions took place at the end of the Mesozoic era about 65 m.y.a., and marked the demise of the dinosaurs and many marine invertebrates (Fig. 3.7). However, its severity was probably low compared to the events which terminated the Ordovician and Permian Periods (438 and 248 m.y.a. respectively).

Adaptations with clear significance can be detected from the fossil record. It is equally clear that some of these adaptations do not enhance survival across the boundary of a mass extinction. One of the better documented cases is the acquisition of anti-predator protection and predatory ability in molluscs. Carnivorous naticid gastropods developed

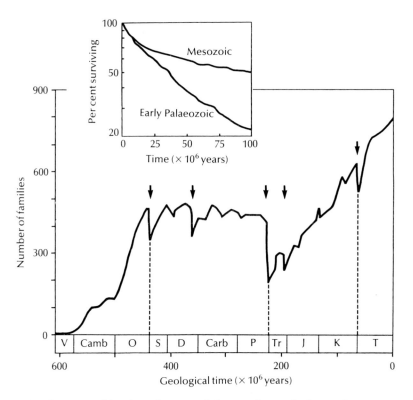

Fig. 3.7 Diversity of families of marine skeletonised animals during the Phanerozoic. Arrows denote unambiguous mass extinctions. The inset compares the survival of families during the radiation that took place through the upper Vendian and lower Cambrian, and the Mesozoic radiation after the Permian. V, Vendian; Camb, Cambrian; O, Ordovician; S, Silurian; D, Devonian; Carb, Carboniferous; P, Permian; Tr, Triassic; J, Jurassic; K, Cretaceous; T, Tertiary. After Erwin *et al.* (1987).

a shell-drilling habit in the late Triassic which would have greatly increased access to food, but the lineage was extinguished in the extinction at the end of the Triassic (Fürsich and Jablonski 1984). The shell-drilling habit re-originated in a related clade about 120 m.y. later. In Jablonski's (1986a) words, the naticids acquired the right adaptation in the wrong place at the wrong time.

Erwin *et al.* (1987) have compared the diversification of marine invertebrates in the first 182 m.y. of the Palaeozoic (late Vendian until the Ordovician extinction), and in the 183 m.y. from the end of the Permian until the Cretaceous extinction. The intervening period was a time of comparative stability in familial diversity (Fig. 3.7). They point out that the Palaeozoic diversification was a time of origination of many new fundamental grades of organisation (*Baüplane*), many of which became extinct and have not been replaced by anything which remotely resembles them (Fig. 3.8). In contrast, the Mesozoic radiation saw few new forms of organisation. Further, the rates of extinction declined dramatically (Fig. 3.7). This topic will be elaborated further in Chapter 9, though a possible explanation for these differences is a progressive increase in genetic resistance to restructuring (Wright 1982), and an increased resilience to extinction (Flessa *et al.* 1986).

Specialise and perish?

The contemporary record of extinction is so dominated by our own actions that our knowledge of natural causes of background extinctions remains extremely rudimentary (Simberloff 1986). Our ignorance not only frustrates attempts to manage natural communities (Chapter 10), but also weakens our ability to comment on background extinctions.

One of the predicted courses of adaptation will be specialisation for the local environment (stenotopy) (see Chapter 5). If the environment changes rapidly it is widely believed that stenotopic species will be more prone to extinction than generalist, or eurytopic species. Vrba (1984) contrasts the evolutionary history of two sister groups of antelope, the Alcephalini (wildebeest, hartebeest, blesbuck and their relatives) and the Aepycerotini (impalas). Impalas are generalists, and have persisted longer the alcephalines, but have shown little diversity in form or number of species. By contrast, alcephalines show high rates of speciation and extinction, and great diversity of form, which in part reflects their tendency to specialise (Fig. 3.9). Vrba suggests that the lack of specialisation for particular foods in impalas increases their resilience to environmental perturbation, minimising the probability of extinction, a species-level character. Nonetheless, there is little evidence that the clade-specific properties of stenotopy and eurytopy have provided either of the sister groups with an evolutionary advantage which can unambiguously be described as species selection.

A related example at a different scale concerns the relative long-term success of parthenogens and sexually reproducing species. There is abundant evidence that parthenogenesis has evolved among animals

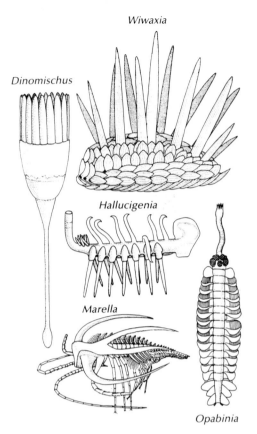

[79]

CHAPTER 3
*The power and
the units of
selection*

Wiwaxia

Dinomischus

Hallucigenia

Marella

Opabinia

Fig. 3.8 Some of the bizarre metazoan fauna discovered from the Burgess Shale. *Marella* is an arthropod, the other groups lack obvious affinities to extant phyla. After Conway Morris (1977, 1979, 1985).

hundreds or even thousands of times (Bell 1982), yet there is only a single example of a taxon above the level of subfamily which is exclusively parthenogenetic (the bdelloid rotifers, a class containing four orders, which Maynard Smith (1986) describes as 'something of a scandal'; Fig. 3.10). In sharp contrast, haplodiploidy, where males are produced from unfertilised haploid eggs and females are produced from fertilised diploid eggs, has arisen only a few times, but has led to several cases of evolutionary success, of which the Hymenoptera are the best example (Maynard Smith 1984). The long-term costs of parthenogenesis may involve both the accumulation of deleterious mutations and a slow response to environmental change (Williams 1975; Maynard Smith 1978), but this pattern does provide some of the best evidence for species selection.

Selection can produce sustainable trends through geological time

Despite these arguments, there do appear to be unambiguous cases of substantial long-term change which can only effectively be explained

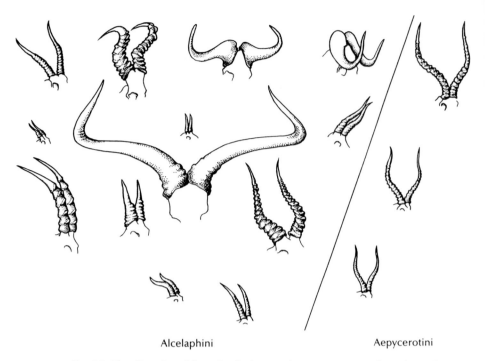

Alcelaphini   Aepycerotini

Fig. 3.9 The diversity of form in the horns of two sister taxa of antelope, the Alcelaphini (wildebeeste, blesbok, etc.) and Aepycerotini (impalas). After Vrba (1984).

by selection. At a small scale, consider the extensive fossil record of trilobites from a shale deposit in central Wales which spans about three million years in the Ordovician. One of the characters which has traditionally been used to separate species and even genera of trilobites is the number of pygidial ribs, unfortunately a trait of unknown function. Sheldon (1987) showed that all eight genera in the deposit showed a net increase in the number of pygidial ribs through time, though there were some temporary reversals in the direction of evolution, and probably several changes of rate (Fig. 3.11). These results are compelling for two reasons. First, they document parallel change in unrelated taxa, minimising the prospect that these changes are not driven by the same directional force, in this case selection on some character correlated with the number of pygidial ribs. Second, they illustrate one of the grave difficulties in using the fossil record as a basis for attributing importance to the process of speciation. Within the lineages depicted in Fig. 3.11, the taxa at the bottom and top of this deposit had been classified as different species or even genera. Yet when the data were greatly expanded it was impossible to detect either speciation or punctuation. It may be that the convenience of the Linnaean system of classification is inappropriate for fossil taxa, and that measurements of the duration of taxa at all levels may be biased greatly by the artificiality inherent in this pre-Darwinian system (Maynard Smith 1987a).

[81]
CHAPTER 3
*The power and
the units of
selection*

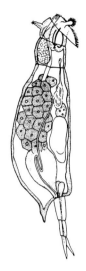

Fig. 3.10 'An evolutionary scandal' — *Philodina roseola,* a member of the only exclusively parthenogenetic higher taxon in the Metazoa, the bdelloid rotifers. After Barnes (1987).

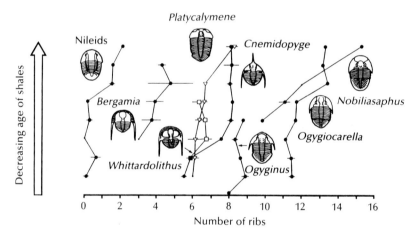

Fig. 3.11 Parallel changes in the mean number of pygidial ribs in eight lineages of trilobites from a fossil deposit in central Wales. Horizontal bars denote 95 per cent confidence limits. After Sheldon (1987).

Across a larger time scale, the best evidence for progressive improvement comes from the vascular plants, a group which has often been neglected in some of the histrionic debate between the neo-Darwinian perspective and the palaeozoology-inspired hierarchical view of evolution. Land plants emerged during the Silurian, increasing in diversity until the mid-Carboniferous. Diversity stabilised until the mid-Cretaceous, when a period of near exponential increase was initiated, caused exclusively by the diversification of a single group, the angiosperms. In sharp contrast to the record for marine invertebrates, there is no evidence of mass extinctions (Knoll 1984). There is,

[82]

CHAPTER 3
*The power and
the units of
selection*

however, repeated replacement of one higher taxon by another, inevitably associated with an improvement of either the vascular architecture upon which plants depend for water uptake and transport, or the efficiency of fertilisation, upon which sexually reproducing species depend for reproduction (Fig. 3.12).

SUMMARY

Natural selection is an obvious consequence of some of the fundamental consequences of life, and will be a potent force in shaping evolutionary history and in promoting adaptedness. Organisms are not designed from scratch, but depend on unique and unpredictable events which occur within existing lineages. These unique events and the design considerations that they impose constrain evolution. Selection will operate wherever the conditions of variation, heritability and fitness differences are satisfied. The plausibility of selection at each of the levels in the hierarchy for life has been satisfied, and there is no

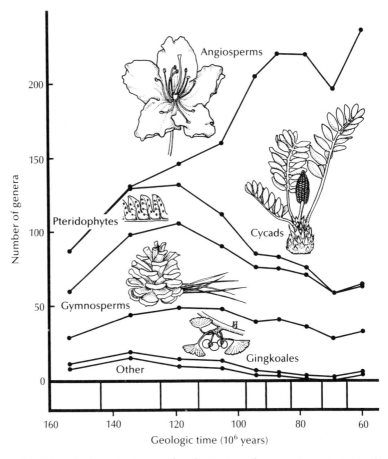

Fig. 3.12 Generic diversity in vascular plants since the upper Jurassic. Note the decline in all groups at the expense of the spectacular radiation of the angiosperms. After Knoll (1984).

[83]

CHAPTER 3

*The power and
the units of
selection*

justification for maintaining that a single metaphor for the unit of selection is biologically appropriate, or of any heuristic value. However, the assumptions which restrict the operation and power of selection above the level of the individual need to be clearly understood.

FURTHER READING

Dawkins (1981) is a forceful advocacy of the view that the gene is the level of selection, and Lewontin (1983) is the most readable criticism of this view. Buss (1987) is a fascinating development of the idea that the cell may be the unit of selection. Levinton (1988) is a comprehensive review of the evidence for selection at the level of the species and the clade, written by an unabashed neo-Darwinist. Raup & Jablonski (1986) is an excellent set of articles from opposing viewpoints on the significance of species selection and the distinction between microevolution and macroevolution.

TOPICS FOR DISCUSSION

1   Gould (1989a) gives the flightlessness of elephants as an example of an historical rather than a formal constraint. Why?
2   Lepidopteran larvae which taste obnoxious often have bright (aposematic) coloration, presumably because it serves to warn predators that they are about to get a nasty surprise. If birds have to learn to avoid bright colours, the caterpillar which is sampled by birds as part of its learning process may die or be removed from its host plant. Discuss why these caterpillars do not 'cheat', producing rapid evolution towards cryptic coloration characteristic of most other caterpillars.
3   Try and find examples of traits which show no intraspecific variation yet show interspecific variation. Could any of these influence the propensity to speciate or to become extinct?

Chapter 4
Studying Selection and Adaptation

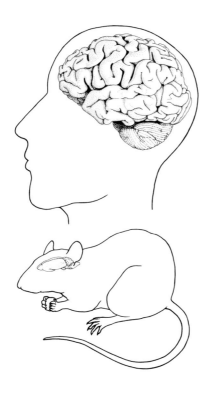

There are numerous useful approaches to the study of selection, adaptation and adaptedness. The best approach will depend largely on the question of interest. Because the remainder of this text will repeatedly make recourse to various techniques, it is crucial to understand the broad classes of approach, the circumstances to which different forms of analysis are well suited, and the pitfalls of each.

As we have already seen, a great deal of confusion arises because analyses of evolution fall into two classes, those which probe process, and those which interpret the results of evolution. Neither has any intellectual supremacy, and both are essential to the progress of evolutionary biology. Nonetheless, certain questions are more amenable to analysis in terms of either process or result. It is of paramount importance to always understand which approach is being used.

For example, adaptedness is a result of natural selection. Plant and animal physiology are good examples of subdisciplines within biology which explicitly study adaptedness, yet almost never study adaptation and selection, even though these processes are implicit in almost all interpretations (for some rare exceptions see Koehn & Hilbish 1987). The debate which arose in the 1970s over adaptive interpretations of social behaviour is illustrative of how tensions may arise between the alternative approaches. Adaptedness was often presumed rather than demonstrated, and process ignored by advocates of genetic determinism, while the critics of sociobiological arguments often assumed that the inability to demonstrate process diminished the arguments about result.

I wish to consider the circumstances in which three classes of technique are useful (the optimality approach, the comparative method and the statistical description of selection). These techniques span the spectrum from concern with adaptedness, and hence the results of selection, to selection in the present, and hence process.

In doing this I do not wish to downplay the importance of the techniques which form the heart of any portrayal of the scientific method — careful observation and experiment. However, I wish to show how these other methods both focus our attention on which of the parameters in the morass that might be measured are likely to reward the investigator, and refine our definition of hypotheses worth submitting to experimental scrutiny.

But first two asides. Throughout this text I will be using the term model, a word which occasionally summons up horrific visions of algebraic madness. In fact the word model has a variety of contexts, all of which are useful to an aspiring ecologist (Box 4.1). It is also important to comment briefly on the place of experiments in evolutionary biology. There are two persistent misconceptions. First, because much of the empirical study of evolution is not based on controlled experiment, it is somehow inferior science or not science at all. Subdivisions of evolutionary biology differ in the extent to which experiments can be applied. For example, palaeontology is largely descriptive, yet is a vibrant science making a profound contribution to evolutionary theory (Gould 1980b, 1980c).

Box 4.1 One of the most confusing words in the ecological literature is model. The term is used to describe four distinct approaches, all of which are immensely useful, but which are based on very different aims, assumptions and knowledge.

Data-free models. We can explore the behaviour of many aspects of nature by examining the mathematical consequences of certain assumptions. Such exploration may be entirely free of relevant data, and rooted only loosely in biological reality. While many empiricists are profoundly sceptical about the conclusions drawn from such analyses, it is important to remember that many of the conjectures we make about nature are implicitly mathematical in content, and mathematics is the appropriate tool for the exploration of these conjectures. However, we must also remember that while the mathematics is likely to be correct subject to the axioms on which it is based, and therefore not open to question, the biological relevance of such theoretical speculation can only be deduced from its ability to provide new insights about the real world, or to suggest limits to the behaviour of systems that could not be deduced empirically.

Data-rich models. By contrast, we often understand or suspect we understand the nature of basic empirical relations (variable y increase linearly with variable x), and wish to see how well our knowledge meshes with theory, or which additional insights can be derived from the data. In this case the use of mathematics is to examine the interplay of the various forces we believe are in operation. The optimisation models described in this Chapter are often of this sort. In is normally necessary to make explicit statements about the data in order to subject it to scrutiny of this sort, and this can be inordinately valuable in helping us examine our belief critically.

Statistical models. On the other hand, ecologists are often faced with a plethora of data but a dearth of theory or concepts with which to interpret that data. We can use statistical procedures to derive useful predictive equations on the basis of that data without understanding the mechanistic basis of that relation. For example, we know that brain weight in mammals is related to body weight according to the formula

log (brain weight) = constant + 0.75 log (body weight)

so we can predict the approximate weight expected for a given mammal, even though it is not immediately obvious why the coefficient is 0.75, rather than say 0.67. We can describe the behaviour of very complex systems like weather in this way.

Indeed, it is extremely instructive to think of most statistical analysis as an exercise in modelling, and to learn to see procedures usually treated separately (e.g. t-tests, linear regression) as part of a general model which can be viewed in a common framework (McCullagh & Nelder 1983). Inevitably the subset of models we use make assumptions about the data. For example, in conventional

(Box 4.1 continued)
linear regression of the brain weight/body weight relationship we
assume that we measure the explanatory variable (body weight)
without error, the variance in the dependent variable (brain weight)
does not change as the explanatory variable changes, the error in the
data is normally distributed, and the data points are independent.
These assumptions are not met, so alternative procedures are
necessary (Pagel & Harvey 1988a, 1988b).

Empirical models. The last common sense in which I will use a
model is to describe organisms whose unique characters and
evolutionary history make them extremely well-suited to probe
empirical questions. Those of you who have tackled the study
questions at the end of the preceding Chapters will realise that
Drosophila was originally chosen as a study organism because the
size of the chromosomes in its salivary glands allowed unusual
resolution of the genome.

Second, it is sometimes suggested that because long-term evolutionary processes like adaptation rely on a series of unique historical
events which cannot easily be reconstructed, controlled experiments
cannot be brought to bear on those processes. A full review of the
experimental approach is beyond the scope of this text, though the
experiments I report should dispel this facile notion. I do wish to
reinforce the lurking suspicion that a well-designed experiment is a
more precise way of drawing reliable conclusions than is possible from
either description, abstract modelling, or the comparative method.

OPTIMISATION

The mechanistic theory of genetics is immensely powerful in allowing
us to model the process of evolutionary change. It says much less
about the outcome of the process of selection, which I have already
argued will be so important in determining the pattern that we observe in
nature. A body of theory has recently emerged which largely ignores
process, but instead inquires what is the best achievable outcome
within the body of constraints which restrict the range of possible
phenotypes. This theory relies heavily on a branch of mathematics
called optimality theory. Optimality theory was developed for application in economics, but has proved useful in biology. This is because
biological problems often reflect the same sorts of trade-offs which
recur in economic decision making. For example, given that two
essential tasks cannot be carried out concurrently, what is the optimum
allocation of resources to each? Or given that a problem can be approached in several ways, which method of tackling the problem produces the greatest rewards?

Optimality models usually rely on three sorts of assumptions
(Stephens & Krebs 1986). First, constraint assumptions, or what are

the range of possible tactics which should be permitted in the analysis? It is at this stage that the range of biologically feasible options is determined. Second, currency assumptions, or how are choices to be evaluated? Do we count a currency of fitness, such as the number of offspring produced, or a currency we presume to be related to fitness, such as energy gain? Third, decision assumptions, which ask what is the problem to be solved, or what is the criterion of optimisation? We run the risk of being outrageously wild of the mark wherever it is necessary to make assumptions in formulating theoretical predictions. However, one of the strengths of the optimality approach is that these assumptions must be made explicit.

Once we have made these decisions it becomes important to decide what aspects of the decision assumptions should be allowed to vary. Simple models (static optimisation) assume that the optimum response of the animal will always be the same. This assumption is justified in many biological contexts, particularly in morphology. For example, the optimum wing shape of a bat or bird will be determined by comparatively simple aerodynamic principles. However, in many behavioural contexts, the state of the animal will change. For example, its knowledge of the environment may improve with learning, or it may be prepared to risk predation if it is starving than if it is has just had a large meal. The appropriate technique is dynamic optimisation wherever the state of the animal is variable.

The optimum solution may depend less on the physical problems of the environment than on the behaviour of other organisms with which an animal interacts. A simple example is the sex ratio. Producing exclusively male offspring may provide extraordinary benefits if the rest of a large population produces female offspring, yet the same tactic may be counterproductive when the population largely comprises other producers of sons. This is an example of frequency-dependence. The appropriate subset of optimisation theory is called game theory, which in evolutionary applications can be used to seek a strategy which cannot be invaded by alternative mutant strategies. Any action which is immune to invasion so long as most of the population follow that action is called an evolutionarily stable strategy, or ESS. As for frequency-independent modelling, game theory may rely on static or dynamic optimisation, though the mathematics used in dynamic game theory are rather intractable, restricting the current utility of this approach.

How long to stay in a patch: static optimisation

Consider an animal which forages in patches of habitat. Its own actions exploit the resources of the habitat and the difficulty in finding food causes the rate of return to diminish, forcing it to seek food elsewhere. This problem was originally treated by Charnov (1976), but has been analysed many times since (Stephens & Krebs 1986). We want to predict the point when it is worth the animal leaving the patch to forage elsewhere (this is the patch residence time, which we assume to be the decision of importance). We make the currency assumption that the goal is to optimise the long-term average rate of energy intake. Then we impose some constraint assumptions to solve for the optimum solution:

1 Expected energy gain within a patch is related to residence time by a well defined gain function $(g\ (T_r))$, where T_r is the residence time within the patch. It is usually assumed and often demonstrated that rewards are initially gained rapidly, but then level off as the patch is depleted (Fig. 4.1). They may even decline if energy expended in searching exceeds the energy obtained from foraging. However, alternative energy gain functions are biologically plausible and confirmation of the energy gain function should not be ignored in empirical testing of the model.

2 Searching for patches and searching for prey within patches are mutually exclusive.

3 Now to the biological assumption. We assume that the animal has complete information. It knows the gain function, the search time, and does not acquire and use additional information as it forages.

Rather than solve this problem algebraically, we will explore the problem graphically by imposing a number of additional simplifying assumptions.

4 There are no costs of searching other than the travel time between patches, so that the rate of energy return on departure from a patch is always given by $g(T_r)/(T_r + T_t)$, where T_t is the travel time between patches.

5 All patches have the same gain function.

The slope of any line which begins at T_t and intersects the gain curve measures the rate of gain to the point of intersection. The greater the slope, the higher the rate of gain, and it is easy to show that the optimal T_r can be found by drawing a line that intersects the gain curve tangentially (Fig. 4.1). Clearly, animals should depart from the patch before it is completely depleted. We can also easily demonstrate that as T_t increases, the optimum T_r will also increase (Fig. 4.1).

One of the most fascinating attempts to demonstrate the accuracy of this model in predicting behaviour quantitatively came not from foraging for food, but from the copulatory behaviour of dungflies, *Scathophaga stercoraria*. Parker conducted a series of experiments with sterilised flies to measure the proportion of eggs of a given female they were able to fertilise if they prolonged copulation. Female flies are fickle beasts, so the male must adopt tactics which maximise the probability that his sperm are not wasted. Copulation would need to last at least 100 minutes to fertilise all the eggs, but the rewards per unit time diminish towards the end of the copulation (Fig. 4.2). In addition, these benefits need to be set against the costs of guarding one female and setting off to find another, which in combination are functionally equivalent to T_t. Parker (1978) found that guarding and searching took on average 156.5 minutes, and so was able to predict

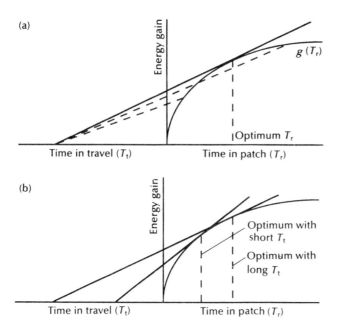

Fig. 4.1 The optimum residence time in a patch (T_r) depends on the gain function $g(T_r)$ and the travel time between patches (T_t); (a) because the rate of energy gain is given by $g(T_r)/(T_r + T_t)$, or the slope of the lines from T_t intersecting the gain function, the returns are maximised by the line that intersects the gain function tangentially; (b) for small T_t the patch should be abandoned earlier.

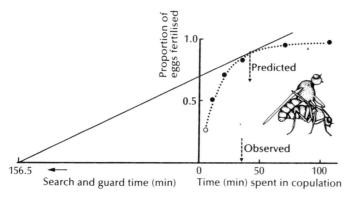

Fig. 4.2 The duration of copulation by a male dungfly (*Scathophaga stercoraria*) predicted as a consequence of the proportion of eggs fertilised as a function of copulation time, and the time it takes to search for and guard a female. After Krebs & Davies (1987).

that copulations should last 41 minutes. The observed time was only slightly less, suggesting that dungflies are following an optimisation rule which maximises their long-term fertilisation rate. This example illustrates one of the strengths of optimisation theory: the logic may be transferred to different contexts provided that the assumptions can be transferred. Indeed, the copulation data are probably more compatible with the assumption of patch invariance than the foraging data for which the technique was developed.

Hunting in groups: understanding dynamic optimisation

Felids usually forage alone. The only exception are African lions *Panthera leo* which live in prides, from which female groups ranging in size from one to eight may cooperate to hunt prey, usually large ungulates such as gazelle, zebra or wildebeeste. Paradoxically, there is little evidence that foraging in large groups enhances the food return to individuals (Caraco & Wolf 1975; Packer 1986; Packer & Ruttan 1988). Observations of foraging show that returns are always highest for pairs, so it seems possible that lions often forage in groups which provide suboptimal rates of return.

Houston *et al.* (1988) and Clark (1987) use the following dynamic model, where the state of the animal depends on its hunger, or the number of kilograms of meat in its stomach. The animal starts each day with a reserve of energy X_t on day t. X has both an upper bound (the capacity of the stomach, about 30 kg in lions), and a lower bound (zero). When X_t reaches this lower level we assume that the lion dies.

Energy reserves are increased by daily food consumption Z_t and decreased by daily energy expenditure a (the minimum daily intake in lions is about 6 kg), so that reserves at the start of day $t+1$ are given by:

$$X_{t+1} = X_t - a + Z_t$$

where Z_t is a random variable whose distribution is determined by the animal's choice of prey (a zebra contains about 164 kg of lion food, a gazelle about 12 kg), the size of the group sharing the prey (n, the food content E is shared equally) and the probability of capture of that prey. If prey H_i is chosen, then Z_t equals E/n with probability p, and zero with probability $1 - p$. The probability of capture of the prey will also be influenced by the group size (corrected for multiple kills, $p_{gazelle} = 0.15$ for singletons; 0.37 for groups; $p_{zebra, dry season} = 0.15$ for one lion; 0.33 for pairs; 0.40 for fours; 0.43 for six or more). We assume that the currencies to be optimised are the maximisation of survival probabilities over 30 days, and wish to solve for the group size at which this probability is maximised. We now assume that the decision is based on risk minimisation rather than energy maximisation. The solution is provided by a tool called stochastic dynamic programming (see Houston *et al.* 1988 for a description). Rather than work through a set of calculations I shall summarise some of the key results which illustrate the unique predictions which arise from a dynamic approach.

First, the optimum group size depends on an animal's current state as well as the prey type and habitat. Where the prey is large and the lion has a fully belly, large groups often provide higher 30 day survival probabilities than small groups (Fig. 4.3). This is because of the low kill probabilities and the inability of a pair to use all the meat on a

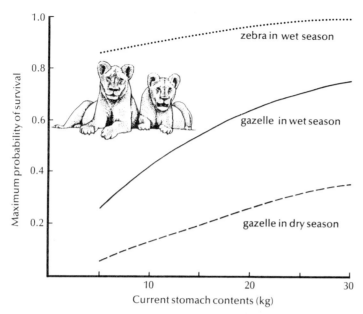

Fig. 4.3 Maximum probability of survival in lions as a function of current stomach contents when feeding on different prey at different times in Serengeti habitat. Optimum group size reaches six for satiated lions feeding on zebra, but is much lower for hungry animals and lions feeding on gazelles (one or two). Lions are lazy. Data from Houston *et al.* (1988).

zebra before it rots. The model also predicts that hungry lions should hunt in smaller groups than well-fed lions. When prey is scarce, prides may break up. Laziness therefore influences the benefits of group hunting (Clark 1987).

Packer (1986) argues that group foraging is a consequence rather than a cause of sociality, and that one advantage of foraging in groups is the enhanced opportunity to scavenge kills from conspecifics, which reduces the variance in food intake, reducing the prospect of starvation. Houston *et al.* (1988) consider the case where a large pride breaks up into smaller groups to hunt, but all the pride share the kill. Thirty-day survival probabilities for a lion which hunts in a small group (pairs) but shares large prey (zebra) with two other such groups is 0.95, while the comparable figure for lions which do not scavenge is only 0.42, strongly supporting Packer's argument.

Of course, this concordance does not mean that lion sociality has been explained in terms of foraging advantage. Although offering more sophistication than previous models of foraging strategies, the analysis of Houston *et al.* (1988) is limited in two important ways. First, it ignores other advantages of group living. One clear advantage gained by female lions living and moving in groups is the prospect of joint defence against infanticide by males, which routinely kill offspring they are unlikely to have sired (Packer *et al.* 1988; Packer *et al.* 1990). Second, long-term maintenance of a territory appears to be vital for reproductive success, yet it is occasionally necessary to leave the territory to forage during periods of prey scarcity. A single lion foraging on a neighbour's territory is much more likely to be wounded or killed than a member of a group (Packer *et al.* 1990).

The hawk–dove game: understanding game theory

Supposedly polite fighting was one of the examples frequently used by advocates of naive group selectionism. Animals equipped with formidable weaponry often fail to use that weaponry in contests. This anomaly can be examined using the classic application of game theory to animal behaviour. Consider the hypothetical behaviour of two animals seeking to use a common resource (e.g. a territory, piece of food, etc.). In the simplest model we compare two alternatives (Table 4.1). Hawks always fight to gain access to the resource. Doves always defer to a hawk but share the resource equally with another dove. First we consider a population of doves. If the value of the resource is B, then the outcome of each dove–dove interaction will be $B/2$, as the resource is shared. Now consider a mutation to hawks. While hawks are rare, they will interact chiefly with doves, and will gain B from the poor doves, who receive nothing from each interaction. Therefore hawks can invade the population, so dove can never be an ESS. Now consider a population of hawks. The payoff from the hawk–hawk interaction will be influenced by the cost of fighting. If one hawk wins and gains B, yet the other loses and suffers a cost (C; e.g. blinded in one eye, an infected scratch, a broken antler), then the average payoff will be $(B - C)/2$. By contrast, a rare dove never gains the resource from a hawk, but suffers no cost of fighting. The payoff to the rare dove remains zero, which will be greater than the payoff to the hawk if $C > B$, or when $(B - C)/2 < 0$. Therefore a population of hawks can be invaded by a rare dove phenotype under certain conditions, and hawk is only an ESS when $B > C$. Under the conditions $C > B$ the populations should equilibrate when dove

Table 4.1 The game between hawks and doves. The winner of any contest gains a benefit B, and the loser of an escalated fight suffers a cost C. When doves meet they share the benefit. Values are provided for one case where $C < B$ ($B = 50$; $C = 100$). As we are interested in the susceptibility of a given strategy to invasion, the opponent may be viewed as the common strategy within the population, and therefore the most likely opponent for all participants, including the invader. When $B > C$ dove does better than hawk in a population of hawks, but hawk does better than dove in a population of doves. After Maynard Smith (1982).

		Opponent	
		Hawk	Dove
Attacker	Hawk	$(B/2) - (C/2)$ -25	B $+50$
	Dove	No benefit or cost 0	$B/2$ $+25$

reaches a frequency given by $p^* = 1 - B/C$. Because both phenotypes are present at this equilibrium frequency we say there is a mixed ESS under this limited contrast, or game. This equilibrium can be obtained in two ways. Either a proportion p^* of animals could always play hawk and another proportion $(1 - p^*)$ could play dove, or each individual should play hawk with a probability of p^*. Sex ratios illustrate these alternatives. Most mammals produce both sons and daughters at the evolutionarily stable frequency $p^* = 0.5$. However, in some isopods and cirripedes the sex ratio is determined by the balance between individuals which produce mainly sons and those that produce mainly daughters (Maynard Smith 1982).

The complexity of the game can be increased by considering alternative strategies in the same way. We first identify a rule which determines when one animal plays hawk and another plays dove. This strategy (bourgeois) is to play hawk when on familiar ground (perhaps a territory) and play dove when on unfamiliar ground (a neighbour's territory). The payoffs are shown in Table 4.2. Bourgeois is immune to invasion from both hawk and dove, and is a pure ESS in this game.

Several insights into fighting behaviour emerge from these models. First, they demonstrate that polite fighting may be evolutionarily stable from an individual perspective. Second, they cast some light on the frequently observed pre-eminence of territory owners in contests over resources. Territory owners may win over intruders for several reasons. The simple explanation would be greater fighting ability (or resource holding potential). Because a territory owner must have some fighting prowess to acquire a patch, it pays an intruder to be deferential.

Territory owners may also have greater motivation than intruders, as they have invested more in learning the intricacies of the territory. For example, when territorial great tits are removed experimentally, the territories are rapidly usurped by intruders, as the population

Table 4.2 The effect of adding a third strategy bourgeois to the hawk–dove game. The payoffs are identical to those in Table 4.1. When bourgeois meets either hawk or dove we assume it is owner half the time and behaves like a hawk, and intruder half the time and therefore plays dove. It therefore derives an average payoff. When bourgeois meets bourgeois it gains the reward if owner and loses it without cost if intruder. Bourgeois has higher fitness than hawk in a population of hawks, and higher fitness than dove in a population of doves. However, neither hawk nor dove do as well as bourgeois is a population of bourgeois, so bourgeois is a pure evolutionarily stable strategy (ESS). After Maynard Smith (1982).

		Opponent		
		Hawk	Dove	Bourgeois
Attacker	Hawk	−25	+50	+12.5
	Dove	0	+25	+12.5
	Bourgeois	−12.5	+37.5	+25

comprises both territory holders and a floating population of birds unable to acquire a territory. When the original owner is released, the outcome depends on the time it has been absent from its territory (Krebs 1982). Rapid release is followed by quick eviction of the intruder. However, if the intruder is given several days to familiarise itself with its new-found home it will usually expel the original owner. In inter mediate cases the outcome is less predictable, and fights are often escalated. In a similar result, Barlow *et al.* (1986) show that the pattern of fights in midas cichlids depends on how long fish have had to familiarise themselves with a tank in which fights are contrived. When fights take place soon after they have entered a tank, aggressive individuals willing to escalate a fight usually win encounters, and the fights are of short duration (Fig. 4.4). When fish have had a long period of ownership body size is much more important in determining out-comes than daring and aggressiveness. Correlated asymmetries of these sorts are easily understood. However, the novel aspect of the hawk–dove-bourgeois game is the prediction that uncorrelated asymmetries are also stable, with the territory owner winning for no other reason than respect for a rule which avoids costly fighting.

Further complexity can be introduced by allowing the state of an animal to vary (dynamic game theory). There is every reason to believe that the state of the organism will also be influential in determining its approach under conditions where frequency-dependence is important. Dynamic game theoretical models should therefore be of inestimable value. Unfortunately the mathematics and computational demands are very formidable, so formal applications of theory are rather restricted. Returning to the example of aggressive behaviour, it is obvious that the cost of abandoning food or some other resource will be greater for a starving animal than a fat one (Grafen 1987a). Houston & McNamara (1988) have modified the hawk–dove game to include a variable that represents the animal's energy reserves, and ask which strategy maxi-mises survival depending on the animal's energy reserves and the time of day. The result of modelling is a pure ESS: play hawk if reserves are below some critical level at a given time of day; otherwise play dove. This critical level increases with time of day, as animals become more desperate to acquire reserves to ensure they survive a period of starvation during the night.

Game theory has been applied to a very wide range of problems, and offers many useful insights (Maynard Smith 1982 provides a very readable review). One reasonably general result is that a pure ESS is a common outcome of a game between two contestants; by contrast a mixed ESS often arises when an individual is playing against an entire population instead of a single individual.

Last, it is worth voicing some caution. Alternative behaviours are common, particularly in mating systems. There is no reason to believe that the existence of alternative behaviours implies a mixed ESS. An important result of game theory is that alternative behaviours should have equal payoffs. The other source of alternative behaviours is making

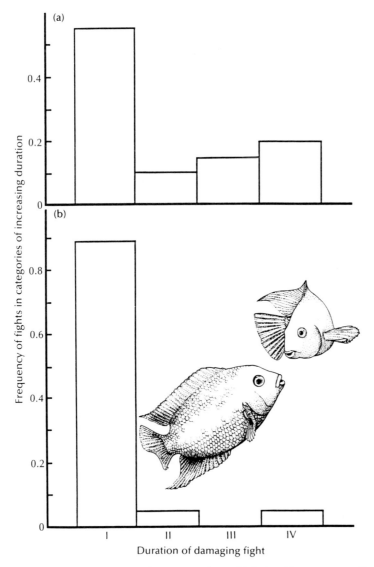

Fig. 4.4 The duration of combats between Midas cichlids (*Cichlasoma citrinellum*) in fights between fish that (a) had been allowed 24 hours to familiarise themselves with a territory, and (b) those which were allowed only one to two hours. After the long exposure, heavier individuals won encounters but aggression did not appear to be important. However, after the short exposure the most aggressive individuals usually won an encounter, though those encounters were of short duration. After Barlow *et al.* (1986).

the best of a bad job'. Consider the mating system of red deer, where males defend females in a harem. Some males will be unable to attract or defend a harem, but may gain small but significant payoffs from attempting to sneak copulations with females within a harem while the dominant male is otherwise occupied. There is no evidence at all that these males do as well as the holders of harems.

Optimisation theory has been the butt of some scathing criticism. Some critics argue that because evolution is dominated by the formal and historical constraints documented in Chapter 3, optimisation is unattainable and scarcely worth investigating. Constraints are undeniably important, yet no optimisation model can be developed without formal definition of a set of assumptions about constraint (the phenotype set). Optimality procedures therefore are unusually explicit in dealing with the problem of constraint. However, as Lewontin (1987) correctly points out, if all the constraints and currencies are correctly known, the evolutionary processes would be fully understood so the optimality approach would be rendered unnecessary.

A much more telling critique of optimality theory concerns the occasions where the model works poorly; how can we interpret a poor or even halfway reasonable fit between prediction and observation? There are several types of explanation. First, the goal, the currency or the phenotype set may be wrong. Second, the population may be behaving suboptimally, either because it is lagging behind environmental change, because it is constrained from reaching the optimum, or because doing well enough is good enough, so precise optimisation is unnecessary. Third, we may be experiencing the difficulties in measurement inevitable in working with the living world. The failure of the model is of great heuristic value if we recognise correctly that either of the first explanations are true. However, we lack a reliable general method drawing us to that conclusion.

Consider one of the most compelling integrations of game theory with real data. Riechert (1986) has used encounter experiments and observations to examine the costs and benefits of territorial defence behaviour by the spider *Agelenopsis aperta* in two quite different habitats (Fig. 4.5). These spiders are always territorial. Spiders living in desert grassland suffer from low prey availability and a limited 'window' of temperatures in which they can forage. There is very little habitat (perhaps only three per cent of that available) which affords conditions permitting reproduction. Unsurprisingly, spiders actively favour these sites, and compete for them fiercely. By contrast, in riparian habitats, there is abundant suitable habitat. These differences in the availability of breeding habitat may lead to very different costs and benefits from fighting over territories. Because there is genetically determined variation in the tendency to fight (Maynard Smith & Riechert 1984), this system seems ideal for application of the game theoretical approach to real data.

Hammerstein & Riechert (1988) have attempted to find an evolutionarily stable strategy for fighting behaviour in the two habitats. Spiders could exhibit three behaviours once they assessed their size relative to their opponent and the value of the site over which they were fighting: withdraw, display or escalate. In order to complete the ESS analysis it proved necessary to gather the following data: 1) the

*Fig. 4.5 Agelenopsis
aperta.*

distribution of body weights in the population; 2) the probability that a
territory owner will win a random contest; 3) the rate of encounter
with conspecifics; 4) the cost of biting; 5) the cost of display; 6) contest
durations in different contexts; 7) predation rate on spiders active on
their web-sheet; 8) total time spent in disputes, building webs and
handling prey; 9) daily rewards in egg production from holding a
particular territory or lacking a territory; and 10) the availability of
sites. The general theoretical result is that spiders should respect
ownership unless the discrepancy in weight exceeds 10 per cent. In
this case the larger individual should win again without serious fighting.

The desert grassland spiders behaved in concordance with the model.
In addition, the costs of escalation in this habitat are rather small, so
fights over high quality territories should occur. Fights are most often
between a smaller owner and a larger intruder, probably because owners
of excellent sites are unusually strong for their size.

By contrast, in the riparian habitat high-quality sites are readily
available, and fighting should be very costly because spiders fighting
on an open web sheet incur a substantial risk of predation. However,
display and escalation occur much more frequently than the model
would predict. Hammerstein & Riechert (1988) argue that it is unlikely
that the predictions of the model err in any significant way. Instead,
they suggest that the apparently suboptimal behaviour is a result of
immigration into the riparian habitat from adjacent arid areas. If this
interpretation is correct, spiders living in the riparian habitat cannot
be studied as an isolated subpopulation, and attention must be paid to
the habitat landscape in interpretation of behaviour. A nice aspect of
this example is that an optimality model of result has produced a
prediction about process which should be testable using the tools of
population genetics. Indeed, we might also expect that the high-quality
habitat should be a net exporter of emigrants rather than a recipient of
immigrants, so direct work on population biology is also suggested.

THE COMPARATIVE METHOD

Biologists usually fall into one of two groups. Some use a single
species as a model and either hope that their results may be generalised,
or alternatively do not care that their results cannot be generalised.
Others look at a diversity of species and compare them. Because the
information available from a variety of species is inevitably weaker
than it is for the chosen models (white rats, *Drosophila melanogaster,
Caenorhabditis elegans, Arabidopsis thaliana, Xenopus laevis* and
Tribolium castaneum spring to mind), the results of cross-species
comparisons often seem as crude to the specialists as the specialisation
seems naive to the generalists. There is obviously validity in both
approaches. Indeed, my aim in this section is to convince you that
when properly applied, the comparative method is not only a powerful
tool for making generalisations about adaptedness, but also one of the

best ways to identify good models for experiment or special attention.

In its simplest form, the way the comparative approach may be used to identify possible adaptedness is by detecting a correlation between characteristics of the environment (possible selection pressures) and a particular trait. For example, the repeated predominance of sex in marine habitats is suggestive of adaptedness, and the repeatable latitudinal clines in allele frequencies of alcohol dehydrogenase and esterase-6 in *Drosophila melanogaster* are indicative of both selection and adaptedness. Comparisons of this sort have been made many times and have had great success in elucidating the pattern of nature. Unfortunately, such comparisons are often bedevilled by difficulties.

Many of the problems can be illustrated with an example, the pattern of brain size among the mammals. Consider the following extreme case. The average brain size of marine mammals is much larger than the average brain size of terrestrial mammals. This observation might be used to argue that the marine environment selects for large brains. This argument is facile, and illustrates two problems with the interpretation of comparative data. First, it is trivially true that the brain size of mammals increases with the body mass of those mammals — it is hardly informative to report that a shrew weighing 2.5 g cannot have a brain weighing a kilogram. Because marine mammals include the largest of all the mammals, it is unsurprising that average brain size is great, and we may be deceived by an evolutionary trend towards large body size.

Second, the taxonomic diversity of marine mammals is not great. Whales and dolphins have descended from an ungulate ancestor, as have manatees and dugongs. The carnivores have given rise to the seals and to the sea otters. It may be premature to judge the direction of evolution until it is confirmed that marine ungulates and carnivores have larger brains than their terrestrial counterparts. The comparative method therefore must take into account two important factors, phylogeny and scaling.

Scaling and adaptationist analysis

The simplest way to approach the problem of scaling is to plot brain size against body size. As for many other parameters of interest (Chapter 3), the relation follows a power function, but can be made linear by logarithmic transformation of both axes. This logarithmic transformation has a second desirable consequence of reducing the variance about large values, helping satisfy the three critical assumptions on which the general linear model used in statistical analysis is based (Box 4.1).

Relations of this sort have been derived for very many traits, and have been used in two ways. First, we can describe the basic relation. Simple allometric functions of this sort often explain 80 to 90 per cent of the variance in life history traits of animals, and are hence regarded as adequate descriptors of the data. Unfortunately, interpretation of these basic relations have not proceeded as rapidly as their documentation. For example, we know that for many life history traits of mammals which are measured with a scale of time (e.g. gestation length, time at maturity, longevity), the exponent (b) falls between 0.1 and 0.4 (Harvey *et al.* 1989). We lack any theory which suggests why this should be so, and why exponents apparently differ. Some authors have argued that they do not, and that the 'pace-of-life' converges on an exponent of 0.25 (Lindstedt 1985). However, this argument appears based on optimism rather than statistics, and even were it true, an explanation for an exponent of 0.25 remains elusive (Harvey *et al.* 1989).

The second use of these relations is to identify significant departures from the basic scaling rule, departures which can be then subject to adaptationist analysis. Returning to the variation of brain weight with body weight in marine mammals, it is possible to show that across all species an exponent of 0.75 describes the basic relation very well (Fig. 4.6). The extreme exaggeration of brain size from the basic regression equation is found in our own species, *Homo sapiens*. The second largest relative brain size occurs in the dolphins. Yet the smallest brain size relative to body mass occurs in some of the baleen whales,

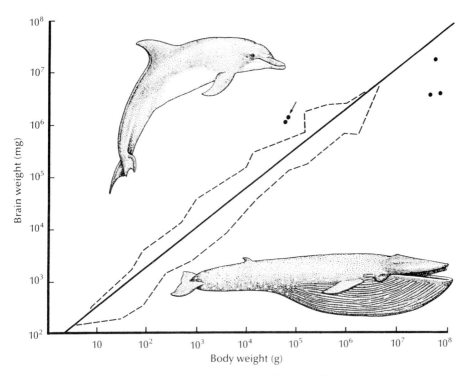

Fig. 4.6 Brain size in marine mammals. The regression line and ellipse are given for all mammals. Humans (arrowed) have the largest relative brain size, and the point nearest to humans is a dolphin. However, baleen whales have the smallest relative brain size. Data from Martin (1981).

probably because these animals have enormous amounts of blubber, rather than that they suffer from congenital idiocy. Clearly we are limited in our ability to draw conclusions about the relative brain size of marine mammals.

Disentangling phylogeny

The second problem of the comparative method is what to count. Closely related species often share characteristics of interest, such as small brain size in tenrecs or carnivory in sundews. Pagel & Harvey (1989a) list three reasons why closely related species are similar. First, unoccupied habitats are likely to be colonised by species that occupy similar habitats elsewhere. If speciation occurs in the new habitat then closely related species may remain similar for adaptive reasons, driven by the common selective environment imposed by their habitat. Second, there may be historical and formal constraints which restrict members of particular lineages from diverging in response to the selective forces operating on the different branches of the lineage. This phylogenetic conservatism is particularly pernicious in assessing the number of origins of a trait. Third, the response to selection will depend critically on the historical constraints on the species in question. Selection for fighting ability in males has lead to the development of antlers derived from bone in deer, horns derived from skin in sheep, and enlarged canines in primates.

We are interested in the number of times a trait evolves in response to particular environmental pressures, and therefore require a method which excludes the overestimation of trait/environment correlations because of common ancestry. In designing a method it is useful to distinguish between categorical characters, where the expression of the trait can be expressed as a series of alternatives, and continuous characters, where the trait is free to vary.

Categorical variables—cooperative breeding in Australian birds

Initially, categorical variables would seem to be treated quite easily. The simplest approach would be examine the correlation between the trait and a particular environmental feature. Cooperative breeding, where young birds fail to disperse and instead assist their parents to rear offspring, is of profound general theoretical interest, but theory has so far proved more effective in accounting for the behaviour of individual species rather than in accounting for differences between geographical areas (J.L Brown 1987). Cooperative breeding is rather common among the Australian avifauna (Rowley 1968; Dow 1980). Recent analyses have identified three important correlates within the Australian avifauna. Cooperation is most common in *Eucalyptus* forest and woodlands (about 30 per cent of all species) rather than rainforest (7 per cent) and desert (15 per cent), is much more common among insectivores than granivores, nectarivores or frugivores, and among insectivores, is most prevalent among ground-feeding species that pursue their prey rather than sit-and-wait predators (Table 4.3; Ford *et al.* 1988). The correlations in Table 4.3 assess the number of species. However, whole radiations are entirely cooperative and exhibit a particular feeding habit (e.g. the corcoracine mud-nest builders), while others do not co-operate and have a uniform feeding habit (e.g. *Rhipidura*) so that a simple count would certainly overestimate the correlation. Ridley (1983) proposes that we count the number of times that a trait has originated in association with a particular habitat condition. Examination of the 55 insectivorous species used reveals seven reasonably well-established clades in which there are cooperative and non-cooperative species (*Sericornis, Acanthiza, Climacteris, Lichenostomus, Pardalotus* and *Artamus*). However, in none of these cases is there a distinct association between foraging style and cooperation.

The picture is further clouded by the recent recognition that the Australian passerines can be divided into two distinct groups, an old endemic radiation (Corvida), and a group of immigrants from Asia (these include the other great radiation of passerines, the Passerida, plus some Corvida secondarily re-entering Australia) (Sibley *et al.* 1988). Cooperative breeding is confined to the old endemic passerines, not having developed in any of the groups which are comparatively recent invaders from Asia (Russell 1989). Indeed, many of the cooperative breeders in other parts of the world are descended from this old Australian stock. We could treat this difference in two ways. First, there may be a phylogenetic tendency towards cooperative breeding

Table 4.3 The distribution of foraging mode among 55 insectivorous birds living in *Eucalyptus* woodland near Armidale in eastern Australia. (a) Foraging sites (some species forage in both sites and are allocated 0.5 to each). (b) Foraging methods according to whether the birds actively pursue their prey (glean or actively pursue the prey), or sit-and-wait (snatch, pounce and hawk). After Ford *et al.* (1988).

(a)

	Breeding system	
	Cooperative	*Non-cooperative*
Ground	10.5	6.5
Leaves	7	13
Bark	3.5	3
Air	2	9.5

(b)

	Breeding system	
	Cooperative	*Non-cooperative*
Pursue	21	17
Sit-and-wait	2	15

amongst these old endemics. Second, the recent invaders have had less time to evolve in response to the Australian environment. This suggests that we need to take the length of time since the clades diverged into account (Pagel & Harvey 1989a). Third, although cooperation is strongly associated with *Eucalyptus* forest, the best records from fossilised pollen of the vegetation history of Australia suggests that the extraordinary prevalence of *Eucalyptus* is quite a recent phenomenon caused by changed fire regimes (Singh & Geissler 1985; Kershaw 1986). We are therefore left to guess whether the current dispersion represents contemporary selective pressures or the preadaptation of birds to new habitats (Heinsohn *et al.* 1990). The other dominant correlation, with insectivory, does not easily explain the paucity of cooperation among North American and Eurasian birds.

Continuous variables dichotomised—sperm displacement in insects

Variables distributed continuously pose even more problems, as the direction of evolution becomes very difficult to discern, though all the old problems of historical contingency remain present. One way to simplify the analysis is to treat the variables as dichotomous.

If female insects mate more than once, male fitness will be profoundly affected by the extent to which the sperm deposited at the first mating will be displaced by the second copulation. We can express the degree of sperm displacement as P_2, the probability that an offspring will be fertilised by the last male. Published estimates of P_2 vary from 0.02 in the mosquito *Anopheles gambiae* to 1.0 in several species, for example the butterfly *Colias eurytheme*. Ridley (1989) identified four hypotheses which try to explain this variation. First, it has been suggested that the predominant influence will be whether the male deposits a plug at the time of ejaculation or not, as the plug could prevent storage of the second male's sperm (Boorman & Parker 1976). Second, Walker (1980) has suggested that the shape of the female sperm storage organ, the spermatheca, will be influential. If the spermatheca is tubular or elongate, the sperm from the second male can push the sperm from the first male away from the opening which is used both for entry by stored sperm, and departure by sperm which will fertilise eggs. By contrast, this would be unlikely to arise in round spermathecae. Third, Gwynne (1984) argues that mating is sometimes costly for males (e.g. because the male feeds the female or guards the female). Such paternal investment is unlikely in species where the assurance of paternity is low. Because those males which cannot be sure that their sperm will have precedence over the sperm from previous matings will receive little benefit from this heavy investment, we might expect a positive association between paternal investment and high values of P_2. Last, it may be that the incidence of sperm displacement is a reflection of the natural tendency to mate and remate. If females typically mate only once in nature, it is unlikely that there would be evolution of male adaptations of sperm displacement, such as the complicated scraper of the male damselfly (Fig. 4.7). The pressures for remating are not one-sided. Females may use sperm and ejaculate as a source of nutrients, or perhaps to improve the quality of their partner, and hence their offspring (Partridge 1980). The latter effect could select for female adaptations to facilitate sperm displacement.

Ridley (1989) has contrasted levels of P_2 according to whether a mating plug is produced or not, whether the spermatheca is spheroid or tubular, whether male investment is high or low, and whether or not females typically remate in the field. If species are counted, three of the four hypotheses are supported quite strongly (Table 4.4). However, when only the independent origin of each trait is counted, only the mating frequency hypothesis is strongly supported. As Ridley (1989) rightly cautions, this does not mean the other factors have no influence, only that they are less important in dictating the evolution of sperm displacement.

Continuous variables—the pace of reproduction in mammals

In many cases it is neither practicable nor desirable to reduce variation to a dichotomy or trichotomy. There are currently two sets of approaches which treat data as a continuous variable, each of which have their difficulties. The first attempt to extract subsets of the data where most interest lies (Clutton-Brock & Harvey 1977; Stearns 1983; Cheverud *et al.* 1985). For example, we might use nested analysis-of-variance to determine the point at which variance is concentrated, decide that the biologically interesting variation is concentrated at the level of the family or order, and assume that such higher taxa have sufficiently different evolutionary histories to be considered independent. Read & Harvey (1989a) have recently considered the relation between life history variables in eutherian mammals. On average 63 per cent of the variance in life history traits was found among orders rather than within them. Across orders life history variables were correlated with body weight, as we would expect from allometric principles. There is therefore a correlation between many of the variables. For example, developmental periods are positively correlated, so animals with short gestation times also mature rapidly (Fig. 4.8a). Interestingly, this correlation persists after the effects of body weight has been removed statistically, so that the orders of mammals can be arrayed along a fast—slow continuum (Fig. 4.8b). We can also identify outliers, such as the Primates and Scandentia.

The problems with these methods are twofold. First, they discard much variation of interest, and are consequently potentially misleading. For example, Mace *et al.* (1981) concluded that nocturnal rodents had larger brains than diurnal rodents, but the level of analysis they had chosen (within families) ignored the families with the largest brains (such as the Sciuridae) which were exclusively diurnal. The second problem is that they make specific and untestable assumptions about the mode and rate at which evolution has proceeded (Pagel & Harvey

Fig. 4.7 The penis of the damselfly *Calopteryx maculata* is modified in shape and morphology to scrape sperm from the female reproductive tract (Waage 1979).

Table 4.4 Levels of sperm displacement (P_2) in insects. Significance tests are probability levels for two-tailed Student t-tests. After Ridley (1989)

Theory		P_2	n	p
Counting species				
Mating plug	Present	0.48	9	0.0056
	Absent	0.75	31	
Spermatheca	Spheroid	0.68	12	0.41
	Tubular	0.75	18	
Male	Low	0.62	22	0.02
investment	High	0.78	31	
Mating	Single mating	0.42	14	<0.0001
frequency	Multiple mating	0.83	39	
Independent trials				
Mating plug	Present	0.46	4	0.1
	Absent	0.73	8	
Spermatheca	Spheroid	0.58	5	0.08
	Tubular	0.77	8	
Male	Low	0.63	5	0.37
investment	High	0.73	7	
Mating	Single mating	0.45	9	<0.0001
frequency	Multiple mating	0.82	8	

1989a). A phylogenetic tree must radiate in a star pattern with all the branches having equal length for this method to be valid.

Far more exciting are a new set of methods which attempt to retain all the relevant data (Felsenstein 1985; Bell 1989; Grafen 1989; Pagel & Harvey 1989a). The latter two methods are particularly valuable because they do not require exact phylogenies and because they make rather few assumptions about the mode of evolution. At the time of writing published applications are rather restricted (e.g. Trevelyan *et al.* 1990), but we can expect to see them used with increasing frequency.

THE DESCRIPTION OF SELECTION

Because selection is a process and adaptedness is a state, there is no reason to believe that understanding the present intensity of selection or the potential for it to operate should tell us how intense selection has been in the past. We certainly expect some stabilising selection to correct mutational and recombinational changes to a well adapted phenotype. However, we cannot assume that asking questions about the intensity of selection will tell us about adaptedness, or about the intensity of selection in the past (Grafen 1988a).

Nonetheless, we also need to understand selection in the here-and-now, so it is important that we reliably distinguish between the per-

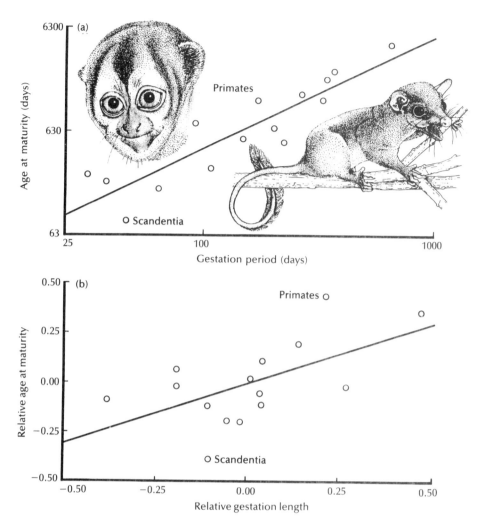

Fig. 4.8 Relation between gestation length and age at maturity across eutherian orders; (a) absolute values; (b) effects of body size removed. The positive correlation remains, though the relative position of some orders changes markedly. Primates and tree-shrews (Scandentia) show the greatest departures from the general relation. After Read & Harvey (1989a).

formance of different genotypes or phenotypes. For single allele differences this is often not very difficult (e.g. Pemberton *et al.* 1988), provided that the difference between the individuals is very marked, as it would be, for example, in the case of a deleterious completely recessive homozygote aa. Let us return to the Hardy–Weinberg equilibrium introduced in Chapter 2. The frequency of AA, Aa and aa before selection are given by p^2, $2pq$ and q^2 respectively. Because fitness is a relative concept we set the fitness of AA and Aa to be arbitrarily equivalent to 1, and define the selective disadvantage of aa as s, so the fitness of aa can be written as $(1 - s)$. For example, $s = 1$ and fitness = 0 when the homozygote recessive causes death before reproduction or sterility. The frequencies after selection of AA, Aa and aa are p^2, $2pq$ and q^2 $(1 - s)$ respectively, so the increase in the frequency of p is $spq^2/(1 - sq^2)$. Note that as the frequency of recessive homozygotes declines, the numerator declines and the denominator increases, so the rate of elimination of the recessive declines dramatically as its frequency declines, even when selection is very strong.

The efficiency of selection is much enhanced when dominance is not complete. Dominance is absent when both alleles contribute equally to the phenotype (we say that $h = 0.5$ in this case, and that h will decline to zero as the recessive makes smaller and smaller contributions). More generally, the fitness of AA, Aa and aa will now be given by 1, $1 - hs$ and $1 - s$ respectively, so the genotype frequencies after selection will be given by p^2, $2pq(1-hs)$ and $q^2(1-s)$ respectively, and the rate of increase of p per generation will be given by $hspq/(1 - hsq)$. Dobzhansky *et al.* (1977) compare the rate that q will change if $s = 0.02$ when $h = 0.5$ (no dominance) or 0 (complete dominance). A reduction in the frequency of q from 0.01 to 0.001 will take only 231 generations in the former case and a massive 90 231 generations in the latter case.

These calculations can be expanded for any case of differing fitnesses between two or more alleles at a locus. Unfortunately, application of the formal theory of population genetics suggests that even very subtle selective differences may have profound impact, and there is therefore reason to believe that important differences may occasionally exceed our capacity for measurement.

We can measure selection coefficients reasonably directly for alleles at a particular locus, but it is also possible to define fitness for any phenotypic distribution. Let the absolute fitness of a phenotype X be $W(X)$, and the relative fitness of phenotype X be definess as $w(X) = W(X)/W$, where W is either the mean absolute fitness or some phenotype against which an comparison is required (as for s, where we define the fittest genotype to have a fitness of 1).

Lifetime reproductive success

A central methodological problem is how do we assign fitness values to a particular phenotype or genotype. Short-term measures of reproductive success are likely to be unreliable indicators of fitness for many reasons. First, short-term indicators will be particularly sensitive to environmental changes unrelated to the quality of the individual. Important differences between individuals may be swamped during temporary conditions of plenty. Second, an individual which debilitates itself by intense reproduction in one year may perform poorly the following year. As an example of a combination of these effects, red deer hinds which have suckled a calf in the preceding year perform worse than those that have not, and this difference is most pronounced at high density (Clutton-Brock *et al.* 1982). Third, reproductive success often depends directly on age. To pursue the previous example, red deer hinds that are old or young are more likely to skip a year of reproduction than those in the prime of life (Fig. 4.9). Measurements taken from animals of different age are often used to reconstruct the reproductive performance of a typical individual. For example, we could measure the fecundity of 1, 2 and 3-year old animals to build up a profile of an animal that normally survives for three seasons. But cross-sectional studies of this sort underestimate the first of the important elements of selection, the change in the trait distribution with age. There are usually fewer old individuals than young individuals, and we should be extremely cautious of assuming that those individuals which do reach a ripe old age are a random sample of those which survive less well. Data on the lifetime reproductive success of animals commonly produce the result that total reproductive success is highly correlated with lifespan (Clutton-Brock 1988; Newton 1989). However, some older individuals make a much greater contribution than would be expected by their age alone (Fig. 4.10).

Even lifetime reproductive success is only a useful index of fitness under certain circumstances. For example, consider the sex ratio, which I used as an example of frequency-dependence. The adaptive benefit of producing equal numbers of males and females derives from its effects on the number of grandoffspring produced by parents, not from the number of their offspring.

Spurious correlations with reproductive success

Grafen (1988a) points out that there are likely to be three additional

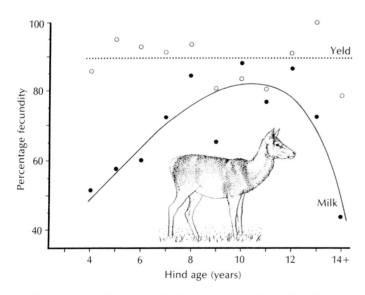

Fig. 4.9 The percentage fecundity of red deer (*Cervus elephas*) hinds in relation to age and to whether they had suckled a calf in the preceding year (milk hinds: ●) or not (yeld hinds: ○). There is only age-dependence for milk hinds. After Clutton-Brock (1984).

cases where correlations between reproductive success and the possession of a trait do not provide evidence for causation. The first is where mutational variance is pleiotropic in effect. Where mutations are pleiotropic, then those that are uniformly disadvantageous are likely to be lost more rapidly than those which have some advantageous effects despite the sum of their effects producing a net disadvantage. The mutations which are eliminated more slowly will therefore tend to produce a negative correlation between characters that are positively associated with fitness.

If some individuals are born in unusually benign conditions, they are likely to be both vigorous and fit (the silver spoon effect). There may therefore be positive correlations among fitness components which obscure any tendency for selection on one trait to have deleterious pleiotropic effects on another trait. We have seen that there is a very strong correlation between lifetime reproductive success and lifespan in many organisms, and that if anything, those organisms which live longest often perform at a slightly greater average rate than those which live for a shorter time (Fig. 4.10).

The third effect which has the power to mislead arises because the phenotypic performance of an organism depends directly on its internal state. Grafen (1988a) conjectures that if asked to choose whether a running or walking child is most likely to catch a school bus, he would bet on the child that walks. The act of running implies that the

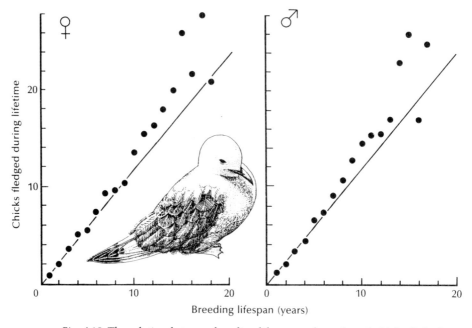

Fig. 4.10 The relation between breeding lifespan and number of chicks fledged during the lifespan of a kittiwake (*Rissa tridactyla*). The lines denote the expected performance based on the average annual output per pair. Note that old individuals also maintain a higher average productivity than expected. After Coulson & Thomas (1985).

child has already assessed that its probability of arriving on time is very low, and has adjusted its behaviour in response to its plight.

Phenotypic description of selection

Measurement of the genetic basis of variance within and covariance between traits is exceedingly difficult (Falconer 1981), and it will only be rarely the case that precise estimates of the appropriate parameters can be made on the species and questions of principal interest to ecologists. However, because the effect of very many loci acting on a trait is to produce a bell-shaped distribution amenable to statistical treatment, and because phenotypic correlations are often good indicators of underlying genetic correlations (Cheverud 1988), we can ask several important questions about selection on particular phenotypes even if we do not understand the genetic regulation of the phenotype. Current effort focuses on two parameters: the shape of the selection curve, the opportunity for selection and partitioning the separate components contributing to selection. The design of statistical procedures which allow these questions to be analysed is a dynamic and exciting area of evolutionary ecology, though there is still no universal panacea available.

The shape of the fitness curve

Understanding the form of the relation between fitness and a trait is of profound importance. Although it is easy in theory to portray directional, stabilising and disruptive selection (Fig. 1.8), there are two problems with their distinction in field data. First, there is no reason other than statistical convenience to assume the neat relations portrayed in the figure, and the incorrect assumption of a normal or quadratic distribution may distort the analysis of other variables. Schluter (1988a) offers a very promising method of fitting curves non-parametrically, and calculating confidence limits for those curves. The results for a range of data are presented in Fig. 4.11.

A second and very general problem arises because often the phenotypic distribution available naturally will be inadequate to assess the fitness of rare and extreme variants. The problem is twofold. First, if rare events are not observed, we may underestimate the strength of selection, or alternatively incorrectly interpret stabilising selection as directional selection, or vice versa. In a classic experiment, Andersson (1982) increased the tail length of male black-tailed widowbirds (with glue) and dramatically improved their success in attracting mates. This is strong evidence that the long tails of these birds are attractive to females, and that the apparently suboptimal tail length observed in nature is restrained by natural selection. We will return to the significance of this observation in Chapter 7. In direct contrast to this result, increasing the length of antlers in red deer caused both males and females to run away from the experimental freaks (R.V. Short, personal communication).

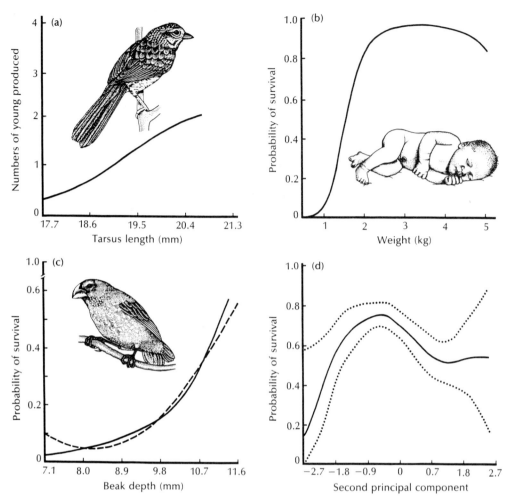

Fig. 4.11 Fitness functions for four traits using a non-parametric cubic spline: (a) reproductive success in female song sparrows, *Melospiza melodia* in relation to an index of body size, tarsus length; (b) survival in relation to birth weight in human infants; (c) survival in relation to beak depth in the Galápagos finch, *Geospiza fortis* (a quadratic function is included for comparison); (d) overwinter survival in song sparrows in relation to a principal component describing several features of morphology (the dashed curves represent ± one standard error, illustrating the difficulty in characterising the ends of the natural distribution of sizes. After Schluter (1988a).

Third, a problem arises in the interpretation of complex selection surfaces. While we can be confident about the surface in the centre of the phenotypic distribution, the confidence limits at the tails of the distribution may be very vague, as the data for overwinter survival of song sparrows indicate (Fig. 4.11d).

The last difficulty has hardly been addressed, and concerns the difficulties in interpreting selection as stabilising, disruptive or directional when more than one trait is selected simultaneously (Mitchell-

Olds & Shaw 1987). Phillips & Arnold (1989) offer a first step in this direction.

The opportunity for selection

By definition, selection on a phenotype cannot proceed without an influence of that phenotype on variation in survival or reproductive success. Although variation in reproductive success in the here-and-now need tell us very little about the extent to which the phenotype has been influenced by selection, it is nevertheless of great interest to observe the opportunity for selection.

Opportunity for selection (I) is defined formally as the ratio of variance in fitness (lifetime reproductive success) to the square of mean fitness (Wade & Arnold 1980). There are some constraints on the validity of the index for comparisons between species (e.g. Downhower *et al.* 1987), but it provides a valuable way of comparing the selection pressures within species. For example, the selection pressures on the sexes may be quite different (Chapter 7). A prominent cause of this difference will be the greater proportion of males that fail to acquire a mate in species where several females may be inseminated by a single male. For example, I may be twelve times higher for male bullfrogs (*Rana catebesiana*) than female bullfrogs (Howard 1983, 1988).

The components of selection

Selection can act at various parts of the life cycle of a plant or animal, and variance in survival or reproductive success will be more pronounced in some stages than others. One aim of all methods is to partition variance between life cycle stages, or between times of functional significance in achieving reproductive success. For example, in Howard's bullfrogs, male reproductive success (the number of tadpoles fathered by a given male) is the product of three components: the number of females that spawn with the male; the average number of eggs laid per mate; and the proportion of eggs that survive to hatching.

Arnold & Wade (1984a, 1984b) present a method for partitioning the total variance into that attributable to these single components and also the covariances between adjacent stages in a temporal sequence, using a technique analogous to multiple regression. In particular, they estimate the selection differentials, or the covariance of the character with a component of reproductive success, and the selection gradient, which is the slope of the best-fitting straight line relating reproductive success to the trait, holding other variables constant. Note that Endler (1986a) advocates the use of coefficient instead of gradient, in order to avoid confusion with geographical gradients in selection pressure. Total fitness of male bullfrogs increases with body size but the majority of this variation (59 per cent) depends on variation in success in obtaining mates, not in the other components of fitness

(Table 4.5; Fig. 4.12). An alternative procedure is presented by Brown (1988), who also enumerates the relative advantages and disadvantages of both methods. Surprisingly, an empirical comparison of the two procedures has not been attempted.

Selection in opposite directions at different life history stages

It is worth briefly noting that considerable complexity arises where there are life history episodes when the direction of selection is reversed. Such reversals can be shown at a variety of levels. For example, Clegg *et al.* (1978) examined allele frequencies for a number of linked esterase loci in inbred strains of barley. For all loci there were large and usually opposing allele frequency changes associated with the transition from zygote to adult and from adult to zygote. The selection could not be simply attributed to heterozygote advantage.

SUMMARY

Any study of either the product of selection (adaptedness) or the contemporary presence of selection requires explicit focus on the type of question and the temporal focus of the question. Several important methods are available to guide the direction of studies of both adaptedness and selection. Optimisation models help generate testable quantitative predictions about adaptedness, and discrepancies between predictions and observations often help focus and redirect hypotheses.

Table 4.5 Partitioning the total opportunity for selection on a reproductive population of male *Rana catebesiana*. After Arnold (1983).

Source of variation in fitness	Symbol	Contribution to the total opportunity for selection (I)	
		Value	Per cent
Number of mates	w_1	1.382	59
Number of eggs per mate	w_2	0.212	9
Offspring survivorship	w_3	0.160	7
Covariance between number of mates and number of eggs per mate	$w_1 w_2$	0.081	3
Covariance between total number of eggs and number of eggs per mate	$w_1 w_2 . w_2$	0.312	13
Covariance between total number of eggs and offspring survivorship	$w_1 w_2 . w_3$	0.032	1
Covariance between total number of eggs that hatch and offspring survivorship	$w_1 w_2 w_3 . w_3$	0.150	6
Total selection	$w_1 w_2 w_3$	2.238	100

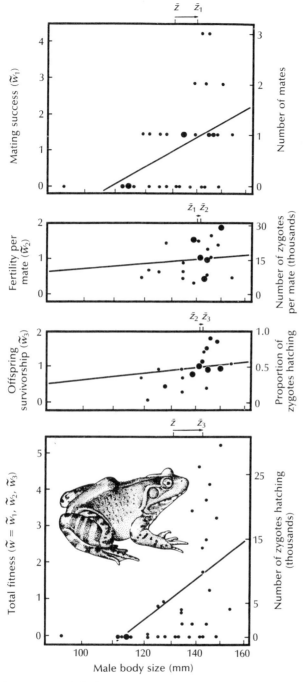

Fig. 4.12 Selection on male body size in bullfrogs (*Rana catebesiana*). Large points represent coincident data. Relative success (mean fitness = 1) is shown on the left vertical axis and actual success is shown on the right vertical axis. The total shift in the mean size $(\bar{z} - \bar{z}_3)$ is shown in the bottom panel, together with the selection gradient (the regression of total relative fitness on male body size). These total values can be partitioned into effects of two kinds of natural selection (average mate fertility, measured as number of zygotes per mate) and offspring survivorship (measured as the proportion of zygotes hatching) plus sexual selection (the number of mates). Sexual selection is dominant in this case. After Arnold (1983).

The comparative method is potentially immensely powerful in identifying correlations between phenotype and environment, and also in
identifying organisms which depart from common patterns and therefore warrant special attention. The statistical study of selection enables
us to examine selection pressures in contemporary populations.

FURTHER READING

Stephens & Krebs (1986) is the most lucid analysis of optimisation
approaches, and Mangel & Clark (1988) do a creditable job of introducing
the difficult topic of dynamic optimisation. Lendrem (1986) clearly
explains how to derive simple models. Maynard Smith (1982) provides
a comprehensive introduction to game theory. Pagel & Harvey (1988a)
is the best review of the problems inherent in the comparative method.
The three final chapters of Clutton-Brock (1988) by Brown, Grafen and
Clutton-Brock are excellent reviews of the problems in analysis of the
potential for selection and the interpretation of data on lifetime reproductive success. Hurlbert (1984) provides a salutary condemnation of
the weak understanding most ecologists have of the principles of
experimental design, and Hairston (1990) is a good overview of experiments in ecology. McCullagh & Nelder (1983) is the standard treatment
of statistical modelling, though most beginners will find Aitkin *et al.*
(1989) more approachable. Krebs (1989) is a good treatment of some of
the special statistical approaches of value to ecologists.

TOPICS FOR DISCUSSION

1 The concordance between the predicted (41 min) and observed
(36 min) copulation times in Fig. 4.2 could be regarded as a success or a
failure. Attempt to list the possible causes of a discrepancy of about
10 per cent in an optimisation model.
2 The exponent of the equation relating brain size to body size in
mammals varies according to the taxonomic level considered, so that
the slope is more shallow for genera than it is for orders (Gould 1975;
Lande 1979). Contrast the explanation for this trend given by Riska &
Atchley (1985) on the one hand, and Pagel & Harvey (1988b, 1989b) on
the other.
3 Try and find a biological example which resembles Grafen's (1988a)
'running to catch a bus' analogy. How do such anomalies relate to the
distinction between static and dynamic optimisation models?

Chapter 5
The Habitat Templet

The environments that determine the selective milieu in which an organism lives are usually extremely complicated. Physical properties vary in space and time, either predictably with the ebb and flow of the seasons, or unpredictably, in response to the vagaries of the weather and other natural phenomena like volcanoes or erosion. In addition, individuals constantly interact with other organisms, both of their own and other species. The diversity and density of those organisms also varies dramatically in space and time. These other organisms pose special problems as they are also subject to the forces of evolution and will change their nature through time.

In this Chapter I examine some of the consequences of this complexity of habitat. When is it best to specialise on a few resources, as opposed to enjoying the advantages of exploiting many resources? When is it best to be tightly programmed to fill a particular role, rather than respond to local conditions? Can inbuilt switches to the developmental programme achieve an optimal response to the environment? How is the distribution and specialisation of an organism related to its abundance? Can habitats be classified according to the selective environment they generate?

HABITAT, NICHE AND ROLE

We can classify the relationship between an organism and the environment it inhabits in three ways. The role of an organism refers to its function, for example, whether it is a detritivore or a sessile filter feeder. The habitat of an organism describes the properties of the environment in which the organism lives. The scale used in ecological investigations depends on the question of interest. For example, we may wish to describe the organisms inhabiting a desert, or the organisms inhabiting a single species of cactus. The other way to approach the question of appropriateness of scale is in terms of the organism itself. A tiny animal with limited vagility perceives the environment as coarse-grained (Levins 1968), and may be able to restrict its activities to a small proportion of the environment. For example, an aphid may feed on a single host plant such as clover for much of the year. However, the same small-scale heterogeneity which dictates the dispersion of the aphids may be all but irrelevant to a large grazing herbivore like a cow, which sees the environment as homogeneous or fine-grained. Nonetheless, the aphid and cow may still be thought of as living in the same habitat, even though they use it in different ways, so long as the grassland is the arena of investigation. By contrast, the third measure of interaction, the niche, describes properties of interaction with the environment which are specific to individual organisms or species. The niche describes the set of conditions which restrict the distribution of organisms, and can be thought of in terms of Hutchinson's metaphor of an 'n-dimensional hypervolume'.

Organisms will have a specific pattern of response along each dimension or axis of the hypervolume. The response of organisms to environmental gradients is usually portrayed as linear (the more the merrier) or Gaussian (Gauch & Whittaker 1972), exhibiting symmetry like a bell. The first simplification will usually only be true over a very narrow range of environmental conditions. A plant may grow vigorously in response to fertiliser but will ultimately be poisoned by excessive application of the growth stimulant. A bacterium may prefer high temperatures but will ultimately succumb at some intense heat. The second simplification has only recently been subjected to direct scrutiny. One of the impediments to critical testing is the use of statistical methods which in themselves assume the Normal (or Gaussian) distribution. The response of some species of *Eucalyptus* to temperature have been modelled by generalised linear models which relax this assumption of normality, and clearly demonstrate that the Normal distribution and statistical procedures which rely upon it need to be treated with considerable caution when statistically modelling response to gradients (Austin 1987; Austin *et al.* 1990; Fig. 5.1).

The fundamental niche of the organism is dictated by the physiological tolerances of the organism. However, hostile species such as predators, parasites and competitors may further restrict its distribution. For example, in North America moose (*Alces alces*) and caribou (*Rangifer tarandis*) are killed by the nematode *Parelaphostrongylus tenuis*, the normal host of which is the white-tailed deer (*Odocoileus virginianus*) (Anderson 1972). The distribution of two of the ungulates is therefore restricted by the worm and its host. This post-interactive distribution is called the realised niche. Niches may overlap, but by definition, no two species have identical niches.

COEVOLUTION: OTHER SPECIES AS FLUCTUATING HABITAT

Before proceeding to discuss the reasons why some species have a broad niche and occupy many habitats while others are both specialised and restricted, it is useful to introduce a particular type of evolutionary interaction which has profound consequences for most of the topics discussed in the remainder of this book. We have already seen that species interact in ways which are both beneficial and deleterious to the partners in the interaction (Table 1.3). These interspecific interactions can exert strong selective pressure. For example, allele frequencies at the human haemoglobin locus are influenced by the presence of malaria, a blood parasite. Where both species are affected by the interaction (for example, in competition, predation, parasitism and mutualism), there is potential for reciprocal evolutionary interactions, with repeated adjustments to improve or reduce the impact of the association. Such reciprocal evolutionary change between interacting species is called coevolution (Janzen 1980). Although true reciprocity will be rare, it has such interesting consequences that it warrants special discussion.

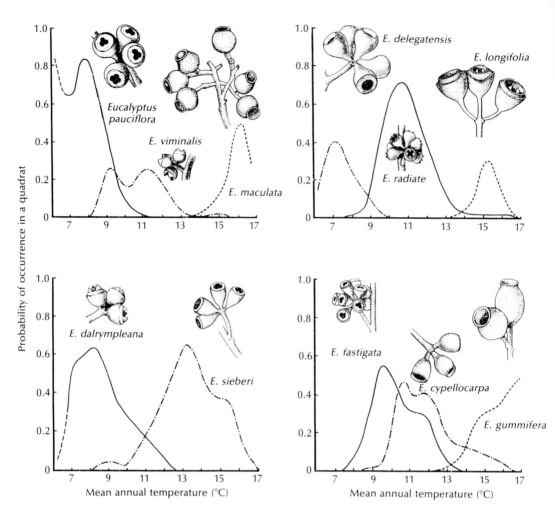

Fig. 5.1 The probability of occurrence of several *Eucalyptus* spp. in south-eastern Australia in response to mean annual temperature. After Austin (1987).

The intricacy of coevolutionary association between species varies greatly. At one extreme, one species may depend entirely on another for successful reproduction, or two species may be utterly interdependent, like the figs and fig wasps described in Chapter 1. At the other end of the spectrum, less rigid associations may occur. A plant may depend on pollinators, but any one of several insects may perform the task equally well. Similarly, a pollinator may be highly adapted to feed on flowers with a particular morphology, such as the hummingbird flowers described in Chapter 3, but any hummingbird flower may suffice. This variation has prompted Thompson (1989) to distinguish three types of coevolutionary interaction.

Gene-for-gene coevolution

Originally proposed by Flor (1956) to account for his observations of patterns of resistance to plant fungal disease, gene-for-gene coevolution describes the case where a particular gene in a parasite which affects virulence has a complementary gene in the host which affects resistance to that parasite. The best examples still come from interactions between plants and pathogenic fungi. In one population of the legume *Glycine clandestina*, eleven different phenotypic resistance patterns and at least twelve genetic resistance factors can easily be identified in response to nine races of the pathogenic rust *Phakospora pachyrhizi* (Burdon 1987a). No individuals were susceptible to all nine pathogen races and half the population was resistant to all of them. However, each of the eleven *Glycine* lines usually possess only one to three resistance genes. The high frequency of resistance genes in this and other plant populations strongly suggests that gene-for-gene interactions are common in plant–parasite interactions (Burdon 1987b; Burdon & Jarosz 1988).

Such intricate genetic associations are unlikely to arise commonly in interactions other than host–pathogen associations. Molecular interactions have a very direct impact on the success or failure of the parasitic infestation. However, most interactions (including host–pathogen associations) will be polygenic in character. Herbivory may very occasionally provide an exceptional example. Dirzo & Harper (1982b) speculate that the two-locus control over the ability of white clover *Trifolium repens* to produce cyanide results from a very specific interaction between the plant and the herbivore, in this case molluscs.

Specific coevolution

Very close coevolution between two species can occur without a gene-for-gene relationship, and indeed less close genetic association is likely to be far more common, and encompass close mutualisms and symbioses, many cases of resistance to parasites, and occasionally competitive interactions. Although there are many instances where such close association are known, such as the interaction between figs and fig wasps, it is only rarely that its genetic basis has been demonstrated, and indeed it seem likely that direct investigation will not prove to be an easy task (Thompson 1989).

Guild (diffuse) coevolution

Vetches (*Vicia* spp.) introduced to California from Europe have extra-floral nectaries which attract ants (*Iridomyrmex humilis*) introduced from Argentina. The ants protect the plants from herbivores, so the interaction is clearly a mutualism (Koptur 1979). In combination with many other studies, there seems no doubt that the nectaries are a trait whose adaptive function is to attract ants, yet the Argentine ants have until recently had no opportunity to contribute to selection for nectar production by vetches. This opportunistic mutualism is made possible by guild or diffuse coevolution, where the reciprocal interaction is between groups of similar species and the benefits are general. Any pugnacious ant can contribute to herbivore defence. Similar examples have arisen from natural dispersal patterns. The scale insect genus

Cryptostigma is known only from ant nests inside tree trunks on the east coast of South and Central America, and from the east coast of Australia and New Guinea (Qin & Gullan 1989). Ants attend the scales and feed from their sugary secretions, and are apparently essential for dispersal of the scales from one tree to another. It seems clear that this mutualism has persisted since the fragmentation of Gondwanaland. Extraordinarily, many different genera of ants and families of host plants are involved—the mutualism is critical for the scale, the partner is not.

Examples of guild coevolution are legion, particularly for mutualisms. Most interactions between plants and their pollinators and between fruits and frugivores are of this sort. The plant depends on an animal vector for the dispersal of pollen or seeds, and provides a reward to the vector. However, association with a specific dispersal vector is often either impossible or undesirable. Some exceptions certainly exist. Some fruits have exceptionally large seeds which cannot be swallowed by most frugivores—specialisation is inevitable, and the elimination of the animal can cause the decline of the plant. For example, the extinction of the dodo caused a dramatic change in seedling recruitment on Mauritius, as the seeds of some plants required the sort of abrasion caused by the crop and gizzard of a large bird in order to germinate (Temple 1977).

Arms races

Where the interaction between species is hostile, as in parasitism, predation and competition, one possible outcome is a persistent 'arms race', with escalating tactics of defence and attack. Despite criticisms of this analogy (e.g. Abrams 1986a, 1986b), it remains a useful way of viewing prolonged coevolutionary interactions between hostile species, subject to some important reservations. We can illustrate the problem with some examples.

Cuckoos

Most cuckoos lay their eggs in the nest of another bird species. Either the cuckoo or its chick then evicts the host eggs, and the parents of the host species raise the young. Not unsurprisingly, this imposes selection on the host species to develop the ability to detect the host egg, and host species vary in their ability to detect and then reject cuckoo eggs (by active ejection, by building the egg into the structure of the nest, or by abandonment of the nest). The ability of hosts to detect cuckoo parasitism can evolve quite rapidly, and directly in response to the presence of parasitism. For example, Soler & Møller (1990) show that the magpie *Pica pica* can detect both mimetic and non-mimetic eggs of the great spotted cuckoo *Clamator glandarius* in an area where they have co-occurred for a long time, but in an area of

recent sympatry detection is much reduced. Outside the range of the
cuckoo the birds do not detect cuckoo young at all.

We can extend this perspective further by examining the unusual
natural history of host use of the cuckoo *Cuculus canorus* available
in Britain. Brooke & Davies (1988) examined historical records of
parasitism, particularly in the classic volume *The Natural History of
Selbourne*, where the fine naturalist White (1770) failed to mention
reed warblers as hosts though they are now very commonly parasitised
by cuckoos, suggesting that cuckoos momentarily have the upper hand.
By contrast, Chaucer reports parasitism of dunnock nests, and this
species still shows no recognition at all (Davies & Brooke 1989). The
species which can recognise eggs impose considerable selection pressure on the cuckoo to deceive the host, usually by close mimicry of
its egg, with individual gens of cuckoos specialising on a particular
host (Fig. 5.2). The genetic basis of this specialisation would be of
extreme interest.

Conflict of interest in mutualism: long-tubed flowers

Even where the relationship is apparently beneficial to both partners
there may be conflicts of interest. In a typically successful piece of
insight, Darwin (1862) concluded from the extraordinarily long (30 cm)
tubes of the flowers of the Madagascar star orchid, *Angraecum ses-
quipedale*, that there must be a hawkmoth pollinator with a very long
tongue. More importantly, he proposed that orchids which compelled

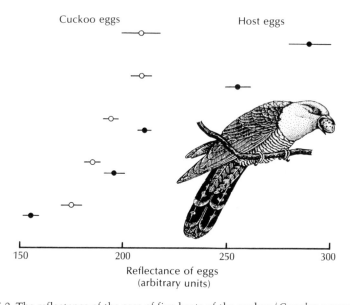

Fig. 5.2 The reflectance of the eggs of five hosts of the cuckoo (*Cuculus canorus*) in
Britain and the reflectance of the cuckoo eggs found in their nests. After Brooke &
Davies (1988).

moths to probe the base of the flower would set more seed than those
which were fertilised by moths with shorter tongues, creating selection
for longer flowers. This would in turn generate selection for longer
tongues, in order that the pollinator would get any nectar at all
(Fig. 5.3). The various components of this hypothesis have recently
been verified (Nilsson 1988).

Changes in the sign of relationships: the evolution of virulence

Prolonged arms races are by no means inevitable consequences of
coevolution between species where the interactions are initially
hostile. Indeed, it is virtually assumed in some parasitological literature
that a well-adapted parasite should not damage its host, as debilitation
and death of the host can cause the death of the resident parasites
(Hoeprich 1977; Alexander 1981). Evolution of the myxoma virus after
it was introduced into Australia to control rabbits provides a famous
example where the virulence of parasites rapidly declined to intermediate levels (Fenner & Ratcliffe 1965). Highly virulent strains kill
their hosts too rapidly for transmission, while strains with low virulence

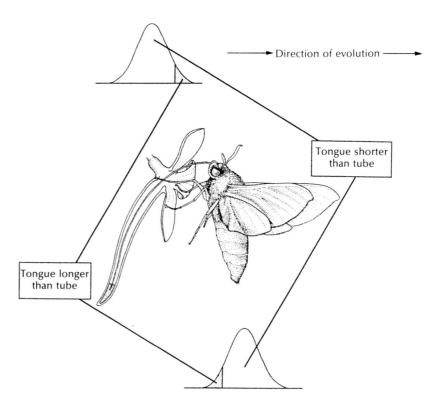

Fig. 5.3 Nilsson's (1988) modification of Darwin's hypothesis for the evolution of
flowers with deep corolla tubes. Long-tongued pollinators need not contact the
sexual organs in individuals with short tubes, so long tubes are favoured by
selection. Short tongues will not reach the nectary, so longer tongues are selected.

have a low rate of further infection. The virulence is also influenced by the evolution of resistance in the host.

This does not mean that host—parasite coevolution will always be in the direction of reduced virulence, and much of this literature assumes group selection is the driving force in the evolution of virulence (Lewontin 1970; see also Wilson 1983). Both theoretical models (Anderson 1982; May & Anderson 1983) and comparative analysis (Ewald 1983, 1987, 1988) suggest that many alternative coevolutionary outcomes are possible.

For example, Knolle (1989) suggests that virulence may often be associated with the extent of tissue invasion suffered by the host. He argues that if host density is high, transmissibility will be high, the life cycle of the parasite can be fast, and there may be selection for specificity to the tissue from which transmission is most easily achieved. By contrast, at low density there is less opportunity for rapid transmission and there may be selection for a slowing of the life cycle and greater tissue penetration.

Despite these exceptions it does seem possible that one outcome of persistent coevolution will change in the sign of the interaction, so mutualists become parasitic, and parasites become mutualistic (Thompson 1982 provides a stimulating review). The ultimate outcome of coevolutionary interactions may differ along geographic gradients (Hairston *et al.* 1987). We have already reviewed some cases where the distinction between negative and positive effects can become blurred, such as the bacteria in the gut of a ruminant (Chapter 2) or the wasps which pollinate a fig (Chapter 1).

The Red Queen

Because any evolutionary adjustments to other species can be countermanded by natural selection on those species, Van Valen (1973a) used the metaphor of the Red Queen to describe biotic evolution; consistent change is necessary, not to increase adaptedness, but to maintain it, just as Alice and the Red Queen had to run as fast as they could to get nowhere.

We will return to the importance of this Red Queen effect in many contexts in the following chapters. However, one implication must be stressed at the outset. Some problems posed by the environment can be resolved by selection. We therefore expect adaptedness to be achieved in time in stable habitats, and therefore evolutionary equilibrium to prevail. However, such equilibrium will be perturbed by evolution in other species, so physical stability need not necessarily promote evolutionary equilibrium.

GENERALISTS VERSUS SPECIALISTS

Some plants and animals are extremely widespread, successfully reproducing in different physical conditions and in different communities of plants and animals which may have no other species at all in common. Others exhibit intense specialisation, being utterly dependent on a particular resource. Alternatively, some species use the resources with their habitat in the same proportion that they are encountered, while others show preference for part of the available resource spectrum. Understanding specialisation or narrow niche width is a central problem for ecologists. It is easy to envisage selection which broadens the range of environments that can be tolerated or effectively exploited, but much less easy to understand decreased niche breadth. Profound macroevolutionary consequences have been proposed to flow from specialisation, yet these extrapolations are often developed with only a rudimentary understanding of the original phenomenon.

Futuyma and Moreno (1988) argue that it is useful to distinguish morphological and physiological limitations to the range of an organism from behavioural specialisation. The former set of limits presumably reflects clear evolutionary pressures towards limited niche width, the latter may be expressed only facultatively. They also draw attention to two ways in which a species may exploit a variety of habitats. Individual phenotypes may have broad tolerance (within-phenotype niche width), or different phenotypes within a population may be capable of exploiting different parts of the habitat (between-phenotype niche width). For example, many herbivorous insects show local specialisation at any site, yet a full list of their host plants from many sites indicates that they are polyphagous, or capable of feeding on many plants (Fox & Morrow 1981; Strong *et al.* 1984).

Prey choice in rich environments

An organism searching its environment to locate prey must decide each time it encounters a potential prey item whether to eat that item or to continue the search in the hope of finding something more rewarding. The spectrum from including all items to eating only rewarding items represents a gradient from generalist to specialist. Optimal foraging theory has been applied to the problem in the following way. We use the following assumptions: the currency to be maximised is the long-term average rate of energy intake, the decision is with what probability should prey type i be attacked upon encounter. The model is constrained for the purposes of simplicity. Searching and handling are assumed to be mutually exclusive, and encounter with prey is assumed to be sequential and a Poisson process. For each of i prey types there is a fixed value of energy gain (E_i), handling time (h_i), and encounter rate (λ_i). The predator is assumed to have complete information, knowing the values of these parameters at the start of the search (it does not learn).

For the sake of argument, consider a predator faced with only two prey types, a profitable prey 1 (high E_i/h_i) and a less rewarding prey 2. If the predator searches for T_s seconds it will obtain

$$E = T_s \left(\lambda_1 E_1 + \lambda_2 E_2 \right) \tag{5.1}$$

by eating both types of prey in a total of

$$T = T_s + T_s (\lambda_1 h_1 + \lambda_2 h_2). \qquad (5.2)$$

where the second half of the equation represents the handling time, so the total rate of intake is given by

$$E/T = (\lambda_1 E_1 + \lambda_2 E_2)/(1 + \lambda_1 h_1 + \lambda_2 h_2). \qquad (5.3)$$

By contrast, if the profitable prey alone is taken then the rate of intake is given by

$$E/T = (\lambda_1 E_1)/(1 + \lambda_1 h_1). \qquad (5.4)$$

Obviously the predator should specialise when the energy gained from eating only the more profitable prey exceeds that for eating both. This is when

$$(\lambda_1 E_1)/(1 + \lambda_1 h_1) > (\lambda_1 E_1 + \lambda_2 E_2)/(1 + \lambda_1 h_1 + \lambda_2 h_2). \qquad (5.5)$$

With a little algebraic gymnastics this can be rearranged to give

$$(1/\lambda_1) < (E_1/E_2)(h_2 - h_1). \qquad (5.6)$$

There are three conclusions which can be deduced from the model. First, the predator should specialise when the more profitable prey is very abundant (h_1 is low). Second, there should never be partial preference. This is only predicted when there is exact balance between both sides of equation 5.5. This condition will be extremely rare, so the predator should switch rapidly from accepting all of both items to only the more profitable prey. Third, and undoubtedly most interesting, λ_2 disappears from equation 5.6, so the abundance of the less profitable prey is not important in the decision to specialise on the more abundant prey. This model can be extended to include any number of prey items, and the results are identical, so prey should be ranked according to its profitability, and only prey above a given rank should be included in the diet. The number of items included will reflect their abundance, but not the abundance of lower ranking food items. Stephens & Krebs (1986) summarise the very large number of attempts to test this model, and conclude that it offers very useful predictions of prey specialisation in both the laboratory and in free-living animals, with the exception that partial preferences are very frequently observed. The ecological implication is clear. We expect greater specialisation in rich environments.

Resource partitioning and competition

As we have already seen, the niche of an organism can be restricted by its interaction with other species. An obvious case is where one organism is so superior at exploiting a particular part of a resource gradient that it precludes another organism from using those resources. This type of interaction (exploitation competition) needs to be distinguished from cases where organisms fight directly for a resource (interference competition). As a consequence of competition, species will appear to be more specialised than would be the case in the absence of the competitor. A popular textbook portrayal of this type of competitive interaction is shown in Fig. 5.4a. Species space themselves evenly along a resource gradient. At the point of interaction the effect of one species upon another is mutually deleterious. It is likely in practice that competitive interactions will often be asymmetrical to the point that they border on amensalism, with one species being capable of displacing others across its entire fundamental niche (Fig. 5.4b; Lawton & Hassell 1981; Connell 1983; Schoener 1983a). For example, two species of goldenrod *Solidago missouriensis* and *S. canadensis* show habitat specialisation in North American prairies, with *S. missouriensis* occurring on ridges, and *S. canadensis* in wetter depressions. When *S. missouriensis* is planted in plots with no competitors it does best in the wetter, more productive sites, but is incapable of competing there with *S. canadensis* (Werner 1979; Werner & Platt 1976). By contrast, *S. canadensis* is a specialist on the wetter sites, and will not expand its distribution if *S. missouriensis* is removed.

There are two distinct ways that competition can influence the structure of communities (Rummel & Roughgarden 1985). First, by selective survival of invading species as a result of competitive exclusion (invasion-structured communities). The mechanism for producing structure is easily envisaged. Similar species compete, and the population size of one or both is reduced, increasing its vulnerability to extinction. The higher probability that one of a pair or group of similar species will become extinct introduces structure to the members of the community which survive extinction. For example, experimentally induced interspecific competition between *Daphnia* spp. will not permit the coexistence of three species (Bengtsson 1989).

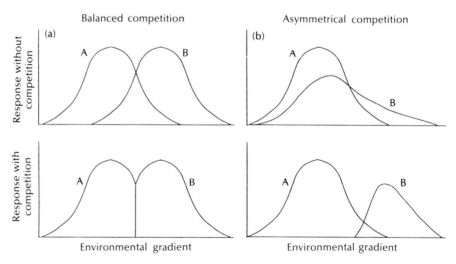

Fig. 5.4 The distribution of competing species in the presence and absence of balanced (B) and asymmetrical (A) competition.

Second, by coevolutionary rearrangement of species that coexist over long periods (coevolution-structured communities). Many potential competitors overlap in only part of their geographic range. If competition is an important determinant of niche width, we might anticipate that the realised niche of both or either species may differ in sympatry compared to the pattern in allopatry (character displacement; Fig. 5.5). Alternatively, two species may typically co-occur. If one finds itself in an environment from which the other is absent, it may be able to exploit a greater range of habitats (character release; Fig 5.6). Aside from methodological difficulties, there are three problems with assessing the relative importance these coevolutionary effects. First, theoretical models suggest that even when competition is present, divergence of characters need not be the only outcome, with convergence of competitors possible under a variety of conditions (Taper & Case 1985),

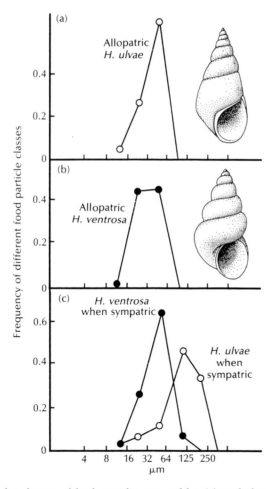

Fig. 5.5 Size distribution of food particles ingested by: (a) *Hydrobia alvae* when allopatric; (b) *H. ventrosa* when allopatric; (c) both species when sympatric. There is a major shift in the diet and shell length of *H. ulvae* in sympatry (Fenchel & Christiansen 1977). After Maynard Smith (1989).

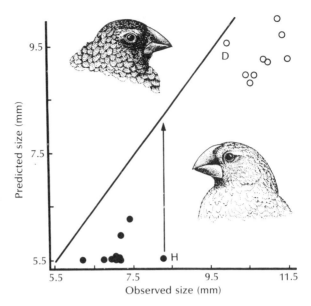

Fig. 5.6 Character release in Darwin's finches. The observed beak size for adult male *Geospiza fortis* (○) and *G. fuliginosa* (●) compared with the mean beak sizes predicted by analysing the beak size which would be expected on the basis of food available on given islands and known fitness responses to the availability of food of given sizes. All sympatric populations depart from the expected size. There are two allopatric populations. The beaks of the single allopatric population of *G. fortis* on Daphne (D) are much closer to the expected size than when the species occur in sympatry. The beak of the single allopatric population of *G. fuliginosa* (H) also shifts size, but becomes much larger than expected, apparently moving in the direction of a second peak in the expected curve. After Schluter *et al.* (1985).

particularly when the two species compete for resources which cannot be replaced (Abrams 1987b, 1987c). Competition should increase with coevolutionary convergence, giving the community the appearance of invasion structuring. Second, even where divergence occurs, it is possible that it reflects other selective forces, such as the prevention of mate recognition necessary to avoid interspecific hybridisation (Taper & Case 1985; but see Chapter 8). Third, and most relevant to discussion of niche width, coevolution may have led to morphological specialisation to the point that the fundamental niche declines in width through evolutionary time. Even if a competitor is experimentally removed, spread by adjacent species into new parts of the resource spectrum may be prevented by this preceding coevolutionary specialisation. These effects have been called 'the ghost of competition past' (Connell 1980).

Enemy-free space

Although competition has played a pre-eminent role in the development of niche theory, it is by no means the only ecological interaction that can influence the volume of realised niches, or the selection towards or away from specialisation. Predation and parasitism play critical roles in dictating the availability of habitat. Indeed, one of the most important forms of competition may be avoidance of these hostile agents, or the occupancy of 'enemy-free' space (Jeffries & Lawton 1984; Bernays & Graham 1988). Jaenike (1985) points out that larvae of some members of the *Drosophila quinaria* species group are tolerant of aminatin, the main toxin found in the *Amanita* group of mushrooms which are so deadly to humans and other animals. Nematodes parasitise large proportions of the larvae on non-toxic mushrooms, so a primary selective advantage in utilisation of the toxic resource may be escape from parasite pressure.

About one-third of the lycaenid butterfly species associate with ants. The relationship is mutualistic: the caterpillars secrete carbohydrates and amino acids which are harvested by the ants; the ants protect larvae and pupae from predators and parasitoids (Atsatt 1981; Pierce & Mead 1981). The butterflies may use ants as a clue for oviposition, so that females will lay eggs on plant species different from their original host species (Pierce & Elgar 1985). Ant-attended lycaenids have a much greater host range than those which are not attended by ants (Table 5.1), but also a strong tendency to feed on nitrogen-fixing protein-rich plants (Pierce 1985). While this seems obviously beneficial for the caterpillar, it does not explain the success of lycaenids not attended by ants on protein-poor plants. Pierce & Elgar (1985) argue convincingly that the real source of the selection pressure towards use of high protein food is the need to supply the attendant ants with a secretion containing not only carbohydrates but also concentrated amino acids. The mutualism can be viewed in terms

Table 5.1 Average number of plant species, genera and families used as hosts by lycaenid butterflies, according to whether their caterpillars are attended by ants. After Pierce & Elgar (1985).

	Ant-attended	Not ant-attended
Number of species		
Australia	4.6	2.9
Number of genera		
Australia	2.7	2.4
South Africa	2.8	1.8
North America		
Theclinae	16.0	1.8
Polyommatinae	5.6	2.0
Number of families		
Australia	1.8	1.7
South Africa	1.9	1.1
North America		
Theclinae	7.7	1.5
Polyommatinae	2.2	1.3

of opportunity and constraint. Ant association affords the caterpillars the opportunity to forage on more species of plants and protection from parasitism. Nonetheless, the need to provision ants with nitrogen-rich food constrains the type of plant which can be utilised.

The analogies with competition can be pursued further. Many plants produce secondary compounds, chemicals which do not contribute significantly to their metabolism, or occur at concentrations much greater than would be anticipated from examining the metabolic pathways of those plants. One of the problems with interpretation of the role of plant secondary substances is the possibility that the chemicals may be detoxified or sequestered to the point that they no longer harm the herbivore at all. *Eucalyptus* trees suffer high and chronic levels of herbivory throughout Australia, their natural environment. Many species have been introduced throughout the world because they grow extremely quickly. The tallest hardwood trees in North America are found in the eucalypt grove on the University of California at Berkeley campus, and the physiognomy of these tall and leafy eucalypts bears no relation to that of their conspecifics in Australia. There is abundant evidence that the profligate growth is at least in part a consequence of predator release. For example, the introduction of a curculionid beetle (*Gonipterus scutellatus*) to South Africa caused widespread defoliation and reduction in growth of eucalypts, until it in turn was controlled by an Australian parasitic wasp (Pryor 1976). It seems certain that their formidable chemical armory is a major deterrent to herbivores with no prior exposure to eucalypts, but it is singularly difficult to show that the same deterrents affect Australian herbivores at all. Another confounding factor in assessing the role of secondary compounds is the possibility that the evolutionary relationship that promoted their occurrence may have ended (Janzen 1979). Some ferns produce mimics of insect moulting hormones (photoecdysteroids), yet these appear to have absolutely no effect on either the fern-feeding insects found on these plants or on polyphagous insects which do not feed on ferns (Jones & Firn 1978). Photoecdysteroids may represent redundant defences against an organism which preyed on ferns at some time in the past, but has since disappeared (Strong *et al.* 1984).

Evidence for trade-offs

The fundamental premise underlying all these arguments is that specialisation has benefits through efficient exploitation of a resource, but imposes costs which limit the range of habitats which can be exploited by such a specialist—a jack of all trades is a master of none (MacArthur & Levins 1967). The intuitive appeal of this suggestion belies the difficulty in testing it adequately. The genotype–environment correlation could occur intraspecifically and interspecifically, provided that there is intraspecific variation in behaviour and morphology.

Most species are polymorphic, with sex the most common correlate of morphological differences. In some species there are distinct foraging morphs uncorrelated with sex. The African estrildid finches, *Pyrenestes* spp. show typical intraspecific variation in most body measurements (wing, tail and tarsus), but extraordinary bimodality in the morphology of their bills. The bimodality occurs in both sexes of three species, and the large-billed morph has a bill 23 per cent larger than the small-billed morph, a greater difference than sometimes occurs between good species exhibiting character displacement (Grant 1986). The foraging ability of the two morphs in *P. ostrinus* vary with diet (Smith 1987, 1990a). On a sedge with hard seeds (*Scleria verrucosa*), the small-billed morph took 25 seconds to crack a seed compared to 18 seconds for birds with large bills. By contrast, the small-billed birds were much more efficient at handling the much softer seeds of another *Scleria* sp. (6 seconds versus 9 seconds). These differences in foraging ability are likely to be a major contributor to maintenance of the polymorphism. Their genetic control is not entirely clear, but application of Schluter's method for portraying selection surfaces does reveal evidence of disruptive selection promoting bimodality (Smith 1990b). Two questions of great interest remain unresolved. First, why is the polymorphism uncorrelated with sex? Second, why in this case has one species accomplished specialisation for a variety of food types, rather than different species exploiting the different resources?

In a related example, bluegill sunfish (*Lepomis macrochirus*) within a single lake tend to show morphological and behavioural specialisation for feeding in littoral or open-water habitat (Fig. 5.7; Ehlinger & Wilson 1988). This is not just a consequence of induction of morphology by early experience, as all fish complete their early development in the littoral habitat. Although much of the prey specialisation observed in sunfish may have its basis in learning (e.g. Wainwright 1986), it appears that part of the foraging specialisation within a single population is genetically determined.

Hybrid zones and pest resistance

Gene/environment correlations are dramatically illustrated in hybrid zones, where two genetically distinct populations interbreed. The hybrids of both plants and animals often bear enormous parasite burdens compared to their source populations (e.g. Whitham 1989; Sage 1986), suggesting that the different populations have different genes for resistance, but the hybrids lack any sort of resistance at all.

Chemical deterrents to phytophagy in insects

Many adult herbivorous insects oviposit, and their larvae feed on the leaves of only a single species of plant, or at least on a group of plants

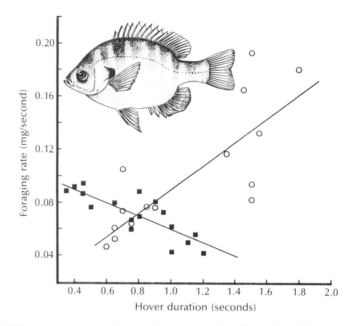

Fig. 5.7 Foraging rates (mg dry weight per second) by bluegill sunfish, *Lepomis macrochirus*, as a function of hover duration for trials when feeding on *Daphnia* in open water (■) and feeding on nymphs in vegetation (○). There is a clear reversal in the rewards associated with increasing hover time. After Ehlinger & Wilson (1988).

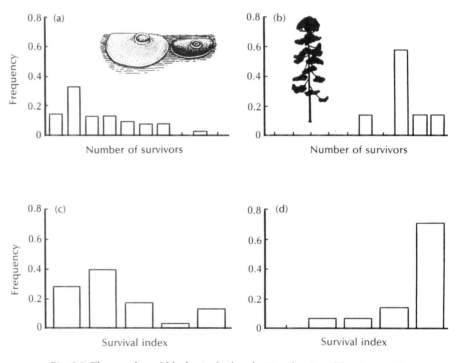

Fig. 5.8 The number of black pineleaf scale *Nuculapsis californica* on *Pinus ponderosa* surviving nine months after being transplanted: (a) between trees; and (b) within trees. The lower graphs show survival indices which eliminate any effect of unequal establishment. After Edmunds & Alstad (1978).

which are closely related (monophagy). Because the food chemistry of plants is usually at least superficially similar, attempts to explain specialisation in terms of nutrition have focused on the numerous deterrents to herbivory found in plants. Secondary compounds do frequently reflect the phylogeny of the plant, and in many cases obviously contribute to deterrence. However, their contribution to the evolution of monophagy is much less clear. The trade-off hypothesis argues that the jack of all trades will not perform as efficiently on host plants as can a monophagous species. Perhaps more importantly, the preference of a monophagous species should be reflected in increased growth or survival on the plant it chooses. There is some evidence for specialisation and local adaptation at the intraspecific level. Natural populations of scale insects and thrips are believed to become differentiated into small demes, each specialised on an individual plant (Fig 5.8; Edmunds & Alstad 1978; Karban 1989). However, perhaps surprisingly, the evidence for gene/environment correlations between phytophages on different species is not very compelling (Bernays & Graham 1988; Futuyma & Moreno 1988). While some well-designed studies demonstrate the predicted correlation (e.g Via 1986), many others do not (e.g. Futuyma & Phillipi 1987). In addition, closely related species with different preferences often fail to show differences in growth and survival on a common host. There are a number of problems with analyses to date, which caution against early dismissal of the trade-off hypothesis. Notably, Rausher (1988) points out that growth experiments are often conducted in the laboratory, where generally benign conditions could promote growth and therefore obscure any negative correlations arising from trade-offs.

Finding a mate versus using more food

Niche segregation does not necessarily imply competition. Communities of parasites within a host often have high tissue specificity. However, high levels of tissue specificity may enhance mate-finding and reproductive output, and many species show extreme specificity in the absence of any interspecific interaction (Rohde 1979). Similar arguments have been advanced to account for host specificity in hummingbird flower mites (Colwell 1986) and butterflies where location of mates takes place on the host plant (Ehrlich 1984).

Learning in bumblebees and butterflies

Foraging social Hymenoptera such as honey bees and bumblebees show behavioural specialisation for a particular nectar source. Naive bumblebees sample the environment but soon develop a preference for a particular species of plant on which they subsequently specialise (majoring). However, they also continue to sample a small range of alternative nectar sources (minoring) as part of their daily foraging routine. On open actinomorphic flowers they show good initial performance, but on zygomorphic flowers with complex morphology and restricted access to the nectar they develop foraging skills gradually, doubling their performance after encountering 60 to 100 flowers (Heinrich 1976, 1979; Fig. 5.9). Why do they spend time in learning a suitable host plant, why do different bumblebees in a single colony specialise on different majors, even when there are great differences in the rewards they provide, and why persist in sampling minors? First, it does not appear that the energetic costs of learning are terribly great (Heinrich 1984). It is clear that bumblebees need to be able to track continuously changing food resources. The rate of temporal change is determined by both seasonal changes in flower availability, and by the rate of depletion of resources by other nectar feeders. Heinrich (1979) has experimentally investigated the effects of decline of food resources. In the field, bumblebees respond first by expanding their search area, but in enclosures where this is impossible they expand their diet breadth, and may switch between host plants. Switching is limited by what other animals in the population are doing. A bumblebee which exploits a flower early in the period over which it is available will encounter a resource not being heavily exploited by other animals. As the number of specialists on that resource increases, these specialists harvest the nectar so efficiently that a naive bee encountering that resource will not receive sufficient rewards on those flowers to encourage learning (Heinrich 1976). By contrast, bees will rapidly learn to specialise on artificial flowers where abundant nectar is provisioned.

When the butterfly *Pieris rapae* learns to exploit a new flower it interferes with its capacity to extract nectar from the first (Lewis 1986). This suggests that one trade-off affecting specialisation is limited neural capacity. Learning may afford the opportunity to deal with a fluctuating environment, but may restrict the number of plants that can be tackled at any time. Alternatively, genetically programmed specialisation may afford the opportunity to reduce costly neural circuitry associated with host selection.

Neural trade-offs cannot be excluded as one explanation for host specificity in phytophagous insects, as a common feature of changes in host-plant preferences from both experience and genetic changes is the

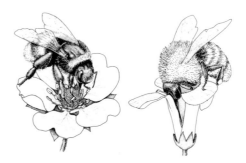

Fig. 5.9 Bumblebees are slow to develop the skills necessary to exploit flowers with complex shapes rapidly, but can exploit simple flowers immediately.

loss of recognition of the original host (Jermy 1987, 1988). The number
of contact chemoreceptors is much greater in unspecialised taxonomic
groups such as the Orthopteroidea than in groups where species exhibit a
high level of host-plant specificity (Chapman 1982). Further, the major
cause of failure to reproduce in many insects is the failure to find hosts
and deposit eggs (Dempster 1983; Courtney 1984; Dixon & Kindlmann
1990). In some circumstances this shortfall may favour generalisation
(Wiklund & Ahrberg 1978), but in some case search efficiency may be
enhanced by a specific search image (Rausher 1978; Courtney 1988),
and a level of learning is important in host selection in some species
(Prokopy *et al.* 1982; Stanton 1984).

The ideal free distribution

One of the difficulties in assessing the performance of individuals in
different environmental conditions is illustrated by the bumblebee
example. The rewards to be derived from a patch may vary according
to the density of animals exploiting the more profitable areas. One
habitat may initially be favoured, but it may ultimately be so heavily
used that individuals may enjoy greater rewards by utilising another
habitat type. Game theoretical models predict that the individuals will
eventually distribute themselves so that their fitnesses are equal. In
this 'ideal free distribution' local density will vary but rewards will not
(Fretwell & Lucas 1970; Fretwell 1972). For example, sticklebacks
Gasterosteus aculeatus, occupy portions of a fish tank so that the
average foraging rewards are equal regardless of the local food density
in the patch (Fig. 5.10). Of course, this pattern will be complicated by
competitive differences between individuals (Fretwell & Lucas 1970;
Sutherland & Parker 1985), and the ability of the organisms to move
freely among patches and assess their quality (Abrahams 1986). However, the theory does suggests that caution should be adopted in
assessing the benefits of living in different habitats.

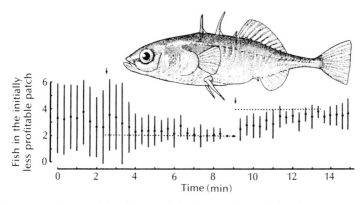

Fig. 5.10 The density of foraging sticklebacks and the ideal free distribution. Food
was added in a way that ensured that one end of the tank had twice as much food as
the other end. At a given point the relative profitabilities were reversed. The dashed
lines represent the predicted density of fish. After Milinski (1979).

The use of learning to facilitate habitat selection draws attention to the widespread occurrence of specialisation in response to some environmental trigger which alters the phenotype. Such changes are an example of phenotypic plasticity (Bradshaw 1965), and represent one aspect of the within-genotype component of niche width. The individual's pattern of phenotypic response to environmental influences is called the reaction norm (Johanssen 1911). It is useful to distinguish two types of reaction norms. First, all genotypes respond similarly to a change in the environment, so no changes in genotypic rank occur along an environmental gradient. Second, genotypes vary in the extent of their response to environmental gradients, so that changes in genotypic rank can occur. This variation in reaction norms is called genotype–environment interaction.

Environmental influences on the phenotype are virtually ubiquitous, but vary enormously in extent. We can recognise a continuum of relative genotypic and environmental influence on the phenotype (Table 5.2). At one extreme, canalisation, the phenotype is buffered from variation in both the genotype and the environment (Waddington 1952). Active buffering is suggestive of selection against variation. Such selection is easy to understand for many properties of the organism. Physiological and developmental homeostasis demand a degree of integration which should enforce a uniform phenotype. However, there are also some ecological circumstances where adaptive responses to environmental circumstances might be precluded. If fluctuations in the environment are more rapid than the generation time, the environment will appear to be fine-grained (Levins 1968). Second, there may be no cues which enable the organism to anticipate future environments (Levins 1968; Lloyd 1984a).

At the opposite extreme to canalisation, environmental cues, sometimes quite subtle, may profoundly influence the course of development of an individual by channeling it along one or more discrete pathways, so that many aspects of the organism change in concert (developmental conversion; Tables 5.2, 5.3). Developmental conversion is presumed to be adaptive. It is important to recognise that alternate morphs may arise in any stage of the life cycle, even if the adult morphs do not differ. In a remarkable example of polymorphism in larvae, caterpillars of the spring brood of *Nemoria arizonaria* develop into excellent mimics of the oak catkins on which they feed. By contrast, the summer brood develop into equally cryptic mimics of oak twigs (Greene 1989). The proximate cause of developmental conversion is the concentration of tannins. The catkins are low in tannins, and induce the catkin morph. Leaves are high in tannins, and induce the twig morph.

Between the extremes of canalisation and developmental conversion, the organism may show a graded variety of responses to the environment, and individual characters of an organism may differ in

Table 5.2 Types of variation. Modified from Lloyd (1984a) to use the terminology of Smith-Gill (1983).

Nature of variation	Characteristic feature	Number of classes of individuals	Number classes of structures on one individual
Uniform phenotype	Canalisation	1	1
Phenotypic modulation	Widely varying adjustment	Numerous, reflects environment, usually continuous variation	
Developmental conversion	Environmentally cued, distinct options	≥2	1
Morphs	Genetic specialisation	≥2	1
Multiple strategy	Simultaneous combinations on single individual	1	≥2

Table 5.3 Comparisons of two types of phenotype plasticity. After Smith-Gill (1983).

	Developmental conversion	Phenotypic modulation
Genetic basis	Specific switch	No specific basis
Developmental programme	Changed	Not changed
Occurrence	At discrete developmental stages	Any time in life cycle; phenotypic results may be stage specific
Phenotype variation	Discrete	Continuous
Adaptive	Yes	Not necessarily
Synonyms		
Lloyd (1984)	Conditional choice	Continuous lability
Schmalhausen (1945)	Autoregulatory morphogenesis	Dependent development

the degree of change. This phenotypic modulation need not necessarily be adaptive (Table 5.3). For example, stunted growth under a poor nutrient regime hardly requires an adaptive explanation. However, many cases of phenotypic modulation may have adaptive consequences, and the capacity to respond to subtle environmental cues may be strongly selected.

Although phenotypic plasticity is of profound importance in all organisms, there has been a special emphasis on the evolution of plasticity in plants. This may be because a sessile life-style eliminates the ability to move in response to fluctuations in ambient conditions (Schmalhausen 1949; Schlichting 1986), or because modular growth is more amenable to the expression of plasticity, with fewer pressures enforcing canalisation towards an integrated phenotype (Bradshaw 1965; Primack & Antonovics 1981). It has been shown that there is heritable variation for plasticity within populations of plants (Jain 1978); indeed it is possible that selection to improve yields in crops often focuses unconsciously on plasticity by selecting genotypes more responsive to improved culture conditions (Simmonds 1981). For example, artificial selection on rice has led to increased plastic responsiveness to weeding, transplanting and fertiliser (Oka & Chang 1964). Analyses of limits to the evolution of plasticity are in their infancy. One promising approach is to treat one character expressed in two environments as two character states, each of which is expressed only in one of the environments (Via 1987). Each character state has a measurable genetic variation within the environment in which it is expressed. The states expressed in the two environments may share some common genetic basis that would produce a genetic correlation (r_G) between them. When r_G is high (± 1), there will be low genetic variation in phenotypic plasticity, and the potential for independent evolution of character states disappears. Preliminary quantitative genetic models using this framework indicate that plasticity can usually evolve to optimum levels provided that $r_G \neq \pm 1$, and that the different environments are encountered frequently by the population, because they are common and there is some dispersal between habitats (Via 1987).

It will ultimately be instructive to seek correlations between the amount of plasticity and the amount of environmental heterogeneity. However, the simultaneous difficulty in measuring heterogeneity and the genetic correlations in Via's model currently clouds our ability to make generalisations. For example, differences between favourable or safe sites for germination by a seed, and areas totally hostile for germination may be expressed at a spatial scale dictated by the falling of a single drop of rain, or a tiny crack in the soil (Harper 1977; Sultan 1987). In addition, because these type of disturbances are so unpredictable in space and time, local adaptation is less likely to be adaptive than is plasticity (Hartgerink & Bazzaz 1984). Useful clues are likely to arise when there are strong differences in the extent of plasticity in different environments. The intertidal whelk *Nucella lapillus* shows great plasticity in the development of the area of its foot in response to the intensity of wave action (Etter 1988). Whelks from a protected site transplanted to an exposed shore form a much larger foot than controls reared at the protected site, no doubt because the increased foot area helps avoid dislodgement by waves. By contrast, whelks from the exposed site differed much less from controls at the exposed site. Etter (1988) argues that where selection has favoured a plastic response to a temporally unpredictable component of the environment, there should be selection for much more rapid change to the stress-tolerant form than to the stress-intolerant form.

It has been suggested that plasticity would inhibit evolution of characters, because where there is broad phenotypic plasticity, no genetic change may be necessary to accomplish morphological modification in response to environmental change (Grant 1977; Stearns 1982), or phenotypic convergence may occur in organisms with very different genotypes, reducing the ability of selection to discriminate between genotypes (Bradshaw 1965). Note that this latter pattern is not the same as canalisation. However, Schlichting (1986) points out that the amount and pattern of plastic response can evolve independently of the character mean (Schlichting & Levin 1986). Evidence for independent evolution of the trait and its plasticity means that a change in environmental conditions could select for both. Characters with high plasticity do not necessarily evolve less.

In modular organisms, developmental conversion can have different effects on modules of the same organism, leading to multiple phenotypes on the one individual. For example, aquatic plants often have greatly different types of leaves on the one plant, with the occurrence of different leaves reflecting the environment affecting the meristems from which they develop (Fig. 5.11). Although many authors have recognised the distinction between developmental conversion, phenotypic modulation and the expression of different structures on the one organism, there have been few attempts to assess their relative importance, or why one might be favoured over the other (Lively 1986). The most comprehensive attempt at classification of sources of variation comes from the New Zealand flora, where Lloyd (1984a) suggests that multiple structures are much more common than either pure developmental conversion or genetically distinct morphs. The same conclusions are unlikely to apply to unitary organisms.

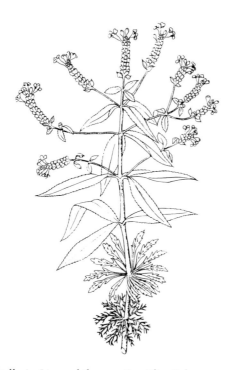

Fig. 5.11 Heterophylly in *Limnophila aquatica* After Subramanyam (1962).

So far we have considered variation and plasticity to be an adaptive (or maladaptive) response to the environment which could alter the mean reproductive success of the organism involved. There has been great interest in the adaptive significance of variance itself since the publication of a pioneering group of papers by Gillespie (1974, 1977). To illustrate this concept with an example we have already discussed, bumblebees prefer flowers that provide constant versus changeable resources, even when the mean rewards are similar (Waddington *et al.* 1981). Perhaps more surprisingly, it has also been argued that there may be trade-offs between the average success and the variance in success. Seger & Brockmann (1987) use the analogy of insurance policies. Over time we spend more on insurance than we ever receive back from insurance companies (otherwise the companies would not exist for very long). However, some of us judge that this slow loss of wealth is worthwhile to minimise the risk of the sudden catastrophic financial failure when we drive our car into the back of our neighbour's Rolls Royce. Our average wealth declines, but the variance in wealth is reduced. These trade-offs have been called bet-hedging (Slatkin 1974; Seger & Brockmann 1987; Frank & Slatkin 1990).

In evolutionary models, variance can be reduced at the expense of mean fitness in two ways. The first, which applies in the insurance analogy, is to avoid risk by reducing the variance at the level of the individual. Phillippi & Seger (1989) call this the 'bird in the hand is worth two in the bush' approach. In the simplest numerical models, it is the geometric mean rather than the arithmetic mean that is maximised during each attempt at reproduction.

The geometric mean is always less than the arithmetic mean when numbers vary, so a bet-hedger should exhibit conservative reproductive effort or foraging strategies when there is temporal variation in the environment. Great tits (*Parus major*) nesting in Wytham Wood lay on average 8.53 eggs, while the most productive clutch size is twelve. There is substantial interannual variation in reproductive success, and years that are poor for survival of young affect individuals laying larger clutches more than they affect those laying smaller clutches. These occasional bad years increase the standard deviation of fitness and hence reduce the geometric mean fitness of large clutches (Fig. 5.12). Geometric mean relative fitness peaks remarkably close to the observed clutch size (Fig. 5.12; Boyce & Perrins 1987).

The second way to reduce variance is also well depicted by an old adage: 'don't put all your eggs in one basket'. The result is the same, as geometric mean fitness over several generations is maximised, but variance in the aggregate outcome over many reproductive bouts may be reduced by increasing the variance of individual outcomes. Kaplan & Cooper (1984) call this adaptive-coin-flipping, and suggest that in environments that are temporally very unpredictable the optimum phenotype may vary from year-to-year. It may be advantageous to

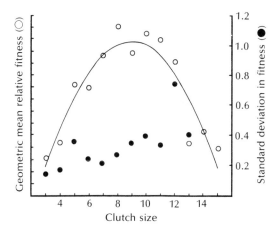

Fig. 5.12 Geometric mean relative fitness as a function of clutch size in great tits, *Parus major*, at Wytham Wood during 1960–1982. The optimum is around 9 eggs and the observed number is 8.5. Although large clutches are most productive, the returns from large clutches are also highly variable. There were insufficient very large clutches to calculate a meaningful measure of variance. After Boyce & Perrins (1987).

produce young covering a range of phenotypes, in order to raise the long-term geometric mean reproductive success. For example, breeding at a temporary versus permanent aquatic breeding site does not affect the clutch size, mean egg size or energy investment of tropical frogs. However, species which breed at temporary sites show more variable egg sizes than those breeding in permanent ponds (Crump 1981). This suggests that variable egg sizes allow female frogs to produce at least a few eggs having the appropriate size for the conditions that prevail.

This diversification of phenotypes caused by this type of bet-hedging must be regarded as ,the variable phenotypic expression of a single genotype, not as a consequence of a genetic polymorphism. Unlike a mixed ESS, which can be maintained by genetic polymorphism, the different phenotypes do not usually have the same expected fitness at equilibrium (Seger & Brockmann 1987).

THE RELATIONSHIP BETWEEN DISTRIBUTION AND ABUNDANCE

Ecology is sometimes described as the study of distribution and abundance of plants and animals. Unfortunately the properties of distribution (biogeography and niche breadth) are often treated separately from abundance (population ecology or demography). Although explicit in at least one approach to ecology (Andrewartha & Birch 1954), it is only very recently that numerous recent studies in evolutionary ecology have revived the view that this separation is unwise. Following Gaston and Lawton (1988a, 1988b), the sorts of relations which might be expected are presented in Table 5.4.

Table 5.4 Some theoretical expectations of the relationship between distribution and local population abundance and variation. After Gaston & Lawton (1988b).

	Dependent variable	
Independent variable	Local population abundance	Local population variability
Proportion or number of sites occupied	Positive correlation	Positive correlation
Body size	Negative correlation	Negative correlation
Feeding specialisation	Uncertain	Positive correlation

Abundance versus range

Widespread species are usually locally abundant relative to species with narrow ranges, at least at the centre of their ranges (Hanski 1982; Brown 1984; Schoener 1987; Gaston & Lawton 1988b). Brown (1984) suggests that widespread species are likely to be more flexible in their use of resources (have a greater niche hyperspace) than regionally restricted species. Because generalists are able to use more resources, they should also be more abundant. In a related argument, Stevens (1989) has demonstrated a repeatable relation between latitudinal range and distance from the centre of the range to the equator. He argues that because species must tolerate greater seasonal extremes of temperatures at any individual latitude, they can also exploit a greater range of latitudes (Fig. 5.13). Attempts to test Brown's model have met with mixed success. Burgman (1989) analysed the distribution of plants in a species-rich region of southwestern Australia and concluded that there was no difference between the habitat volumes of rare and ubiquitous plants, but suggested that rare plants may fill habitat space more sparsely.

Abundance versus diet breadth

As a corollary of Brown's argument, species which feed on a greater variety of foods should be more abundant than specialists, because they can exploit a greater variety of the available resources. Specialisation could in turn affect population variability in two ways (Fig. 5.14). MacArthur (1955) suggested that species with a broad dietary breadth should be less susceptible to population declines associated with the loss of their food source, and hence should also show less variability. By contrast, Watt (1964) suggests that because generalists can feed in a greater proportion of the total habitat volume, they will be able to reach high numbers more rapidly during rare conditions favouring rapid population growth. The most comprehensive attempt to distinguish these hypotheses comes from long-term surveys of British insects by the Rothamsted Experimental Station, where careful analysis

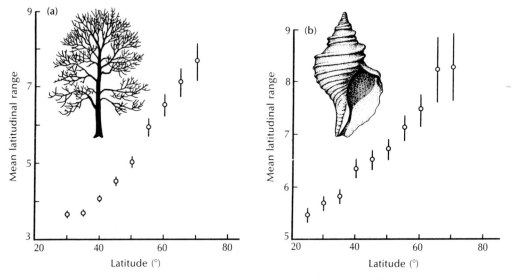

Fig. 5.13 Mean latitudinal range of North American trees (a) and marine molluscs (b) with hard body parts native to various latitudes. After Stevens (1989).

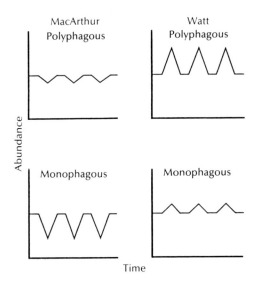

Fig. 5.14 MacArthur (1955) suggested that specialist species may have more variäble populations because of their susceptibility to population crashes, while Watt (1964) suggested that generalists will be more variable because of their potential for population explosions in favourable periods. After Redfearn & Pimm (1988).

tends to support MacArthur's argument (Redfearn & Pimm 1988). For example, there is a negative relation between variability and number of food species for aphids, and this relation is more extreme for host-alternating species than species which do not alternate (Fig. 5.15).

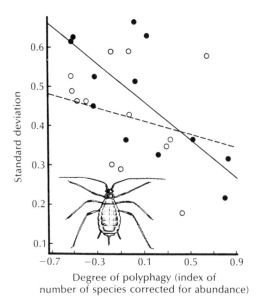

Degree of polyphagy (index of
number of species corrected for abundance)

Fig. 5.15 Standard deviation of the log-transformed yearly counts in the
Rothamsted Insect Survey from 1970 to 1983 for aphids which show host-plant
alternation (○) and which lack host plant alternation (●), plotted against an index of
polyphagy. The lines are the result of an analysis of covariance. The slope is most
steep for non-alternating species. After Redfearn & Pimm (1988).

Abundance versus body size

Big organisms are usually thought to occur at lower population densities
than small organisms, and Damuth (1981, 1987) has suggested that
log Density $\propto$ log Mass$^{-0.75}$. As the metabolic rate (MR) increases with
body size according to log MR $\propto$ log Mass $^{+0.75}$, the total energy used
by a population would be independent of body mass as log D.log MR $\propto$
log Mass$^{+0.75}$ log Mass$^{-0.75}$ = 1. However, several recent studies on
different types of organisms have concluded that very small species
also occur at low densities (Brown & Maurer 1986, 1987, 1989; Morse
et al. 1988; Lawton 1989; Fig. 9.5), though Blackburn *et al.* (1990) have
suggested that this relation is a statistical artefact. Species diversity
also peaks at intermediate body sizes (Fig. 9.5). The smallest range of
densities is typically found in the size categories containing few species,
so the body size category containing the most species is also likely to
contain the particular species with the highest population density.
Resolution of this problem awaits analysis of a larger data base.

Despite the problems of interpretation, it does seem that the slope
relating population density to body size is much shallower than estimated by Damuth, so large species may utilise a disproportionate
share of the resources available within an ecosystem (Brown & Maurer
1986). Why this should be the case is poorly understood, but Brown &
Maurer (1986) suggest that dominance in antagonistic encounters,
greater mobility and more efficient homeostatic mechanisms may all
contribute to the pattern.

The problem of devising a 'periodical table' for the classification of habitats has been a central task for ecologists, and one over which there remains considerable discussion. Southwood (1977, 1988) has recently shown that the various schemes attempting to classify habitats agree that conditions to which organisms are exposed can loosely be divided along three axes: 1) an axis of disturbance, or the extent to which habitats become disturbed; 2) an axis of adversity; and 3) an axis of biotic interactions.

The axis of disturbance

First, there is the question of the extent to which the habitat is filled with members of the species of interest. More precisely, would the addition of further individuals be inimical to the presence of those already present, or do residents prevent the addition of conspecifics either passively or actively? At first, this question seems of little relevance. We are often told that nature abhors a vacuum, so there is a common expectation that populations will grow rapidly to fill any vacancies. In fact, several types of habitat are rarely saturated. The two commonest cases are sites of very frequent disturbance (so that colon-isation is possible but the habitat decays or is destroyed before saturation is possible), or those where pressure by enemies such as predators and parasites restrict population growth. For example, corals growing on reef edges and platforms are often destroyed by cyclones. Algae growing on rocky shores are grazed so heavily by molluscs and other herbivorous invertebrates that we are hardly aware of their presence except during seasonal blooms of growth, or when something affects their predators.

There is every reason to believe that selection pressures in the two environments should differ. If habitat is vacant (either because of a recent disturbance or because the next door neighbour has just been eaten), then population growth is possible and contribution to that growth is strongly selected. Following the familiar symbols of the logistic equation for population growth (Box 5.1), selection in a popu-lation which is at low density and therefore approaches exponential growth is called r-selection, and is presumed to maximise r, the rate of increase attained without any density effects. This can be achieved by a number of demographic effects such as increased brood size or de-creased time to maturity. By contrast, selection at the carrying capacity is often called K-selection, and as population growth under these con-ditions approaches zero there will be no obvious selection on r. Instead, anything which increases the carrying capacity, K, should be selected. For example, *Drosophila melanogaster* populations maintained at high density show improved larval survival to adulthood when reared under dense conditions than do populations maintained at low density (Fig. 5.17). It is often presumed that the outcome of K-selection will be selection for interference behaviour (Pianka 1970), but in fact a variety

Box 5.1 One of the fundamental parameters that can be derived for a population is the reproductive rate, R_0. This can be expressed simply in terms of life table statistics (Lanciani 1987). Consider a number of females (S_0) born as part of a single cohort. We measure the lifetime survival $(S_x$ is the number of survivors at age $x)$ and reproduction $(F_x$ is the total number of female offspring produced by all females of age $x)$ of members of the cohort. From these parameters we can estimate the proportion of the original cohort surviving to age x $[1_x = S_x/S_0]$; the proportion surviving to age x but dying by age $x + 1$ $[q_x = (S_x - S_{x+1} + 1)/S_x]$; and the average number of female offspring produced per female of age x $[m_x = F_x/S_x]$.

$$R_0 = \sum_{x=0}^{\infty} 1_x m_x$$

or more conveniently by substitution

$$R_0 = \frac{\sum_{x=0}^{\infty} F_x}{S_0}.$$

In species with discrete or non-overlapping generations (like some annuals), R_0 measures not just the average number of female offspring produced by a female in its lifetime, but also the overall extent by which the population increases or decreases. Populations will increase when $R_0 > 1$, and decrease when $R_0 < 1$. In populations with overlapping generations the situation is a little more complicated, because population growth depends both on the birth of new individuals and on the survival of existing individuals. We therefore need to know R, the fundamental net per capita rate of increase for a population which contains N_t individuals at time t where

$$R = \frac{N_{t+1}}{N_t}$$

so over t time periods, a population which comprised N_0 individuals will grow or decline such that:

$$N_t = N_0 R^t$$

Because R_0 is the multiplier which converts one population size to another in one generation $(T$ time intervals), so $N_T = N_0 R_0$, and therefore:

$$N_T = N_0 R^T$$

and $$R_0 = R^T$$

The rate of population increases in size $(r$, the intrinsic rate of natural increase) is given by $\log_e R$ (or $\ln R$), so that the population increases when $r > 0$, and decreases when $r < 0$, and is approximated by:

$$r = \frac{\ln R_0}{T}$$

(Box 5.1 continued)

[149]

CHAPTER 5
*The habitat
templet*

We can derive a reasonable approximation for the generation time from life table statistics by computing the average length of time between the birth of an individual and the birth of its own offspring:

$$T = \frac{\sum\limits_{x=0}^{\infty} x 1_x m_x}{\sum\limits_{x=0}^{\infty} 1_x m_x}$$

so by substitution

$$T = \frac{\sum\limits_{x=0}^{\infty} x F_x}{\sum\limits_{x=0}^{\infty} F_x}$$

The net rate of increase of a population (dN/dt) can also be defined as

$$\frac{dN}{dt} = rN$$

while the average rate of increase per individual is given as

$$\frac{dN}{dt} \cdot \frac{1}{N} = r$$

so long as the population continues to grow without check. Of course it is unlikely that a population will increase in size indefinitely $(r > 0)$, despite the best efforts of our own species to prove that this is indeed possible. As a habitat is filled this possible intrinsic growth rate will not be achieved (Fig. 5.16). In the simplest case the average rate of increase will decline linearly until the carrying capacity of the habitat (K) is attained when:

$$\frac{dN}{dt} \cdot \frac{1}{N} = 0$$

For any population density

$$\frac{dN}{dt} \cdot \frac{1}{N} = \frac{-r}{K} \cdot N + r$$

or alternatively

$$\frac{dN}{dt} = rN \frac{(K - N)}{K}$$

or through integration

$$N_t = \frac{K}{1 + \left(\dfrac{K}{N_0} - 1\right) e^{-rt}}$$

where t is the time from the point where the population grows most rapidly, which in this logistic model occurs when $N = K/2$. This is called the logistic equation.

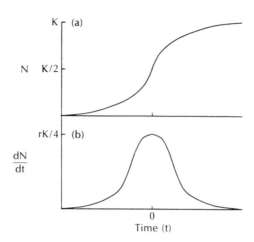

Fig. 5.16 The logistic curve: (a) the increase in population size over time: (b) the rate rate of change in population size through time. Population growth is symmetrical around the inflection point, when growth is maximal. After Hutchinson (1978).

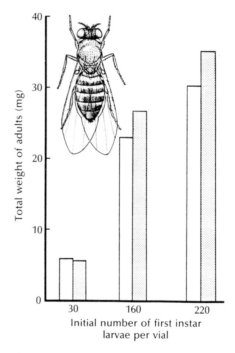

Fig. 5.17 Mean numbers of *Drosophila melanogaster* first instar larvae surviving to the adult stage from populations with different initial density. Data are the averages from lines maintained at high density (*K*-populations; stippled) and those maintained at low density (*r*-population; unstippled). After Bierbaum *et al.* (1989).

of outcomes are possible (Boyce 1984), including selection for tolerance to density (Gill 1978), even to the point of favouring amicable behaviour (Cockburn 1988). Presuming that the phenotypic outcome can be arrayed along a single gradient trivialises a complex process, a point I will stress repeatedly in this discussion of life history evolution.

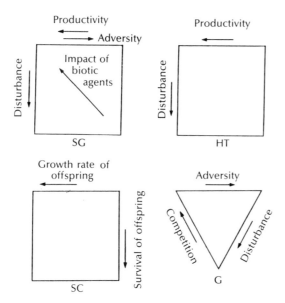

Fig. 5.18 Four attempts to classify habitats oriented so that their axes have approximately the same orientation (SG, Southwood/Greenslade; HT, Hildrew/ Townsend; SC, Sibly/Calow; G, Grime). After Southwood (1988).

Attainment of the carrying capacity will depend directly on the stability of the habitat. In particular, the habitat must remain available for sufficiently long to allow reproduction by the early colonisers. The stability of the habitat must therefore be interpreted in terms of the reproductive span of the organisms which inhabit it. A fig lasts quite long enough for fig wasps to complete their life cycle, and to compete for oviposition sites. However, to a spider monkey, a fig is only an ephemeral part of its daily diet. Southwood (1977) defines the durational stability of a habitat as H/t, where H is the time that the habitat is suitable, and the generation time of the organism is t. H/t must be at least one in order for reproduction to occur.

The generation time of an organism is also important in interpreting responses to environments when there are breaks in the periods in which reproduction may occur, or changes in the availability of breeding opportunities (these gaps may be regular, perhaps caused by seasonality, or less predictable, perhaps caused by changes in weather conditions). These responses include diapause (or escape in time) and dispersal (or escape in space), and will be discussed in Chapter 6.

Adversity

The limitations to population growth occurring in the logistic model are density-dependent; they are a reflection of the pressure imposed by intraspecific competition for limited resources, or habitat. A second type of environment is inimical to both individual growth, population growth and population size, and the limitation to density is the adversity of the habitat, which may impose its own selective pressures. Many environments are so inclement that special adaptations are necessary to allow their occupation. For example, members of the Australian ant genus *Melophorus* are typically only active when soil temperatures exceed 45°C. These ants face few competitors from other harvesting species under these conditions yet their very existence presumably reflects novel biochemical and physiological adaptation.

The exploitation of these novel habitats presumably imposes costs (a jack of all trades is a master or none), and limits the ability of the organism to do other things. For example, plants on a Canadian lake subject to high levels of exposure are poor competitors (Wilson & Keddy 1986a, 1986b).

The habitat templet

There have been a number of efforts to combine the effects of relative levels of stress and disturbance into a general classifications of the selective environment generated by different habitats. These include both general models (one based explicitly on the concepts of adversity and disturbance by Southwood (1977) and Greenslade (1983); SG; and a second describing the opportunities for growth of young and survival by Sibly & Calow (1985); SC) and classifications focused on a particular taxon (e.g. plants, Grime (1977, 1979); G), and habitat (e.g. aquatic benthic communities; Hildrew & Townsend (1987); HT).

Southwood (1988) suggests that the differences between the models are more apparent than real, and it is worth briefly reviewing their similarities. For example, they all predict three or four basic circumstances (Fig. 5.18):

1 Physically favourable and stable habitats (*K*-habitats). In these habitats biological interactions will be intense. Survival of adults may be high but survival prospects of young are poor. Possible outcomes are selection for reduced niche breadth, low mobility and increased competitive ability and investment in resistance to pathogens and predators in both adults and young.

2 Physically favourable but frequently disturbed habitats (*r*-habitats). Selection will be for the rapid production of large numbers of small offspring as soon as possible, coupled with high dispersal capacity of parents or offspring. Specialisation to particular physical conditions is less likely, though organisms can be specialised for disturbed habitats. This is the weedy or ruderal habit (Grime 1977).

3 Adverse and stable habitats (*A*-habitats). Biological interactions will be less important than tolerance to the prevailing conditions. The growth rate of young will be low and it is likely that a few large young will be produced. Longevity of established organisms might be quite high.

4 Adverse unstable habitats. Grime (1977, 1979) assumed that life under these conditions would be impossible, and restricted his classification to the preceding three alternatives. However, it is not incon-

ceivable that if mobility was exceptionally high such species could persist.

We will return to the problem to detecting suites of life history variables that have evolved in response to habitat pressures at the end of Chapter 6.

SUMMARY

The external environment is the ecological stage in which selection takes place. Environments are never completely stable, because even if physical conditions are invariant, interactions with other species also undergoing selection will place additional selection pressure on organisms for change. Reciprocal evolutionary interactions between species are called coevolution, and may take a variety of forms, and have a variety of outcomes. It is widely believed that habitat specialisation will arise because a jack of all trades is a master of none. Specialisation is likely to be best developed in environments rich in resources, and where there is pressure from competitors and other enemies. One way of achieving a level of specialisation yet coping with variable environments is through phenotypic plasticity, adopting alternative morphology and behaviour in response to environmental cues. Variable environments exert another sort of selective pressure; in certain cases it may be advantageous to hedge bets and accept a lower mean return in order to minimise the variance of return, and hence the prospect of catastrophe. The degree of habitat utilisation is intimately related to patterns of abundance. Attempts to derive a general classification of habitats emphasise the importance of disturbance, adversity and biological interactions.

FURTHER READING

Futuyma & Slatkin (1983) is an excellent series of papers on coevolution, and Thompson (1982) an interesting account of the diverse outcomes that can arise as a consequence of coevolution. Futuyma & Moreno (1988) provide a good review of the evolution of habitat specialisation. Cohen *et al.* (1990) is a compilation of papers describing some of the generalities that have arisen from the analysis of food webs. McArdle *et al.* (1990) document some of the problems that arise in comparative analysis of abundance patterns. Southwood (1977, 1988) discusses the problem of classifying habitats.

TOPICS FOR DISCUSSION

1   Consider the explanations offered by Brooke & Davies (1988) for the failure of the dunnock to evolve recognition of cuckoo eggs in the light of the evidence by Soler & Møller (1990) that host recognition can evolve extremely rapidly (see also Davies & Brooke 1989). How might these explanations be tested?

2   Room (1990) reports the extrarodinary success of a programme of biological control of the aquatic weed *Salvinia*. Discuss the types of scientific expertise required to mount a successful programme of this sort, and its cost-effectiveness. Relate your findings to the discussion of habitat specialisation in this Chapter.

3   Make a list of the bird species that occur in your immediate area. Attempt to classify them according to their abundance and geographical range as determined by a regional field guide.

Chapter 6
Reproductive Effort

Life histories can be usefully divided into four phases: growth and development prior to reproduction; recruitment into the reproductive population; the lifetime distribution of reproduction itself; and life after reproduction. In many species the demarcation between these phases is clear. For example, species with complex life cycles often complete their pre-reproductive development as larvae which lack the capacity to reproduce. By contrast, in many social vertebrates a large proportion of the males in population acquire the form of an adult (e.g. antlers and active testes in red deer), but are never recruited, as they never acquire a harem. Extensive post-reproductive life is very rare in free-living species other than humans, though it does occur in some whales (Kasuya & Marsh 1984; Marsh & Kasuya 1984).

Selection has the opportunity to act on all these phases if parents continue to influence the fitness of their offspring or relatives after they become reproductively senescent. However, in general the first three phases will be most important. Even then, field studies are only rarely able to partition all three adequately. For example, odonates have proved very simple models for measuring mating and oviposition success in the wild, but are quite unsuitable for examining the potential for selection prior to reproduction. By contrast, birds are exceptional study organisms for following young throughout both their early development and reproduction, provided that the study can be sustained for adequate duration. The frontispiece of Clutton-Brock's (1988) compilation of studies of lifetime reproductive success graphically shows an illustration of George Dunnett at the start and towards the end of the life of an individual fulmar, *Fulmarus glacialis*. Unfortunately, the 34 year study has wearied the biologist more than the bird.

In this Chapter I will first concentrate on the problem of how trade-offs could influence reproductive tactics, and how those trade-offs could be measured. I will then review (a) pre-reproductive tactics, with a special emphasis on co...p.ex life cycles; (b) the way in which reproductive effort is distributed through the life-span of an organism, and what is the determinant of life-span; and then consider two special cases of trade-offs within a reproductive bout, the latitudinal gradients in brood size in birds and mammals, and the determinants of seed size and number in plants.

TRADE-OFFS: THE COST OF REPRODUCTION

The most influential development in the theory of life history suggests that costs incurred at one stage of the life cycle must be repaid elsewhere. Because resources are used in reproduction, effort devoted to reproduction may reduce growth or survival. More formally, the Demographic Theory of Optimum Reproduction states (Williams 1966b) that:

1 When both reproductives and non-reproductives have the same limited resources available for investment, an increase in reproductive effort inevitably results in both an increase in current reproductive

output and a reduction in somatic investment (somatic costs —
e.g. reduced growth, disturbances to homeostasis, decreased time for
predator avoidance).

2   When reproduction takes place at the expense of somatic investment,
the somatic costs reduce the probability of surviving to breed again
(survival) and/or reduce future reproductive output (fecundity).

3   When reproduction results in survival and/or fecundity costs,
there is a trade-off between current reproductive output and residual
reproductive value.

4   As a corollary, this trade-off is optimised by natural selection.

Begon *et al.* (1986) present a simple algebraic definition of reproductive value. The reproductive value of an individual at age x (RV_x)
can be defined in terms of life table statistics (Box 5.1):

$$RV_x = \sum_{t=x}^{\infty} m_t \cdot p_{x \to t} \cdot \frac{N_x}{N_t}$$

where m_t is the birth rate of the individual at age t, and $p_{x \to t}$ is the
probability that the organism will survive from stage x to stage t, and
therefore produce offspring at that age. In life table terms this is
equivalent to l_t/l_x. N_t is the population size at time t. The reproductive
value generally rises then declines (Fig. 6.1). The initial rise occurs
because as an individual gets older it has a greater chance of surviving
to reproductive age. We can divide this reproductive value into current
reproductive output (m_x) and residual reproductive value (reproduction
from age $x = 1$ until death).

$$RV_x = m_x + \sum_{t=x+1}^{\infty} m_t \cdot p_{x \to t} \cdot \frac{N_x}{N_t}$$

We are interested in the reproductive effort which maximises RV_x
for all x. If the population is stable the formula is simplified because
$N_x/N_t = 1$, and so cancels out. If the population is expanding or
declining the situation is rather more complex. In an expanding population $N_x/N_t < 1$, so the value of current reproductive output relative
to residual reproductive value increases. This suggests there should be
an advantage to early intense reproductive effort among colonising
species. In a declining population $N_x/N_t > 1$, the converse is true.

The effects of current reproduction may be of two sorts. First,
current reproduction (m_x) may affect future fecundity (m_t). Second, it
may affect the probability of survival ($p_{x \to t}$).

Three broad approaches have been adopted to the measurement and
analysis of the costs of reproduction. First, we can measure phenotypic
costs as negative correlations between current reproductive effort and
future survival and reproduction. Second, we can measure the physiological investment in reproduction and maintenance. Third, we can
examine the genetic covariance between traits in order to detect
antagonistic pleiotropy between traits. Much discussion has been
devoted to the relative virtue of these approaches (Calow 1979; Reznick
1985; Bell & Koufopanou 1986; Rose *et al.* 1987). As Stearns

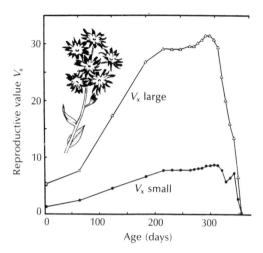

Fig. 6.1 The change in reproductive value (V_x) in relation to age in large and small *Phlox drummondii*. After Leverich & Levin (1979).

(1989) points out, '... each of these levels makes an essential contribution to our understanding. It is not a question of *either* genetic correlations *or* phenotypic correlations *or* physiological trade-offs but of how such measurements combine to deliver information about potential evolutionary responses. A study conducted at just one of these levels is likely to be as little use as the information on the nature of the elephant delivered by one blind man holding its tail.'

Phenotypic correlations

The key trade-offs important to Williams' model have been demonstrated using experimental or descriptive measures of phenotypic correlations many times. A direct relation between resource availability and reproduction has been shown many times. A famous example is the predatory ciliate protozoan *Tokophrya*, which produces a single offspring every time it eats a *Paramecium* (Kent 1981).

There are many examples of a trade-off between somatic and reproductive investment in the presence of limiting resources. For example, the onset of reproduction is often correlated with a decrease in the rate of growth and diversion of resources from somatic to reproductive functions. Indeed Bell & Koufopanou (1986) comment '... the drain on somatic tissues associated with reproduction is so widespread it may be appropriate to mention that it is not universal ...'. It is less often demonstrated that these changes in investment influence future reproductive attempts. Bell & Koufopanou point out that this will commonly be the case where reproductive performance is correlated with size, so any decrease in growth impairs future reproductive performance. Such correlations are often found in invertebrates and lower vertebrates. For example, in the grassland-dwelling isopod *Armadillidium vulgare*, reproductive females grow much less rapidly than non-reproductive females, particularly in the summer, apparently because allocation of energy to reproduction takes place at the expense of allocation of energy to growth (Lawlor 1976; Table 6.1). Because reproductive performance is extremely strongly correlated with body size (Paris & Pitelka 1962; Fig. 6.2), present reproduction should inhibit future fecundity. Trade-offs are likely to be even more conspicuous in modular organisms such as plants. A meristem committed to a reproductive role will often preclude further increase in the amount of photosynthetic tissue and development of other reproductive meristems (Watson 1984).

A trade-off between reproduction and the ability to breed again has also been repeatedly demonstrated. For example, male *Drosophila melanogaster* supplied with virgin females have lower longevity than males kept without access to females (Partridge & Farquhar 1981). The cost is an instantaneous one. Stopping the reproduction by males which have previously had access to females results in those males enjoying the same life expectancy as flies of the same age which have never mated. Further, giving unmated males access to females late in life quickly produces a life expectancy the same as in males of the same age kept with females throughout their life (Partridge & Andrews 1985). Increased exposure to males or increased egg production will also decrease the subsequent survival of female *D. melanogaster* (Partridge *et al.* 1987). This result is apparently almost universal among invertebrates, but may be reversed among mammals, for reasons which remain poorly understood (Bell & Koufopanou 1986).

Of course both future survival and future fecundity may be affected by increases in reproductive effort. Experimental manipulations of avian clutch sizes suggest that fecundity trade-offs are more common than survival trade-offs, and that trade-offs affecting the condition of

Table 6.1 Growth and energy allocation by female *Armadillidium vulgare* (Isopoda). After Lawlor (1976).

Postmoult size (mg)	Female size class (mg premoult live weight)		
	20–39	40–59	60–99
Growth rates (weight increase at ecdysis in mg)			
Spring			
Non-reproductive	4.71	6.53	7.29
Gravid	3.25	3.62	4.35
Summer			
Non-reproductive	3.91	3.13	3.10
Gravid	0.96	1.01	0.46
Energy allocation (joules expended during one moult cycle)			
Reproductive growth		41.8	49.8
Reproduction		66.9	110.5
Total		108.7	160.3
Non-reproductive growth		100.8	127.6

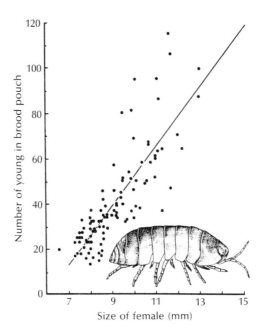

Fig. 6.2 The number of young in the brood pouch versus size of the mother in
Armadillidium vulgare. After Paris & Pitelka (1962).

offspring are more common than those affecting the parents themselves
(Lindén & Møller 1989).

The silver spoon and positive phenotypic correlations

Despite these encouraging results, there are many instances when the
correlation between current and future reproduction are positive rather
than negative, and it has been occasionally argued that these instances
are the rule rather than the exception (Bell 1984a, 1984b). First, there
are differences between individuals in the capacity to reproduce. These
might arise because of genetic differences in fitness, because of age
effects, or because of environmental effects which cause the 'silver
spoon effect', leading to some individuals born in favourable conditions
enjoying a reproductive advantage throughout their lives. Unless this
individual variation is controlled for, there will be a tendency for
positive correlations between traits to arise, although we might have
anticipated a priori that these traits would be negatively correlated.
Several examples are illustrative. A puzzling result that arose from
Lawlor's (1976) analysis of *Armadillidium* life histories was a positive
correlation between growth and reproduction for individual females,
despite the overwhelming evidence of a trade-off between early reproduction and subsequent fecundity.

In birds, there is growing evidence that individuals differ in their
individual ability to rear a clutch (Högstedt 1980, 1981; Smith 1981;
Gustafsson & Sutherland 1988; Pettifor *et al.* 1988). For example,

Pettifor *et al.* (1988) showed that number of offspring recruited from individual broods of great tits depended on the size of the brood in two ways. First, recruitment increased with the number of eggs laid. Second, recruitment declined if eggs were added to or removed from the clutch for all clutch sizes (Fig. 6.3). This implies strongly that each bird lays its optimum clutch size, but that the more fecund tits are in some sense superior to the less fecund tits. This superiority need not have a genetic basis (Chapter 3).

Correlations of this sort are likely to be reflected in the considerable variation in the magnitude of costs according to variation in the conditions to which organisms are exposed. Organisms raised in benign laboratory cultures may provide weak evidence of the magnitude of costs and trade-offs (Reznick *et al.* 1986). Field studies are not immune to effects of this sort. Experimental manipulation of clutches of the blue tit, *Parus caeruleus*, provided evidence of costs (decreased survival of adult females) in two years of a study, but not in a third (Nur 1988). In Chapter 5, I showed that rare bad years can influence the evolution of clutch size in the great tit, as birds producing large clutches are affected more in time of hardship (Fig. 5.12). Studies which ignore temporal variation in the conditions to which organisms are exposed are potentially very misleading.

However, as Partridge (1989) points out, the repeated evidence from correlational studies for costs associated with reproduction in spite of this difficulty is comforting for theory, as costs are likely to be generally underestimated.

Allocation of energy and other resources

What should be the currency with which we assess reproductive effort? A common tendency has been to assume that the energy expended in producing reproductive propagules is an adequate descriptor of investment, however, this correlation has only rarely been assessed. The results that are available are not very comforting, particularly for homeotherms like birds and mammals. For example, increased litter size in the golden-mantled ground squirrel, *Spermophilus saturatus* is not associated with great increases in daily energy expenditure (Kenagy *et al.* 1990), again probably because most squirrels work close to the maximum tolerable physiological level, but some mothers are better at converting increased energy expenditure into offspring.

Measurements of resource allocation should be particularly straightforward in plants. The most comprehensive attempt to compare the relative utility of different measures of investment concerns the allocation of carbon by plants (Reekie & Bazzaz 1987a, 1987b, 1987c). It is certainly not possible to assume that the ratio of the biomass of the inflorescence to the above-ground biomass is a realistic measure of effort. Among other reasons, reproductive structures may contribute to photosynthesis (Bazzaz *et al.* 1979). However, it does appear that the allocation of carbon is reflected in the allocation of other resources,

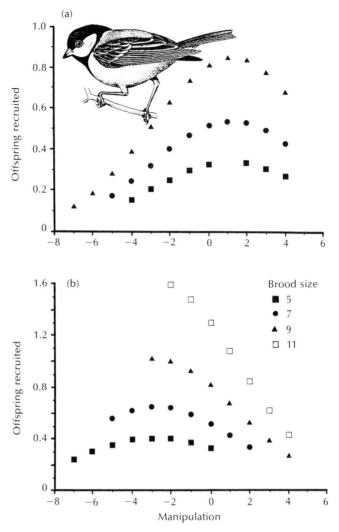

Fig. 6.3 The predictions of log-linear models describing the recruitment of *Parus major* chicks from nests which have been manipulated by the addition or removal of eggs. (a) Offspring recruited from clutches of five, seven and nine. Recruitment is maximised at or close to zero manipulation; (b) offspring recruited from final broods (original size plus manipulation) of 5, 7, 9 and 11. Except for the extreme removal nests, the broods which originally had larger clutches result in more offspring being recruited for any final brood size. After Pettifor *et al.* (1988).

and will often be a reliable indicator of investment (Reekie & Bazzaz 1987b).

Once again, a note of caution is necessary in considering modular organisms which grow partly by vegetative proliferation. In contrast to unitary organisms, where the resources that can be converted to offspring are obtained through a single mouth, the developmental plan of modular organisms means that they can generate new 'mouths' by proliferation of ramets. If a particular meristem is committed to the production of a flower, it may be at the cost of further vegetative proliferation, and hence the harnessing of new resources which could ultimately be devoted to the production of inflorescences (Watson 1984; Watson & Casper 1984).

At the opposite end of the spectrum of developmental flexibility, in animals such as nematodes and rotifers the number of mitoses in different cell lineages is strictly determined. Bell (1984b) argues that his success in detecting a negative correlation between early fecundity and later reproduction in the bdelloid rotifer *Philodina* sp. may be an inevitable consequence of a finite number of eggs, rather than strong support for the cost hypothesis. A related problem could arise in many invertebrates with complex life cycles, where individuals gather the resources they need to produce eggs during the larval stage, and metamorphose into adults with a fixed complement of eggs (Godfray 1987). When and where trade-offs will arise is not always clear.

Genetic correlations

We have already seen examples where phenotypic and genotypic correlations work in opposite directions (Chapter 3). Despite some positive results (Cheverud 1988), there is no good evidence of a general rule relating genetic and environmental correlations (Bell & Koufopanou 1986). This has led some authors to advocate restricting discussion of costs to cases where the genetic correlations between traits are known, as only where genetic correlations are present is there likely to be a response to selection (Reznick 1985). However, as Stearns (1989) points out, there will be cases where there are genetic correlations between two traits, but so much of the variation within and between the two traits has an environmental basis that the trade-off will not be present in the phenotype and selection will not occur.

There are only a few measurements of genetic correlations for life history traits. Such correlations can be measured in two ways. First, we can use the breeding designs of quantitative genetics to measure the correlations between a suite of life history variables (Falconer 1981). Using this procedure, many correlations between life history traits are positive, a result that at first sight seems unsupportive of the cost hypothesis (Bell & Koufopanou 1986). The most likely explanation is that these data come from inbred lines (Rose 1984). The effect of both inbreeding and the origin of new inbred lines as a consequence of mutation will be deleterious, so lines formed in this way will usually have inferior fecundity and lower survival. Most genotypes will express low values of both fitness components, and only in a very few cases will both be improved. For example in a study of mutants of barley, all lines produced heads earlier than the source line, and the overwhelming majority had a lower kernel yield (Gaul 1961; Bell & Koufapanou 1986). Thus the procedure of genetic investigation increases the prospect of positive correlations between fitness components. It is also likely that rich or novel conditions will affect all fitness components in the same way, so only in outbred populations in stressful conditions will we see the anticipated correlations. Of course, it is in these latter conditions that most ecologists are interested. A more pernicious problem arises when we detect weak positive correlations between a pair of life history traits of interest. Charlesworth (1990) shows that this does not necessarily mean that traits are free to respond independently to selection, because they may well be imbedded in a larger set of characters that are subject to constraints. Under these circumstances, traits may reach an equilibrium a considerable distance from their individual optimum. Second, we can select for increase or decrease of a particular trait and observe the correlated changes in other traits. A decline in one component of fitness in response to selection for another component is good evidence of genetic trade-offs through antagonistic pleiotropy (Chapter 2).

Whether antagonistic pleiotropy represents a permanent restraint on evolution is weakly understood. Correlations can arise through both linkage disequilibrium and pleiotropy. Disruption of negative correlations caused by linkage will ultimately occur because of crossing over. In plants pollinated by the wind or by unspecialised insect pollinators, all morphological traits are highly correlated (Berg 1960), presumably because all traits respond to general mechanisms of growth regulation, or respond in a similar way to the environment (Riska 1986). By contrast, in plants with specialised insect pollinators, vegetative traits are correlated with other vegetative traits and floral traits are correlated with other floral traits, but correlations between vegetative and floral traits were usually close to zero (Berg 1960). The necessity for close matching between the flower and the pollinator has led to strong selection for close developmental integration of flowers and reduced their susceptibility to the factors that cause variation in vegetative traits (Riska 1986).

GROWING UP: LIFE BEFORE REPRODUCTION

Reproductive performance is intimately linked to acquisition of condition and resources in the pre-reproductive phase, as we have already seen for *Armadillidium* (Fig. 6.2), and in development of wood frogs (Table 2.6). In considering variation in reproductive performance we can now return to one of the questions raised in Chapter 1: can we explain the extraordinary diversity of pre-reproductive patterns observed in plants and animals? Starting to breed as soon as possible appears to be highly advantageous in expanding or constant populations. The earlier young start to spread through the population the sooner they start to have grandchildren, so fitness accrues in a manner analogous to compound interest. For these reasons, decreasing the time to maturity by a given proportion has a much more important effect on fitness than increasing fecundity by the same proportion (Lewontin 1965a, 1965b). Indeed it is difficult to see why time to reproductive maturity should ever be delayed, all other things being equal (Sibly & Calow 1986).

Nonetheless, phenotypic variation in time to maturity is commonly reported, and it is clear that a proportion of this variation has a genotypic basis (e.g. in insects—Dingle *et al.* 1977; Istock 1983, and amphibians—Berven 1987).

Diapause and dormancy

Many organisms undergo either a facultative or obligate diapause, or delay in development. This may be associated with a distinct life cycle stage, such as a dormant seed, or the pupa of an insect, or through more subtle delays in reproduction, such as arrest of development in the embryo of a kangaroo. Diapause is an adaptive response to temporarily unfavourable conditions, and may represent an escape from either high density, or poor conditions. The commonest environmental cause of delay will be the relatively predictable changes to breeding conditions associated with seasonality, though theory suggests that delayed reproduction can be adaptive in environments that fluctuate randomly (e.g. Tuljapurkar 1989, 1990).

The diversity of responses to seasonality reaches its pinnacle in the arthropods, where it is the dominant element of life cycle evolution (e.g. Bradshaw 1986; Tauber *et al.* 1986). The response to seasonality will be dictated directly by the ability of an organism to mature within the season favourable for reproduction. For example, organisms as diverse as voles and wasps will mature rapidly and attempt to breed if born early in the season, but will adopt a slower growth trajectory or enter diapause if born late in the season (Cockburn 1988). Despite continuous variation in time to maturity, there is a strong negative correlation between the time to maturity in parents and offspring in the collembolan *Orchesella cincta*. This has the effect of ensuring a positive correlation between offspring and their grandparents, which experience similar environments, in contrast to parents and their offspring, which experience very different environments (winter and summer) (Janssen *et al.* 1988).

In many circumstances the population will comprise a mixture of individuals born in the current season and those emerging from reproductive diapause. One interesting consequence of dormancy will be the tendency of the dormant pool to influence contemporary genetic constitution because the genetic constitution of the dormant seed pool

Table 6.2 The effect of extent of habitat disturbance on the number of different genotypes of *Capsella bursa-pastoris* determined by enzyme electrophoresis. From Bosbach *et al.* (1982).

Number of different genotypes	1	2	3	4	5	6	7	8	9	10
Number of populations										
Less disturbed	12	5	1	—	—	—	—	—	—	—
Highly disturbed	—	4	3	2	4	2	1	1	—	1

will reflect past rather than present selection pressures (Templeton & Levin 1979; Brown & Venable 1986; Levin 1990). There are only a few empirical investigations of this phenomenon. For example, populations of the extremely phenotypically variable weed *Capsella bursa-pastoris* show much higher genotypic diversity in disturbed habitats than in areas where soil disturbance is limited (Table 6.2; Bosbach *et al.* 1982). A large proportion of the seed is buried by a variety of processes, including transport by earthworms (Hurka & Haase 1982). Seed may be stored for as long as thirty years (Salisbury 1964). In undisturbed soils most of the crop will be derived from the seeds near the surface which are those most recently falling to the ground. By contrast, soil disturbance will bring old seeds to the surface and may contribute to very high genetic diversity among the seedling population. In addition, it appears likely that there will be more intense selection in undisturbed populations, possibly restricting the number of seedling genotypes that can persist. Most important, however, is the observation that the seed and seedling genotype pool can differ.

The most convincing evidence that dormant propagules can retard evolution comes from the copepod *Diapotomus sanguineus* (Hairston & DeStasio 1988). These copepods produce two sorts of eggs; eggs which hatch immediately, and eggs which can enter diapause, largely to avoid predation by fish (*Lepomis* spp.). Most diapausing eggs hatch soon after they are produced, and make up the winter population, but some eggs do not hatch because they become buried in sediments. A drought in 1981 in one pond (Little Bullhead) exterminated the fish, leading to a much later date of switching from the production of immediate eggs to diapausing eggs (Fig. 6.4). There is evidence that the trait is genetically mediated. This supports a direct role for predation by fish in timing of diapause because a nearby pond (Bullhead) showed no such effect. The second consequence of the drought was a population explosion during 1983 to 1985 of the predatory fly larvae *Chaoborus americanus*, which are normally controlled by the fish. The *Chaoborus* killed most of the emerging copepod population for two consecutive years, until it in turn crashed in the summer of 1985, presumably as a result of starvation. The following winter the copepods successfully bred, apparently from eggs laid in 1982–1983 and before. However, the copepods maturing from this dormant propagule pool did not match the behaviour of the eggs in 1983. Instead there was a broad distribution of time of switching from intermediate eggs, suggesting that both pre-drought eggs and drought eggs were contributing to the adult population (Fig. 6.4). The return to the 1983 behaviour had started to occur in 1987, but evolution was clearly retarded by dormancy.

Complex life cycles

Despite the pervasive myth that humans are the pinnacle of evolution towards complexity, our own species has a rather simple life cycle. We are born as small versions of our adult form, we are diploid throughout

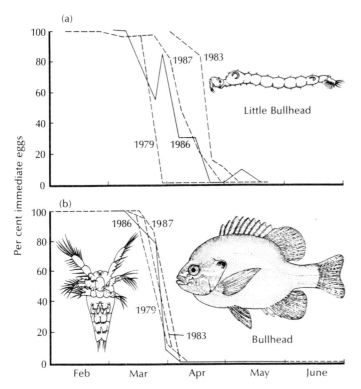

Fig. 6.4 The timing of switch from production of immediate to diapausing eggs by the copepod *Diaptomus sanguineus* in two ponds. (a) Little Bullhead pond, where fish were killed by a drought in 1981, and there was a population explosion of *Chaoborus americanus* from 1983 to 1985. (b) Bullhead pond, where fish were present throughout the study. After Hairston & DeStasio (1988).

our lives after the formation of the embryos from which we develop, and we always reproduce sexually. A very large proportion of plants, animals, protists and fungi have more complex life cycles. Complexity can arise in two ways. An individual may pass through two or more very different morphological forms to reach adulthood. For example, tadpoles metamorphose into adult frogs. Such metamorphosis is found in the great majority of phyla of animals (Werner 1988). Second, the offspring of one morphological form may differ from the offspring of another form either in ploidy, as in the alternation of generation of plants, in mode of reproduction, as in the cyclical parthenogenesis of aphids, or in habitat requirements, as in the various hosts required to complete some parasite life cycles.

The evolution of these complex life cycles has not been well treated in the theory of life history evolution, and it has proved even more intractable to empiricists. For example, association of adult and larval forms is often too difficult to allow individuals to be followed throughout their lives in the field.

Trade-off between dispersal and growth

Most theoretical treatments of complex life cycles have emphasised that opportunities for dispersal, growth, and numerical expansion might differ between habitats. Because morphological specialisation for growth and numerical expansion are potentially inimical to specialisation for dispersal, organisms use one life history stage to achieve growth and the other stage to achieve dispersal (Wilbur 1980). For example, larvae of anurans and holometabolous insects usually have little capacity for dispersal, but grow rapidly prior to metamorphosis. Adults may disperse from their natal site to reproduce. By contrast, many marine invertebrates such as barnacles metamorphose from a dispersive larval form and only grow very rapidly on their recruitment into the adult population (Fig. 6.5).

Optimising growth schedules

The major alternative to this view has been provided by Earl Werner, who argues that almost by definition, metamorphosis within a life cycle implies a shift in the resources used by the separate forms (Werner & Gilliam 1984; Werner 1986, 1988). These changes are almost always more dramatic than the resource shifts that occur during the growth of individuals that do not metamorphose. For example, tadpoles are usually aquatic herbivores and adult anurans are terrestrial carnivores. They cannot easily be classified as adapted for growth or dispersal as in some species post-metamorphic growth accounts for 80 per cent of the body mass of the adult. Werner's central argument is that problems of scaling are an inevitable consequence of increase in size

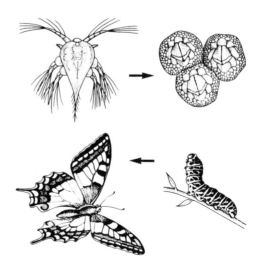

Fig. 6.5 The transition from larval to adult form often reflects change from a form specialised to dispersal to a form specialised for growth; however, the direction of the transition differs for barnacles and butterflies.

during development, and often necessiate a shift in the ecological niche. This shift forms the impetus for the evolution of complex life cycles. For example, consider the changes in the physical parameters most important to an organism as its size increases (Table 6.3).

Werner (1988) follows Cohen (1985) in suggesting four solutions to the problems of change in size

1 Remain very small as an adult, as do nematodes, rotifers and protozoans.

2 Modify or multiply the basic trophic unit so that the organism grows larger but the basic relationship between organisms and the resources they consume stays the same. Tapeworm segments, coral polyps and some plants are examples.

3 Heavily provision the egg with yolk or provide parental care to the young so that the young can take up the parental life style. Birds and mammals are the quintessential parental provisioners, even though the egg at fertilisation is exceedingly small. The kiwi depicted in the Chapter vignette is remarkable for producing a single enormous egg.

4 Develop a complex life cycle and use different resources throughout development. Werner (1988) points to an almost continuous gradation in the effects of changing food resources on the morphology of the feeding form. For example, some fish change diet as they develop with no morphological change, while postlarval flatfish like flounder exhibit a massive metamorphosis from a typical cod-like larva to a flat bottom-dwelling fish with the neurocranium and one eye rotated through 90°.

The optimum size at transformation. The time of transition from larva to adult varies within and between species. For example, some frogs metamorphose within the egg and some are even born in an adult morph, while others undergo prolonged tadpole phases. Once again, the appropriate theory is in its infancy. Werner (1986) contrasts organisms occupying two habitats which provide different age-specific growth rates (g_{larval} and g_{adult}), and mortality rates (μ_{larval} and μ_{adult}). Hypothetical values are provided in Fig. 6.6, where the transition point that will maximise lifetime growth rate is given by s^*. However, in this

Table 6.3 Crucial physical parameters and structures for small versus large organisms. After Horn *et al.* (1982) and Werner (1988).

Function	Small organism	Large organism
Swimming or flying	Viscosity	Inertia and streamlining
Locomotion and support	Surface tension, molecular and electrostatic cohesion	Gravity
Structure (nonaquatic species)	Exoskeleton	Endoskeleton with tensile elements
Physiological transport	Diffusion	Convection, stirring, and circulation

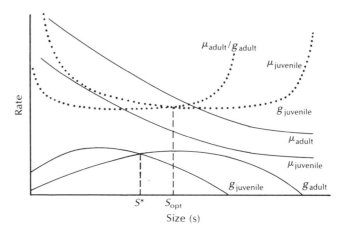

Fig. 6.6 The optimum size at metamorphosis (s_{opt}) in species with complex life cycles. Two habitats (juvenile and adult) offer different size specific-growth rate (g) and mortality rates (μ). The optimum size for metamorphosis is given when $\mu_{adult}/g_{adult} > \mu_{juvenile}/g_{juvenile}$. For this case $(\mu_{adult} > \mu_{juvenile})$, s_{opt} is delayed relative to the optimum predicted on the basis of size alone (s^*). After Werner (1986).

case there is a trade-off between growth and mortality, and mortality is greater in an adult (the values are hypothetical, the situation could easily be reversed). If we now seek the optimum trade-off between growth and mortality (maximise g/μ) the switch of habits is delayed. Werner (1986) elaborates this model with a list of predictions and possible further developments of this theory (see Ludwig & Rowe 1990 for a recent dynamic optimisation model).

Performance during one phase may influence fitness in another. One of the problems associated with analysis of complex life cycles is the influence that variance in performance in a cryptic phase may have on fitness in another phase. These effects may be environmental or pleiotropic in origin. Because field studies only rarely follow individuals throughout a complex life cycle, it is easy to underestimate the importance of this covariance. For example, adult fitness in the salamander *Ambystoma talpoideum* is strongly dependent of larval performance (Semlitsch 1987; Semlitsch *et al.* 1988). The effects could be quite subtle. Larval amphibians have been shown to discriminate kin from non-kin (Blaustein *et al.* 1987), and some evidence suggests that this ability is retained through metamorphosis (Blaustein *et al.* 1984; but see Waldman 1989). Because kin recognition appears to depend on phenotype matching, or comparison with a model learned early in development, the retention through a major metamorphic shift is rather surprising.

Alternatively, there is recent evidence of complex reversals in behaviour associated with metamorphosis which suggest that behaviour of adult and larval form can be decoupled. For example, the larval form of the salamander *Ambystoma talpoideum* is aggressively superior to larvae of its congener *A. maculatum* (Walls & Jaeger 1987). However, post-metamorphic *A. maculatum* are superior competitors to *A. talpoideum* (Walls 1990). Such reversals could conceivably facilitate species coexistence (Wilbur 1980).

Costs of complex life cycles

Istock (1967) proposed that complex life cycles would be unstable if the rates of evolution in each habitat was not the same, or if both habitats were not saturated. He reasoned that if one stage provided a bottleneck which constrained the other there would be strong selection for its elimination. In fact, we know rather little of the patterns of recruitment, evolution and population regulation in organisms with complex life cycles (see Strathmann 1985; Roughgarden *et al.* 1988; Olson & Olson 1989; Hughes 1990 for some recent approaches). Nonetheless there is often evidence of great stability in such life cycles without much evidence of the balance suggested by Istock.

A case study; host-plant alternation in aphids

The problems of habitat specialisation, complex life cycles and seasonality interact in a fascinating way in the use of host plants by aphids, in an example which also sounds a caution against adaptive interpretations of habitat switching and life history evolution (Moran 1988). Some aphids have a complex life cycle which entail seasonal switching between host plants which are often taxonomically and physiognomically quite unrelated (heteroecy). For example, one species specialises on both a shrub (sumac, *Rhus*) and mosses. Aphids exhibit cyclical parthenogenesis, the alternation of a sexual generation with parthenogenesis. In these heteroecious life cycles mating and the production of sexual eggs take place on a primary host, which is also utilised by the fundatrix, a parthenogenetic female which emerges in the spring from a sexual egg. Aphids have a rapid generation time, aided by parthenogenesis, which can even allow some young to develop inside their mothers before their mothers are born. Many generations are therefore possible in each growing season. Some of the subsequent parthenogenetic females descending from the fundatrices utilise a secondary host. Cladistic analysis suggests that heteroecy has evolved from the use of a single host (monoecy) several times. Monoecy has occasionally been reacquired secondarily, though an alternative reduction of the complex life cycle involves dispensing with the sexual generation and becoming fully parthenogenetic (e.g. Moran & Whitham 1988).

The evolution of heteroecy has a number of correlates (Moran 1988). The range of secondary hosts is usually much greater than the range of primary hosts, both within and between closely related species. Heteroecious species more frequently induce galls or cause tissue damage, and they also show greater morphological specialisation or poly-phenism. In particular, the fundatrix is often highly modified, charac-terised by a loss of wings and reduction in the sensory apparatus, presumably to allow increased abdominal size and hence the production of embryos.

Several hypotheses have been developed to account for the evolution of heteroecy (Moran 1988). These fall into two classes. The first hy-potheses suggest that the switching between hosts is adaptive, and represents optimisation of available resources, an option better available to aphids than most organisms because of the telescoping of generations. For example, because host-plant switching often (though not always) involves movement from a woody to a herbaceous host, it has been suggested that the decline in phloem quality throughout the season forces movement to a herbaceous host. Alternatively, it has been suggested that the secondary hosts are best for growth, but the primary host represents high quality sites for oviposition and egg survival. Because some woody plants will selectively abscise leaves with galls, it has even been suggested that movement from one host to the other represents escape from defences induced in the plant.

The second hypothesis suggests that specialisation in the fundatrix prevents complete switching to alternative hosts, even if such switching is adaptive. This can only be achieved with parthenogenetic females, which therefore lose any benefits associated with sexual reproduction (Chapter 7). Several lines of evidence provide strong support for this latter hypothesis (Moran 1988). First, there is the greater specialisation on the primary host. Second, the repeated evolution of heteroecy, which is rare in other groups, indicates selective pressures peculiar to aphids. Third, the fundatrix of secondarily autoecious forms tends to resemble the summer generation of the heteroecious forms rather than the winter fundatrix (Fig. 6.7), suggesting that heteroecy involves a quantum step from past host-plant affiliation. Fourth, host-plant associ-ations of heteroecious species are occasionally very ancient, and hardly consistent with optimisation of varying resources (Moran 1989). Last, most of the large and species-rich groups of aphids specialising on herbaceous angiosperms appear to be derived from the rare occurrences of life cycle reduction (Moran 1990). Collectively, a complex life cycle both arises from a constraint and constrains further evolution in aphids (Moran 1988).

THE TEMPORAL DISTRIBUTION OF
REPRODUCTIVE EFFORT

Maximising the age-specific distribution of reproductive effort once reproduction has commenced is the second central problem of life history evolution. In the ensuing discussion I analyse three problems. First, what causes the deterioration of survival in organisms as they grow old (senescence)? Second, when should reproductive effort be sufficiently intense to hinder further survival prospects, with particular

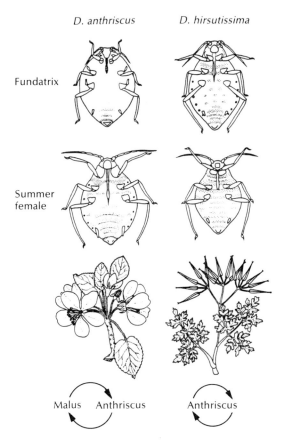

Fig. 6.7 Acquisition of secondary autoecy in *Dysaphis* aphids. The fundatrix of the autoecious *D. hirsutissima* resembles the summer morph of the heteroecious *D. anthriscus* rather than the specialised fundatrix. After Moran (1988).

attention to the extremes of breeding only once (semelparity) and many times (iteroparity)? Third, does an iteroparous organism late in its life start to behave like a semelparous organism because its residual reproductive value declines?

Senescence: why grow old?

The bodies of animals senesce, or deteriorate as they grow old, leading to a decline in fertility or life expectancy late in life (Fig. 6.8). The majority of animals do not die from senescence, which is thus most conspicuous in animals protected from other natural hazards. Members of affluent human societies with well developed medicine serve as an excellent example. The mechanistic basis of senescence is surprisingly not well understood, but involves attacks on the normal cellular machinery and DNA by radiation, highly reactive products of metabolism such as oxygen radicals, and other toxic products which accumulate through time (Gensler & Bernstein 1981). Failure in the control of gene regulation is probably very important. Some deleterious mutations

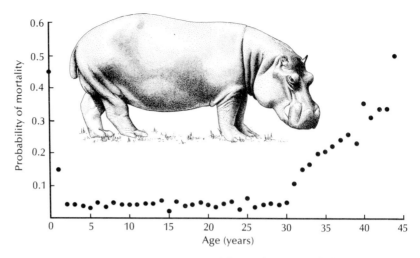

Fig. 6.8 Senescence in *Hippopotamus amphibius*. After a period of low juvenile survival characteristic of all mammals, the age-specific survival probability stabilises until thirty years of age, when senescence becomes evident. After Harvey *et al.* (1989).

such as Huntington's chorea are also expressed late in life. Alzheimer's disease, a conspicuous accelerator of ageing in humans, is increasingly thought to have a genetic basis.

Evolutionary versus non-evolutionary hypotheses?

Damage to cells appears to be almost inevitable, unless they can be sequestered in some manner from exposure to hostile agents. The non-evolutionary view of senescence would suggest that the rate of decay is simply correlated with exposure to hazards. Australians of European origin usually have drier, more wrinkled skins than their same-aged relatives in their ancestral country, presumably because of their greater exposure to solar radiation. Yet this cannot be the complete story.

Bell (1984c) compared the pattern of fertility with age among cultures of asexual species reproducing by fission (two oligochaetes) and those reproducing by eggs (two rotifers, and ostracod and a cladoceran). Survival rate declined with age in the egg-producing taxa but not in the oligochaetes. Although this comparison is necessarily confounded phylogenetically, the absence of senescence in cultures where there is no reproductive trade-off (because adults always split in two) is compatible with evolutionary theories but incompatible with theories attributing senescence to natural hazard, which is constant across both groups.

Do plants senesce?

Just as analysis of metazoans reproducing by fission is revealing, it is of obvious interest whether senescence can be detected in the age-specific fecundity of modular organisms. Because the germ line is not sequestered, and reproductive modules are often derived from very old meristems, individual cell lines may be of great antiquity, and therefore have extreme exposure to natural hazards without the same recourse to recombinational repair found in unitary organisms. Age-specific reproduction in plants will depend on whether plants have a single apical meristem or many meristems (Watkinson 1988). Maximum re-productive effort in plants with a single meristem is strongly constrained because leaf number remains constant as size increases, so reproduction typically increases rapidly then stabilises. Any decline in reproduction late in life may be a consequence of increased difficulties in vascular transport as the plant gets taller. The same constraints do not apply where there are many meristems. Fecundity typically increases with the age of the plant, but this is a consequence of the confounding effect of increased size and number of meristems (Watkinson & White 1986). Unfortunately, the increase in size also hampers detection of senescence. Survival may ultimately decline with age, but this may also be confounded by size, as large plants are more subject to natural hazards such as lightning and windfall (Watkinson 1988). Concurrent mechanical damage may increase susceptibility to disease. Watkinson (1988) concluded that the best evidence of senescent decline in fecundity came from the shrub *Acacia suaveolens*, where fruiting declines expo-nentially with age. There is strong emphasis on early reproduction in this species because of the high risk of death of the adult by burning (Auld 1987).

Another factor confounding the detection of senescence in plants with many meristems is the possibility of selection between those meristems (or modules). In Chapter 3, I expressed scepticism that somatic mutations are an innovative evolutionary agent promoting survival in variable environments (e.g. Whitham & Slobodchikoff 1981; Gill & Halverson 1984). However, Klekowski (1988) correctly points out that the irregular distribution of mutations among cell lines may lead to accumulation of greater mutational load in some modules. Death of these modules may not be detected in measurements of senescence. Further experimental investigation seems warranted.

Antagonistic pleiotropy versus mutation−accumulation

There are two evolutionary hypotheses which attempt to account for senescence in terms of the age-specific effects of mutations. The mutation−accumulation hypothesis suggests that the mutation/ selection equilibrium will change according to the age of the animal (Medawar 1952; Edney & Gill 1968). This is because a deleterious gene affecting reproduction will have a greater cumulative effect if expressed early in the life-span than if expressed late in life. Selection will be more potent against these early acting deleterious alleles. The antagon-istic pleiotropy hypothesis suggests that alleles with beneficial effects early in life may have deleterious effects later in the life-span, and vice versa. Following the same logic as the mutation−accumulation hy-pothesis, the greater selective value of early favourable mutations will cause selection against prolonged life, and lead to senescence (Williams 1957).

Unsurprisingly, most of the experimental tests of these hypotheses come from *Drosophila*. Some evidence is broadly consistent with both hypotheses. For example, strong artificial selection for early reproduction (killing animals soon after they attained reproductive maturity) shortens life-span in both *Drosophila* (Mueller 1987) and *Tribolium* (Sokal 1970). This may be because selection can no longer restrain the accumulation of mutations, or because the selective tension generated by antagonistic pleiotropy swings further in the direction of early reproduction. Rose & Charlesworth (1980) argue that the mutation–accumulation hypothesis predicts an increase with age in additive variance for age-specific fitness components, as selection should depress additive variance, which would depend for its maintenance on the relative importance of mutational input. There is no evidence of the predicted effect on egg-laying rates (Rose & Charlesworth 1980), but variance in mating activity does increase with age (Kosuda 1985). Selection for postponed senescence is effective in *Drosophila*, but depresses fecundity early in life (Fig. 6.9), partly because of the pleiotropic consequence of reduced ovary size (Rose *et al.* 1984). Flies selected for postponed senescence show increased resistance to stressors including starvation, desiccation and ethanol toxicity (Service *et al.* 1988). Selection on these flies in the opposite direction reversed the effects of resistance to starvation and early fertility, compatible with the antagonistic pleiotropy hypothesis. However, there was no reversal of the effects on resistance to stress from desiccation and ethanol. This suggests that additive variance in these traits had been depleted

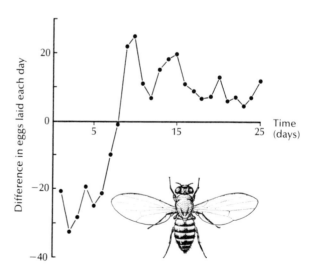

Fig. 6.9 The difference in the number of eggs laid each 24 h by *Drosophila melanogaster* between individuals selected for their ability to survive for 21 days and control individuals. The good late performance reflects the poor survival of control lines to that age. After Rose & Charlesworth (1980).

by selection for long life, and is compatible with the mutation–accumulation hypothesis. In summary, the *Drosophila* data support a conjoint influence of antagonistic pleiotropy and mutation–accumulation on senescence.

Even more convincing data for the antagonistic pleiotropy hypothesis come from the study of *Caenorhabditis elegans*, a nematode that the herculean efforts of Sydney Brenner and his colleagues seems likely to ensure will be the first metazoan whose DNA sequence will be completely known, as is its development (Sulston *et al.* 1983). Individual alleles can be identified which extend life-span. These are of two sorts: those which express their effects in animals which are already post-reproductive, and this cannot enhance fitness (Johnson 1987), and those which lengthen life but dramatically reduce fertility. Friedman & Johnson (1988) showed that a mutation to the *age-1* gene greatly increased life-span but led to a 75 per cent decrease in hermaphrodite fertility.

Because human genetic disease is subject to exhaustive clinical analysis, it is curious that evolutionary hypotheses have not been applied more often to human data, though perhaps the hopes for a 'cure' for senescence do not fit comfortably in this framework. The long post-reproductive life of humans would be compatible with mutation–accumulation. However, Albin (1988) suggests that a variety of genetic diseases are also consistent with the antagonistic pleiotropy hypothesis. For example, there is some evidence that Huntington's disease, a neurological degenerative disease with simple inheritance, is correlated with enhanced fecundity. Other candidates are idiopathic haemochromatosis and peptic ulcer disease.

Semelparity versus iteroparity

The dichotomy between breeding once and breeding many times has been a central preoccupation of life history theory. Cole (1954) was the first to realise that a useful way to approach this problem theoretically is to ask how much more fecund must a semelparous organism be in order to compensate for the abandonment of future reproduction. His analysis provides the following paradox. In a population of annual semelparous organisms which all survive to breed but die thereafter, the rate of increase λ_{annual} is:

$$\lambda_{annual} = N_{t+1}/N_t = b_{annual}$$

where b_{annual} is the yearly fecundity per individual, and N_t is the number of individuals reproducing at time t. If there is no death after reproduction, the individuals present at time t will also breed in the next generation. We can calculate the rate of increase of an immortal perennial or iteroparous organism with $b_{perennial}$ offspring as:

$$\lambda_{perennial} = (N_t \cdot b_{perennial} + N_t)/N_t = b_{perennial} + 1$$

The rate of increase is equal when $\lambda_{annual} = \lambda_{perennial}$, which is true when $b_{annual} = b_{perennial} + 1$, so an annual which produces one more offspring than the perennial is as fit as the immortal perennial. For many organisms, this increased fecundity seems easily achieved, so elaborations of the model have been sought which explain the persistence of iteroparity. Charnov & Schaffer (1973) allowed the probability of surviving to breed ($p_{juvenile}$) and surviving to breed again (p_{adult}) to differ and be less than 1. Under this conditions fitnesses are equal when:

$$b_{annual} = b_{perennial} + p_{adult}/p_{juvenile}$$

Because adult survival is very much higher than juvenile survival for many organisms such as the hippopotamus (Fig. 6.8), semelparity is unlikely to be favoured relative to iteroparity. These mathematical arguments have been extended to allow for the effect of time to maturity, time between reproductive episodes, and senescence (Young 1981), and more recently, the effect of variation in survival imposed by a stochastic environment (Orzack & Tuljapurkar 1989; Tuljapurkar 1989).

Testing these hypotheses is confounded by the problem of circularity. It is certainly true that the proportion of the energy budget devoted to reproduction by semelparous animals and plants greatly exceeds that of comparable iteroparous species (Calow 1979). It is no accident that we derive an enormous proportion of our staple foods from annual grain crops (Table 6.4). However, once a semelparous life history has evolved, further reproductive commitments may take place in response to this evolution. For example, males in the dasyurid marsupial genus *Antechinus* typically die shortly after they mate. Spermatogenesis fails some months before the annual rut (Kerr & Hedger 1983). Males can be nursed through the period of mortality, but are sterile for the rest of their lives because spermatogenesis does not resume, and any sperm they had left at the end of the mating is lost in the urine months before females become fertile again. Why spermatogenesis fails is unclear, but it may be that there is no selection for its maintenance because semelparity is inevitable in any case. This has the evolutionary consequence of locking *Antechinus* into an evolutionary dead-end of semelparity, because restoration of spermatogenesis is unselected, yet is necessary before the reacquisition of iteroparity, which might be selected, is possible. Exactly the same may be true of the exaggerated reproductive investment of most semelparous plants and animals. Therefore it is not possible to claim that a comparison of the investment

Table 6.4 Approximate proportion of annual net assimilate devoted by herbaceous plants to reproduction. After Harper (1977) and Calow (1979).

Type of plant	Approximate range (per cent)
Herbaceous perennials	1–15
Wild annuals	15–30
Annual grain crops	25–40

by semelparous and iteroparous organisms is a definitive test of theory, as reproductive effort may become progressively exaggerated as commitment to semelparity increases.

One compelling example where the genetic basis of life history evolution and the principal selective agent have been established is the evolutionary response by guppies to predation pressure (*Poecilia reticulata*). Differences in predator pressure are known to influence male coloration, morphology and behaviour. Reznick & Endler (1982) have contrasted the behaviour of free-living populations which were exposed to the different levels of predation which result from variation in the population density and foraging behaviour of other predatory fish. At a site where adults were subject to heavy predation by *Crenicichla alta*, but juveniles generally escaped predation, guppies allocated a higher proportion of their body reserves to reproduction, had a shorter interval between broods, produced smaller young and matured at smaller weights than guppies at a site where *Rivulus hartii* imposed moderate predation, but predominantly on juveniles (Fig. 6.10). At the third site where there were low levels of predation on all age classes, largely by *Aequidens pulcher*, the population exhibited reproductive patterns that combined elements of both these patterns. In the case of the guppies, all these traits have high heritabilities, which led Reznick & Endler (1982) to speculate that introduction experiments would rapidly produce the optimum phenotype. In one of the most

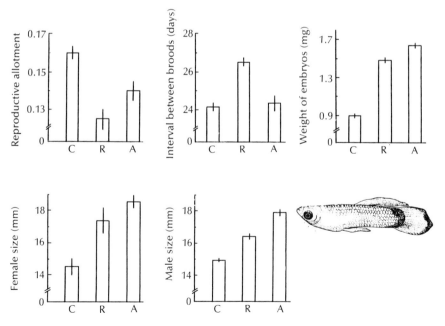

Fig. 6.10 Life history variation in guppies in three streams with differing predation risk, C, high predation of adults; R, moderate predation of juveniles; A, low predation on all age classes. Reproductive allotment is the proportion of female biomass allocated to reproduction. Vertical lines denote standard errors. After Reznick & Endler (1982).

exciting direct experimental tests of predictions concerning life history evolution, changing the age-specific predation on a population from an adult bias to a juvenile bias led to dramatic life history evolution in the predicted generation in only eleven years (as few as 30 to 60 generations; Reznick *et al.* 1990).

However, such predator-induced shifts may on occasions be a reflection of phenotypic plasticity. For example, Crowl & Covich (1990) have shown that the freshwater pulmonate snail *Physella virgata* subsp. *virgata*, exhibits very different life histories according to the presence or absence of a predatory crayfish *Orconectes virilis*. Populations free from the predator showed rapid growth until the shell length reached about four millimetres, at which time reproduction began, and the growth rate declined. Longevity ranged from about three to five months. In the presence of the predator, an increase in the age and size at maturity was observed, and egg production did not begin until the shell were seven to ten millimetres long. Individuals lived from 11 to 14 months. These patterns are consistent with a system where juvenile mortality is greater than adult mortality. Crayfish prey mainly on small snails, so it is advantageous to delay the onset of reproduction and grow to a larger size. However, in this case the life history switch can be induced by a chemical cue from the crayfish.

Big-bang reproduction

An interesting special class of semelparous life histories occurs where a single bout of reproduction leads to the death of the adult, but the onset of reproduction is delayed for many years. This has been called 'big-bang' reproduction (Gadgil & Bossert 1970). Although rare, life histories of this sort have arisen independently on several occasions, for example in periodical cicadas, salmon, bamboo, agaves, and some palms. These organisms appear to completely debilitate themselves by a massive burst of reproduction. For example, in *Agave deserti* a rosette 60 cm in diameter can produce a flowering stalk 4 m tall (Schaffer & Schaffer 1979; the sugar stored in the stalk of a related species is used in the production of tequila). A number of plausible explanations exist for the evolution of big-bang reproduction.

One advantage of temporally spaced but synchronous reproduction is the opportunity to swamp predators. Predator satiation influences seed release in a variety of ways. For example, multi-stemmed mallee *Eucalyptus* release seeds throughout their lives, but there are virtually no seedlings in old stands of mallee. Recruitment takes place predominantly after fire. Wellington & Noble (1985) have shown that seed-harvesting ants remove as much as 100 per cent of newly-fallen seed of *E. incrassata*, and that an individual seed has a half-life as little as 5.2 days. Fire induces a mass release of seeds stored in woody capsules, and this causes both seed storage and germination. Germination comparable to that found in recently burnt stands can be induced in unburnt stands by experimentally adding seed. However, soil conditions in the unburnt stands did not appear to support subsequent recruitment of the seedlings.

In *E. incrassata* an external cue (fire) is used to synchronise reproduction. In other systems such cues may not be available, leading to programmed mortality. Whereas predators could conceivably track annual emergence of a prey species, very long periods between emergence could prevent any tracking. Under these circumstances there would be very strong selection for reproductive synchrony as well as delayed reproduction, both to ensure pollination and to ensure that the few seeds released at the wrong time were not consumed by predators. Predator satiation may also be a plausible explanation for the life histories of the bamboos described in Chapter 1 (Janzen 1976), and hence a source of the unfortunate problems suffered by pandas. There is some evidence that the problem of tracking by predators continues to impose difficulties even for very long periods between reproductive bouts. For example, different varieties of periodical cicadas emerge every thirteen or seventeen years. Both 13 and 17 are prime numbers, so there is no simple period around which the life cycle of a predator could revolve.

The closely related monocotyledon genera *Agave* and *Yucca* both contain some species which exhibit big-bang reproduction and some which do not. Yuccas are all pollinated by moths of the genus *Tegeticula*, in a remarkable relationship in which female moths push pollen down the stigmatic canal, even though they do not derive any nutrition from the plant. They do benefit from injecting eggs into the immature ovary of the plant. When the larvae develop, they consume a proportion of the developing seeds. By contrast, agaves attract a variety of invertebrate and vertebrate pollinators. Schaffer & Schaffer (1977, 1979) have proposed that selective behaviour by these pollinators drives the evolution of big-bang reproduction. Where seed-set is limited by the availability of pollinators, and pollinators prefer tall flowering stalks, it may pay the plant to invest heavily in increasing the size of its inflorescence, even at the expense of future reproduction. For five yuccas and seven agaves, Schaffer & Schaffer (1977) measured pollinator preference as the slope (m_f) of the regression equation relating the percentage of flowers developing into fruits to inflorescence height. This relation is only strong in big-bang reproducers (Fig. 6.11).

An interesting behavioural polymorphism occurs in male coho salmon (*Oncorhynchus kisutch*), in which all animals die after spawning. All females mature at three years, but in males there are two classes of individuals: enormous hooknose males which spawn at three years of age, and smaller jack salmon which mature at two years of age (Fig. 6.12). The large hooknose males use their elongated snout and large teeth to fight for access to females, while the small jacks sneak behind rocks and debris in order to ensure that they are in close proximity to a female when she spawns. Large males are poor at sneaking, and small males are poor at fighting. The difference between the two sorts of males appears to be genetic, and the benefit of the two

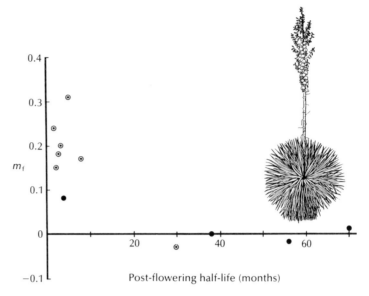

Fig. 6.11 The relation between m_f (the slope of the regression line relating seed set to flower height) and post-flowering half-life in *Agave* (⊙) and *Yucca* (●). In both genera, species where tall flowers do best have short lives after flowering. After Schaffer & Schaffer (1977).

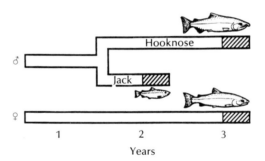

Fig. 6.12 Development in big-bang reproducing coho salmon (*Oncorhynchus kisutch*). While females all mature after three years, males have two morphs. Large hooknoses mature at three, but small jacks mature at two. The divergence in time to maturity appears to be a mixed ESS. After Gross (1985).

behaviours appears to be frequency-dependent. Sneaking should be easier when sneaks are rare, and least profitable when there are large numbers of sneaks, forcing some of the sneaks to fight. Gross (1985) provides evidence that the pay-offs derived from both tactics in natural populations are approximately equal, suggesting the alternative morphs represented a mixed ESS.

Terminal reproductive investment

Because mortality is inevitable, the expectation of future reproductive success for an iteroparous animal will decline with each reproductive bout. Because annual fecundity will not change dramatically in animals with unitary growth, the value of current reproduction to residual reproductive value will increase with age, and this increase will be very pronounced towards the end of its life. This poses the question: should an animal close to the end of its life increase its reproductive investment to reflect this declining residual reproductive value. Clutton-Brock (1984) has tried to test this hypothesis with the very comprehensive data from his long-term study of red deer. The convincing argument he develops is worth considering at some length, as it illustrates many of the difficulties in measuring the cost of reproduction. In red deer, females live for 10 to 15 years, conceiving for the first time as 2- or 3-year olds. Reproduction does impose a cost. Mothers which have reared a young one year are much less likely to be successful in rearing a calf the following year. This difference in fecundity between milk hinds (those that have just suckled a young) and yeld hinds (those have not) is confounded by two factors. First, it is density-dependent, being most pronounced at high breeding densities. Second, the difference is most pronounced for young and old hinds (Fig. 4.9). Further, old mothers produce somewhat lighter calves which in turn survive worse than calves of young and middle-aged mothers. As the pregnancy rate depends on the weight of the female, it is possible to compute the pregnancy rates that would be expected for milk and yeld hinds. While young hinds conform to expectation, old females (> 9 years) have pregnancy rates lower than would be expected on the basis of their weight.

These last four results are hardly indicative of intense reproductive effort in very old females. Nonetheless, two separate lines of evidence are supportive of increased reproductive effort in old females. First, the duration of sucking by a calf is longest for the offspring of both old and young females. This ability to increase investment in young mothers probably reflects their superior body condition, but this explanation will not suffice for the older mothers, which are usually in poor condition. The second telling comparison comes from measures of the kidney fat of calves and their mothers as a index of condition. Calf condition relative to maternal condition increases with age (Fig. 6.13), and it is only in very old females that the kidney fat of calves exceeds that of their mothers, suggesting that although old mothers are debilitated by declining condition and senescence, they are prepared to take greater reproductive risks in each reproductive bout than do younger animals.

CONSERVATION OF REPRODUCTIVE EFFORT?

Thus far we have been primarily concerned with the distribution of reproductive effort throughout an organism's life-span, and the various trade-offs this may involve. Of course, it is also possible that there will be trade-offs within each reproductive bout. I will consider two examples, latitudinal gradients in brood size of birds and mammals, where hypotheses differ in interpreting variation as a consequence of

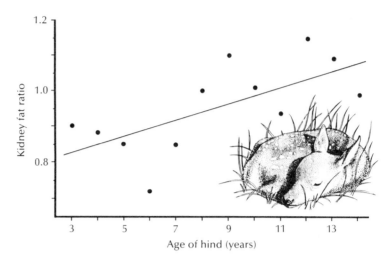

Fig. 6.13 The ratio of condition in calf to condition in mother. Condition is assessed with an index of the amount of kidney fat. With increasing age mothers are more likely to produce young in better condition than themselves. After Clutton-Brock (1984).

between-season or within-season trade-offs, and some of the patterns in seed size in plants, a topic ripe for much greater investigation.

Latitudinal gradients in brood size

If intrapopulation level variation may reflect individual optimisation of clutch size, it is important to ask whether geographic trends can be ascribed to the same factors. In many vertebrates the clutch size declines with proximity to the tropics. Explanations for this decline fall into three classes:

Cody (1966) attributes greatest importance to trade-offs within a single breeding season. He argued that extra investment in reproduction would firstly decrease investment in predator avoidance, and secondly increase the risk of nest predation directly, as large clutches are more noisy, and require more visits to the nest by their parents, making them much more conspicuous. This expectation has never been confirmed empirically, nor is it immediately clear how it could be measured. The hypothesis relies heavily on the expectation that some birds escape from predators by migration. It is true that some island populations of birds which lack nest predators have higher clutches than their mainland counterparts. The hypothesis is of uncertain relevance to mammals, which show the increased litter size regardless of any tendency to migrate, and where increased litter size does not necessarily make the nest more conspicuous, as food can be supplied to all young simultaneously by a lactating mother.

An alternative hypothesis implicates trade-offs between reproductive bouts (Charnov & Krebs 1974), in a direct application of Williams Demographic Theory of Optimum Reproduction. According to this view the costs of migration and the rigors of living in temperate and arctic environments reduce the prospects that birds will survive to breed again in contrast to their counterparts in the more equable tropics. Although some birds in the tropics appear to enjoy long lives, the appropriate detailed studies of the life table have not been made for a large range of populations of any species, frustrating our attempts to test this hypothesis directly.

According to Lack (1947a, 1948, 1954), food availability is the predominant determinant of clutch size. He suggested that in the bird species in which the latitudinal trend was originally detected, daylength increases with increasing latitude and therefore so does the time available for foraging. The ability to harvest more food allows birds to increase their clutch. This hypothesis is limited in its generality. Increases of brood size with latitude occur in nocturnal mammals and owls, and the clutch size of the snow bunting (*Plectrophenax nivalis*) increases across the Arctic Circle, although increases in daylength cease. In a far more general alternative, Ashmole (1963) proposed that the resources which can be devoted to breeding depend not only on the gross availability of food, but also the way that seasonal fluctuations in resource availability affect the number of breeding individuals. In highly seasonal environments there may be regulation of population levels during the non-breeding season, determined by the decline of food availability at this time of year. Any animal which survives this period of hardship until the burst of productivity with which reproduction is associated should enjoy conditions of comparative plenty, and be able to devote the surplus food available to reproduction. By contrast, if the environment is comparatively aseasonal, there will be few extra (net) resources to devote to reproduction, even though the gross food availability is very high (Fig. 6.14). Ashmole's argument is an appealing explanation for a variety of cases. For example, it predicts the parallel altitudinal increases observed in some mammals, and the differences between habitats at the same latitude such as savannah and rainforest (Lack & Moreau 1965).

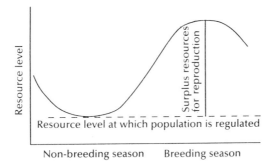

Fig. 6.14 Ashmole's hypothesis for the effect of seasonality on brood size. Brood size is set not by the gross availability of food, but the surplus food which becomes available to the animals that survive periods of food shortage in the non-breeding season. After Ricklefs (1980).

It is difficult to distinguish the proximate effects of food availability and an evolutionary response to seasonality. In the dasyurid marsupial genus *Antechinus*, the litter size is proximately regulated by the number of teats. Supernumerary young which fail to obtain a teat at birth (presumably because the teats are usually saturated), soon die but represent a negligible investment by their mother, as they weigh only 16 mg at birth. Teat number shows evidence of genetic control and the influence of strong repeatable selection pressures. For example, there are parallel clines in sympatric species (Fig. 6.15; Cockburn *et al.* 1983). *Antechinus* mothers are always semelparous or biparous, yet if anything, the clines are most pronounced in the species where semelparity predominates, suggesting that trade-offs between the first and second year of life are not influential in determining the extent of the cline. Teat number is highest in seasonal environments (the alps), and lowest where seasonality is dampened (low latitudes, and on the capes and promontories of southern Australia, where the climate is strongly influenced by the roaring forties).

Seed size and number

Many taxa show a direct trade-off between the number of offspring and size of the offspring within a brood. For example, the kiwi depicted in the Chapter vignette produces a single enormous egg which X-ray photographs suggest must be close to the maximum which can be accommodated in its abdomen. By contrast, other birds of similar size produce many smaller eggs in each clutch.

In no group is diversity in reproductive output within a brood more bewildering and challenging than in plants. Harper (1977) comments: 'In the course of its life a desert annual may produce 10^{0-1} seeds, a weedy annual grass (e.g. *Alopecurus myosuroides*) about 10^2 seeds, about the same number as the double coconut tree (*Lodoicea seychellarum*); and a coastal redwood tree will produce 10^{9-10} seeds, about the same number as a small herbaceous orchid!.'

Seeds have a variety of functions. First, they are one agent whereby plants produce copies of their genes, and usually the stage at which recombination to produce new genotypes occurs. Second, they are very often the stage of the life cycle where dispersal and colonisation occur (but see Chapter 5). Third, the seed is very often an agent of persistence (diapause) through periods of hardship such as winter, drought, and fire. Fourth, the seed usually contains food reserves to assist the developing embryo, and is therefore a target for predators which can utilise those reserves. Harper (1977) points out that these functions involve a variety of trade-offs. A large seed may provide a huge advantage for early growth, but will be unsuitable for diapause because it is likely to attract predators, and will be poor at dispersal that does not rely on animal agents.

The size of seeds often exhibit strong correlations with particular habitats, provided that the obvious phylogenetic constraints are taken

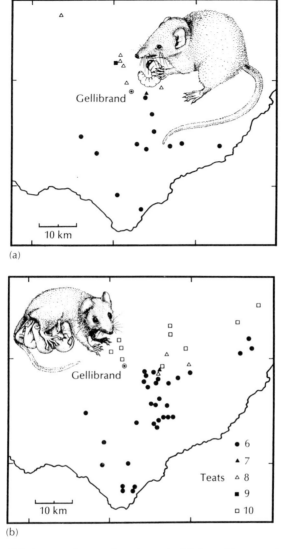

(a)

(b)

Fig. 6.15 Parallel clines in the number of teats in (a) *Antechinus swainsonii* and (b)
A. stuartii. Teats are usually saturated, and hence reflect litter size. After Cockburn
et al. (1983).

into consideration (e.g. Salisbury 1942; Harper *et al.* 1970; Baker 1972;
Mazer 1989). For example, species in closed habitats typically have
larger seeds than those in open habitats, perhaps because the light
available to support development of the seedling is more restricted.
These correlations could come about in two ways (Silvertown 1989).
First, seed size could undergo selection *in situ*. Rapid response to
selection does appear possible in some cases. For example, seeds of
Camelina sativa (Brassicaceae) mimic those of the cultivated flax
crops they infest, so they are harvested and resown with the crop

(Wiens 1978). The alternative view is that although there is considerable intrapopulation (and intra-individual) variation in seed mass, much of the variation is largely environmental or maternal in origin, and will therefore show limited response to selection. Hence much of the variation may reflect constraints on the ability of the plants to occupy particular habitats rather than evolution of seed size. For example, Baker (1972) showed that species introduced into California show the same correlations between seed size and habitat that native species do, suggesting that those species have colonised habitats for which their seed size is appropriate. Any adaptive variation may therefore lie in the evolution of plasticity rather than seed size *per se*. For example, experimentally transplanted rosettes of *Prunella vulgaris* (Labiatae) produced larger seeds in deciduous woodland than in the middle of old-fields, regardless of the point of origin (Table 6.5; Winn 1985). Plants lacking this plasticity may be unable to occupy more than one habitat.

However, at least part of the variation appears to inexplicable simply in terms of plasticity. The seeds produced by a single plant and even flower often exhibit great and discrete variation in morphology (Venable 1985; Fig. 6.16), or more subtle and continuous variation in size and germination behaviour (Silvertown 1984; Michaels *et al.* 1988). Heteromorphism of this sort is more common in short-lived annuals than in perennials, as indicated by the compilation of reproductive habits in the genus *Crepis* (Asteraceae) in Table 6.6. What then are the special pressures on annuals? The most likely explanation is the difficulties inherent in putting 'all your eggs in one basket'. Total reproductive failure is possible if the seeds are all eaten or germination conditions are unfavourable in the following year. The absence of a second chance at reproduction imposes selection pressure for diversified reproductive behaviour (Schulz 1989).

COVARIATION OF LIFE HISTORY TRAITS REVISITED

Chapter 4 included a discussion of covariation in life history traits in mammals, and Chapter 5 concluded with the suggestion that habitats,

Table 6.5 Mean weight (mg) of seeds produced by *Prunella vulgaris* reciprocally transplanted between a deciduous woodland and an old-field site in south-western Michigan. After Winn (1985).

Transplanted from		Transplanted to	
		Deciduous woodland	Old-field centre
Transplanted from	Deciduous woodland	0.50	0.34
	Old-field centre	0.43	0.34

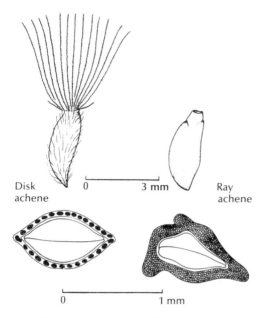

Disk
achene

0   3 mm

Ray
achene

0   1 mm

Fig. 6.16 The disk and ray achene of *Heterotheca latifolia* (Asteraceae). Note the
hairs and pappus of capillary bristles in the wind-dispersed disk achene, and the
thick fibrous pericarp of the ray achene. Disk achenes do not contribute to the seed
bank, but 24 per cent of the ray achene seed bank is still viable after one year
(Venable & Levin 1985b). After Venable & Levin (1985a).

Table 6.6 Number of European taxa in the genus *Crepis* with monomorphic or
polymorphic achenes, divided according to whether the species follows an annual or
perennial lifestyle. There are 71 species in 21 sections of *Crepis* in Europe. To avoid
phylogenetic artefacts, multiple occurrences of a particular habit within a section
have been ignored. After Silvertown (1985).

	Life-span	
Number of achene types in a brood	Annual or annual/biennial	Perennial or perennial/biennial
1	5	15
2 or 3	6	0

although showing enormous variation, could be classified as to the
selection pressures that they impose on life histories. The results of
this Chapter should emphasise that while trade-offs are perhaps not so
pervasive as has been commonly believed, the costs and intricate links
between development and reproduction mean that all aspects of the
life history need to be considered to fully understand the life cycle of a
plant or animal. In Chapter 7 we shall see how the mode of reproduction
adds even greater complexity to that complex web of interactions.

SUMMARY

If resources are limiting, devoting resources to reproduction is likely to limit the amount of resources that are available for other functions. Similarly, pleiotropy means that improvement in one aspect of fitness may lead to a deterioration in another aspect of fitness. Therefore trade-offs should be widespread in the evolution of life history traits. Indeed, the absence of what Law (1979a) called an optimum 'Darwinian demon' which performs superbly at all aspects of growth and reproduction confirms this expectation. Nonetheless such trade-offs should be demonstrated rather than assumed, and positive correlations between fitness components are commonly reported. Common causes of positive correlations are the positive effect of extremely good environmental conditions on all aspects of fitness, and because the great majority of mutations leading to genetic variants phenotypically distinguishable in laboratory cultures are likely to be deleterious. The problems that dominate life history evolution include the acquisition of size and condition necessary for reproduction, accommodating the massive changes in size and morphology typical in development, and disposition of reproductive effort within a lifetime.

FURTHER READING

Stearns (1976, 1977) contributed two very influential reviews of the early literature on life history evolution. Partridge & Harvey (1988) is a good recent summary. Bell & Koufopanou (1986) is the most detailed analysis of the difficulties involved in measuring the costs of reproduction.

TOPICS FOR DISCUSSION

1 Martin & Simon (1988) provide an unusual molecular perspective on the duration of diapause in periodical cicadas. Discuss shifts between thirteen- and seventeen-year cycles from an ecological perspective.

2 Dispersal is an alternative to diapause as a way of escaping from unfavourable local conditions. Identify the sorts of environments where dispersal might be favoured.

3 Construct a list of other trade-offs likely to be important within a single bout of reproduction. In what organisms will each be influential, and how might such trade-offs be examined.

Chapter 7
The Ecology of Sex

This Chapter is all about sex. By sex I mean the formation of a new organism containing genetic material from two parents. Bell (1982) has commented that: 'Sex is the queen of problems in evolutionary biology. Perhaps no other natural phenomenon has aroused so much interest; certainly none has sowed as much confusion.' Sex is of profound interest in two different ways. First, as will become clear, there appear to be extensive costs associated with sexual reproduction. The high incidence of sex in the face of these costs has been one of the most troublesome areas for evolutionary biology. Second, sex has some fascinating consequences. This chapter uses adaptationist analysis to deal with two of these: the way in which selection pressures may be very different on the two sexes (sexual selection), and the way in which parents invest differently in male and female offspring (sex allocation).

WHY BOTHER?

Recombination and genetic diversity

The consequences of sex are more obvious than its causes. Sex involves meiosis and syngamy, generating segregation and recombination. Recombination has several interesting effects. First, it disrupts linkage, breaking up favourable gene combinations. Second, it allows the transfer of new alleles derived from mutation or gene flow throughout the population. As a corollary, the chance that new alleles will become homozygous is greatly increased. This exposes recessive alleles to selection. Because mutations are often deleterious, heterozygosity within individual progeny of asexual species will often be very high, because deleterious mutations are protected more from selection in asexual species than in sexual species (Table 7.1). Selfing both exposes recessives to selection and hinders gene flow, so heterozygosity will be very low, though at least initially progeny will be variable because of recombination. By contrast with heterozygosity within individuals, the genetic differences between the progeny produced by a single individual will by definition be limited in asexual species, depending on the mutation rate in each generation. The diversity of progeny in sexual species can be large, though will vary between species according

Table 7.1 The consequences of different modes of reproduction. After Bell (1982).

	Progeny genomes		
	Diversity	Heterozygosity	Costliness
Asexual	Low	Very high	Minimal
Sexual			
Self-fertilisation	Intermediate	Low	Low
Sex	Very high	High	Very high

to the extent of recombination. For example, in many taxa recombination is confined to zygotes of only one sex (Bell 1982; Trivers 1988), and as we have seen from the discussion of the major histocompatibility complex in Chapter 2, recombination is much more likely in some regions of the genome than in others. Intermediate levels of recombination also occur in the species which alternate or combine sex and parthenogenesis (Hebert 1987).

The low diversity of progeny among individual parthenogens does not mean that there cannot be intrapopulation variation. Most parthenogenetic populations of both plants and animals contain several clones (Ellstrand & Roose 1987; Hebert 1987; Weider *et al.* 1987). Very widespread clones found in all subpopulations are exceptional.

Species selection

In Chapter 3, I pointed out that with the single exception of the bdelloid rotifers, the absence of sustained adaptive radiation in parthenogenetic lineages among the Metazoa was the best evidence for the operation of species selection. The same is true to a lesser extent in plants. The Euglenophyta are parthenogenetic, but most other parthenogens are patchily distributed. Because parthenogenetic descendants can often coexist with their sexual antecedents for long periods of time, it may be that some long-term sorting process is necessary to discriminate between the alternative modes of reproduction. It is less obvious why parthenogenesis predisposes clades to extinction. It has often been argued that sex facilitates response to environmental change by generating new gene combinations, which in turn allow populations to track a dynamic environment. This is because adaptive favourable mutations can be combined horizontally through a population (the Fisher–Muller hypothesis; Fig. 7.1). However, the advantage declines in small populations, weakening the appropriateness of models of group selection. Where there are epistatic interactions, sex can actually slow the pace of evolution by breaking up co-adapted groups of genes (Maynard Smith 1987b).

Muller (1964) advanced a more plausible model of group selection (Muller's ratchet), which suggests that the advantage sexuals enjoy over parthenogenesis at the group level reflects the ability to remove deleterious mutations. Both sexual individuals and parthenogenetic clones vary with respect to the number of deleterious mutations. Sampling error will occasionally lead to the failure of the best adapted genotypes, particularly in small populations. These best-adapted genotypes are easily recreated in sexual populations, but can only be recreated in clones by favourable mutation, which is less probable than the acquisition of further harmful mutations. Each loss of the best-adapted clone acts like a ratchet, slowly increasing the average mutation load, and the load suffered by even the best-adapted clone. Extinction ensues when the load becomes intolerable. This hypothesis fits comfortably the assumption of small population size which underlies most group

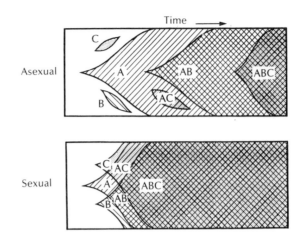

Fig. 7.1 The Fisher–Muller hypothesis. Sex speeds evolution because favourable mutations A, B and C can be combined horizontally through the population. After Maynard Smith (1978).

selection models, because extinction of the least favoured genotype is most probable in small populations. It is also broadly concordant with the patchy distribution of clones and their restricted geographic spread. It is quite possible that Muller's ratchet could set an upper limit to the size of the genome in groups where recombination is rare, such as many prokaryotes (Maynard Smith 1988). This is because the mutation rate should be proportional to the size of the genome.

The paradoxical cost of sex

There are two problems with this argument. First, there is the general dissatisfaction with models of group advantage. Second, attempts to model short-term advantages of sex have revealed several very severe costs relative to parthenogenesis, so all other things being equal, sexual forms should easily be outcompeted by parthenogens long before the ratchet could exert its influence. Lewis (1987) points out that there are five costs of sex for most organisms.

1 Recombination scrambles genotypes, disrupting favourably adapted gene combinations. For example, parthenogenesis could preserve the advantages of overdominance at the haemoglobin locus in a human population exposed to malaria, without the production of young homozygous for the sickle cell allele (Chapter 2).

2 Meiosis and syngamy appears to take longer than double mitosis, slowing the pace of reproduction. These cellular–mechanical delays are potentially of enormous importance in very small organisms where they may be the principal determinant of the rate of population growth.

3 In higher organisms, courtship and mating may be risky, because of the risk from predation or even venereal disease. There may also be wastage of gametes, and costs associated with the maintenance of sexual dimorphism and sexual competition. For example, plants must

rely on pollinators to fertilise all their ovules. Pollinators prefer to visit the male flowers of sexual *Antennaria*, greatly reducing seed set relative to parthenogenetic conspecifics (Fig. 7.2; Michaels & Bazzaz 1986; Bierzychudek 1987a).

4 At low population densities, sexual reproduction may be improbable or difficult to coordinate with favourable environmental conditions. Parthenogenesis ensures that reproduction will be possible at any time and place.

5 Most important of all, sexual females suffer from genome dilution. Consider parthenogenetic and sexual females producing eggs at the same size and rate. Because the sexual female contributes only half the genetic material to each egg, she suffers a twofold cost relative to the parthenogen, at least in the case of anisogamy, where zygotes are formed from the fusion of tiny gametes derived from the male and large gametes produced by the female.

The geographic and taxonomic distribution of sex

The relative importance of the costs of sex will vary between organisms, but they are collectively so severe that some explanation for the short-term maintenance of sex is clearly required. Sex cannot just be some phylogenetic legacy against which selection is impotent. The persistence of sex in species which exhibit both sex and parthenogenesis clearly attests to its resistance to invasion by parthenogenetic clones.

Hope for an evolutionary explanation of the maintenance of sex in the face of these apparently severe costs have emphasised the patchy ecological distribution of sex. In a monumental review of the distribution of sex in the Metazoa, Bell (1982) shows that there is a consistent

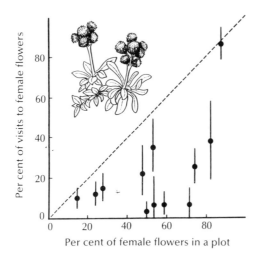

Fig. 7.2 The percentage of visits by pollinators to male and female flowers in sexual populations of *Antennaria parvifolia* in relation to the frequency of female flowers in a plot. The line represents no preference for male or female flowers. Pollinators prefer male flowers. After Bierzychudek (1987a).

trend for sex to be most prevalent where the habitat is not subject to dramatic physical fluctuations. Parthenogenesis is common in fresh-water where nutrient concentrations and temperature fluctuate, but sex predominates in more constant marine habitats (Chapter 1). Par-thenogens are more common early in succession and at extreme lati-tudes and altitudes in both animals and plants (Bell 1982; Bierzychudek 1985). Furthermore, species like aphids and some parasites which alternate sex and parthenogenesis tend to coincide the sexual phase with high density (Bell 1982). Sex occurs in stable, highly competitive environments.

Unfortunately these very strong correlations can be interpreted in two ways. They might indicate that sex only enjoys an advantage over parthenogenesis in stable environments, so we should seek an expla-nation for sex in the selective conditions which prevail in stable populations. Why does recombination provide benefits in these en-vironments which outweigh both its own costs and the costs of genome dilution (Bell 1982)? Alternatively, the advantages of sex may be wide-spread, but parthenogenesis may be favoured in hostile environments because density is low and colonisation opportunities frequent, so that the costs of mate or gamete location may tip the balance in its favour (Lloyd 1988). Or, of course, both factors may be important.

THE MAIN HYPOTHESES

Hypotheses for the short-term maintenance of recombination can be classified by a variety of criteria (Bierzychudek 1987b; Stearns 1987; Maynard Smith 1988). Many hypotheses have waxed and waned in popularity, deriving strength less from their own virtues than from the failure of their competitors. Four important groups of ideas currently enjoy popularity.

1 Segregation and recombination at meiosis is generally advantageous because it facilitates some other process. The possibility that genes for recombination could increase in frequency (hitchhike) because they facilitate selection on other functions has been well explored theor-etically, and may provide an explanation for the initial spread of recombination (Maynard Smith 1978). However, the models do not explain the maintenance of sex, because continued benefits are required. Most current attention focuses on the role that recombination plays in the mechanics of DNA repair (DNA repair hypothesis).

2 Recombination is generally advantageous because of direct benefits. These hypotheses rely on the diversifying effect of recombination. The most commonly invoked benefit is the purging of deleterious mutations (Feldman *et al.* 1980; Kondrashov 1988; the mutation/elimination hypothesis).

3 Segregation at meiosis is generally advantageous because of direct benefits. This interesting recent model suggests that recombination is of less importance than is commonly assumed. Instead, the direct

advantage arises because segregation allows a single advantageous mutation to become homozygous (Kirkpatrick & Jenkins 1989; the segregation hypothesis).

4 Recombination is directly advantageous, but only under some conditions. These hypotheses also rely on the diversifying effect of recombination, which is thought to provide benefits when there is a strong interaction between genotype and local environment. Because particular genotypes are not consistently successful across microhabitats or through time, the disruptive effect of recombination is beneficial rather than costly (Bell 1987; gene/environment hypotheses).

Bierzychudek (1987b) points out that there are two ways of examining hypotheses of this sort, particularly as the basic plausibility of all models can be confirmed with simple algebraic analysis. Direct examination of unique predictions is obviously valuable, but it is also important to examine the assumptions on which the models are based. The advantage of generally applicable hypotheses is that they do not demand any specific environmental conditions which are difficult to imagine operating across a great range of organisms (Grafen 1988b; Kondrashov 1988). Their assumptions are to be sought in aspects of genetic organisation and cellular machinery which are increasingly amenable to direct measurement. Their disadvantages are that they do not explain the geographic distribution of sex.

DNA repair

Bernstein and his colleagues have suggested a function for recombination which does not depend directly on the generation of variation (Bernstein, *et al.* 1987, 1988). DNA is often broken by radiation, oxygen radicals and other agents. Repair is only possible if a suitable template exists. In the early stages of meiosis, four homologous chromatids are produced which are then able to recombine in pairs. The homologues provide the template, and repair involves crossing over. There is abundant molecular evidence that such double strand repair occurs (Bernstein *et al.* 1987). It is less clear whether DNA repair is the principal function of sex (Maynard Smith 1988).

There are two problems. First, crossing over is absent in some circumstances (e.g. it is often absent in one sex), yet there is no obvious increased mortality associated with the absence of repair (Maynard Smith 1988). Second, although the benefits of double-stranded DNA repair should be present in all habitats, the premium on repair should be greatest in environments where damage to DNA is high. Stearns (1987) points out that radiation damage is likely to be greatest at high altitudes, yet parthenogenesis predominates in these habitats. It therefore seems preferable to regard DNA repair as a beneficial consequence of recombination and crossing over, rather than its principal cause.

Kondrashov (1988) offers a model implicating deleterious mutations as the major problem affecting the success of parthenogens. His model differs from Muller's ratchet both in its capacity to affect short-term success, and because it views the effect of mutations as deterministic, emphasising the elimination of unfavourable mutations rather than the stochastic loss of well adapted genotypes. Kondrashov suggests that most mutations are deleterious and the accumulation of many deleterious mutations may have fitness effects that are synergistic rather than additive. In each generation new mutations increase the average genome contamination (Fig. 7.3). In the extreme case, more than a certain number of deleterious mutations prevents reproduction, leading to truncation selection. In an asexual population, this threshold is passed when a clone close to the boundary acquires a new mutation, so deleterious mutations are eliminated separately. By contrast, in a sexual population, there are more very bad genotypes, and the genotypes that are eliminated may contain many deleterious mutations. When the populations are reconstituted after the truncation, the sexual population contains many more very good genotypes. Sex increases the efficiency of the selection.

The assumptions of this hypothesis are a high rate of deleterious mutations per generation, synergistic interactions between those mutations, and the operation of a mode of strong selection similar to truncation selection. Kondrashov (1988) argues strongly for a high mutation rate per generation (though see Chapter 2), but concedes data on the mode of selection are lacking. The significance of this model is therefore difficult to evaluate.

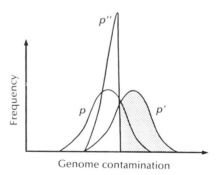

Fig. 7.3 Kondrashov's hypothesis of the action of mutation in a sexual population at equilibrium. During one generation, mutation, selection and reproduction all occur. Before mutation the distribution of genome contamination is given by p. After mutation the distribution becomes p'. As a result of truncation, selection individuals in the shaded area die, leading to distribution p''. Reproduction restores normality. In an asexual population there are fewer individuals in the shaded area in each generation but all individuals lie close to the truncation point. After Kondrashov (1988).

The Fisher−Muller hypothesis relied on the ability of sex to spread favourable mutations between individuals (Fig. 7.1). Parthenogenetic clones cannot acquire favourable mutants that occur elsewhere in the population. Kirkpatrick & Jenkins (1989) point out that the same argument applies to the acquisition of favourable homozygous mutations within an individual. Once a favourable mutant A1A1 $\Rightarrow$ A1A2 arises the adaptive transition A1A2 $\Rightarrow$ A2A2 is not possible without segregation. Kirkpatrick & Jenkins (1989) calculate the advantage enjoyed by a sexual population as a result of the segregational load suffered by an asexual population according to the selection coefficient s, degree of dominance h, and the number of homozygous loci L (see Chapter 4). The twofold effect of genome dilution in the sexual population is compensated for even when selection coefficients and the number of mutable loci are at moderate levels.

There are a number of problems with this appealingly simple idea. First, the heterozygous state is assumed to be of intermediate fitness. This will not be the case for complete dominance. More important, any loci with overdominance will reduce the advantage enjoyed by sexuals greatly. Next, it cannot explain the maintenance of sex in species that are haploid for much of their lives, such as some protists, fungi, algae and bryophytes. Furthermore, it does not provide any help in understanding why self-fertilisation is apparently rather rare and clearly selected against in some populations (Bull & Harvey 1989).

Habitat heterogeneity and gene/environment correlations

Discussion of advantages for sexual reproduction initially focused on the apparently rapid changes facilitated by recombination — so changes in time seemed an appropriate explanation for the advantages of sex. This hypothesis is no longer popular, because its central prediction is clearly false; sex is not more prevalent in physically extreme or fluctuating environments. Indeed, the exact opposite is true.

The Tangled Bank and spatial heterogeneity

The famous concluding paragraph to Darwin's (1859) *The Origin of Species* begins: 'It is interesting to contemplate an entangled bank, clothed with many plants of many kinds, with birds singing on the bushes, with various insects flitting about, and with worms crawling through the damp earth, and to reflect that these elaborately constructed forms, so different from each other, and dependent on each other in so complex a manner, have all been produced by laws acting around us.'

This image of a locally complex environment (a Tangled Bank) has been used by Bell (1982) to name a group of hypotheses which suggest that spatial heterogeneity is profoundly important in the maintenance of sex (e.g. Ghiselin 1974; Williams 1975). The arguments rely on local spatial heterogeneity so that a genotype favoured at one location may perform poorly elsewhere. Because the fitness of a successful adult may not be transferred to its offspring dispersing to random sites nearby, it may be advantageous to break up genotypes regardless of their local success. In addition, sexual groups may achieve a greater exploitation of all of the microsites in their local environment than can a limited number of parthenogenetic clones. The total productivity associated with sexual reproduction may be higher both because of this greater penetration of microsites and because sexual siblings compete less as they can specialise in different microsites.

The Red Queen and biological heterogeneity

A related hypothesis attributes environmental heterogeneity not to physical problems posed by the environment, but the species with which they interact. This is the Red Queen effect described in Chapter 5. The importance of hostile species as selective agents is believed to depend on their relative generation times. Predators and competitors often have similar lifespans, while parasites sometimes have lifespans much shorter than those of their hosts, and should therefore experience much greater rates of selection. Most variants of the Red Queen hypothesis, therefore implicate parasites as a strong selective agent (Levin 1975; Jaenike 1978; Hamilton 1982; Bremermann 1987; Seger & Hamilton 1988). Selection on the parasite will be frequency-dependent; common host genotypes will exert a stronger selective pressure. Sex in the host would be favoured because it generates diverse progeny, some of which may have novel resistance genotypes, and be able to withstand disease. There is therefore the same type of gene/environment correlation as in the Tangled Bank, but the environmental heterogeneity is temporal rather than spatial. The fitness of a genotype is dependent on its frequency in the immediate past. Theory suggests that these models work best with organisms with low mutation rates for defence mechanisms, and also where the predominant mode of selection is soft truncation selection, as for the mutation–elimination hypothesis (Hamilton et al. 1990).

Contrasting the assumptions and predictions

The high incidence of sex in physically stable environments where species diversity is high is consistent with both the Tangled Bank and the Red Queen hypotheses, as both intraspecific competition and interspecific interactions should be most intense in stable habitats. Therefore comparative approaches do not directly distinguish between the importance of spatial and biotic heterogeneity. Several recent attempts have been made to test the predictions of the hypotheses directly.

The most thorough experimental analysis of the assumptions underlying genotype–environment interaction hypotheses come from a

series of studies on sweet vernal grass, *Anthoxanthum odoratum*
(Antonovics & Ellstrand 1984; Ellstrand & Antonovics 1985; Schmitt
& Antonovics 1986a, 1986b; Kelley *et al.* 1988). Because this grass
reproduces both sexually by seed and parthenogenetically by tiller,
transplant experiments can be designed which simulate any combination of variable or invariant offspring. A number of experiments
reveal higher reproductive rates for sexual offspring than parthenogenetic offspring. While some studies report improved total performance of mixtures than of uniform groups (e.g. Antonovics & Ellstrand
1984; this also occurs in *Drosophila*; Martin *et al.* 1988), other
competition experiments using a variety of plant species do not
(Schmitt & Erhardt 1987; Willson *et al.* 1987; Kelley 1989a, b; McCall
et al. 1989; Karron & Marshall 1990). Any advantage that does occur
appears better associated with frequency-dependence, as required by
the Red Queen hypothesis, than with density-dependence, as suggested
by the Tangled Bank model, though occasionally small fitness differences between individuals are greatly exaggerated at high density (Karron
& Marshall 1990; Schmitt & Erhardt 1990). In one case the frequency-
dependent advantage to minority types was associated with aphid
attack (Schmitt & Antonovics 1986a). Further, the advantage to sexual
progeny persists over the entire natural dispersal distance of the seeds,
suggesting that there is no local microsite variation which could
dramatically affect the outcome of competition (Fig. 7.4).

The genetic assumptions of the Red Queen hypothesis are clear and
have been much better investigated than those of the Tangled Bank
hypothesis. There should be a close interaction between the genotypes

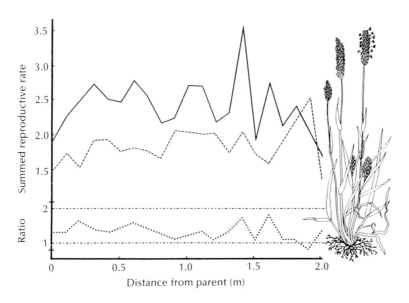

Fig. 7.4 Mean reproductive rate for sexually (unbroken line) and asexually (broken
line) progeny of *Anthoxanthum odoratum* at different distances from the parent.
The ratio is shown as a dotted line. After Kelley *et al.* (1988).

of host and parasite, and variation in the extent of resistance. Substantial variation for disease resistance within populations is now well documented for both animals (O'Brien & Evermann 1988) and plants (Burdon 1987b). The best evidence for intricate association between genotypes of free-living hosts and their parasites comes from plants (Chapter 5). Dramatic local variation in susceptibility to disease and in resistance genotype suggests that the presence or absence of particular resistance genes will reflect the selective environment in the immediate past, rather than any long-term solution to disease (Burdon & Jarosz 1988; de Nooij & Van Damme 1988; Burdon *et al.* 1989).

There is also some evidence that parthenogens are more susceptible to attack by parasites than sexually reproducing organisms. Biological control of weeds is much more likely to be successful when the target is parthenogenetic (Burdon & Marshall 1981). Genetically uniform crop species or zoo populations which are genetically impoverished can be devastated by disease (Lupton 1977; O'Brien *et al.* 1985). Indeed, effective defence in parthenogens may depend on development of permanent morphological barriers to infection rather than local resistance *per se* (Parker 1988).

Burt & Bell (1987a) argue that the level of recombination in mammals is indicated by the frequency of excess chiasmata, or sites where crossing over has occurred. They suggest that if sib competition is important, as suggested by the Tangled Bank hypothesis, then recombination should be most prevalent when litter size is high, when within-brood variation is possible. However, if parasites are important, as required by the Red Queen hypothesis, then recombination should be more important in long-lived species where the number of parasite generations will be greater than the number of host generations, as the slope relating parasite life cycle to host life cycle in mammals is much less than one (Fig. 7.5). Because litter size and generation time in mammals are inversely correlated, the predictions are therefore inimical. In both eutherians and marsupials there is a positive correlation between chiasmata frequency and generation time (Burt & Bell 1987a; Sharp & Hayman 1988).

Lively (1987) also found support for the Red Queen hypothesis in a study of the occurrence of sex in the New Zealand freshwater snail *Potamopyrgus antipodarum*. The proportion of males varied from 0 to 40 per cent, and indicated the importance of sex within a population. Sex was much more prevalent in stable lakes than in variable streams, supporting both the Red Queen and Tangled Bank hypotheses. However, the variation within both stable and variable habitats was much better correlated with incidence of trematode infection than with any habitat variable. This relation does not arise because males are infected more heavily, and strongly supports the tenets of the Red Queen model. Furthermore, parasites show strong local adaptation to their hosts (Lively 1989; Fig. 7.6), suggesting that there should be strong selection for escape from this adaptation.

In summary, available tests lend support to the importance of the

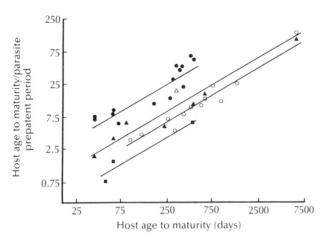

Fig. 7.5 The ratio of mammalian host age to maturity to parasite prepatent period increases with host age to maturity in Coccidia (●), Cestoda (▲), Schistosomatidae (○) and Acanthocephala (■). After Burt & Bell (1987b).

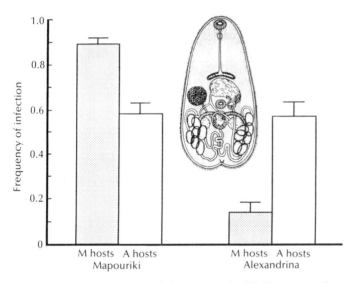

Fig. 7.6 The frequency of infection of the New Zealand freshwater snail, *Potamopyrgus antipodarum*, from two different lakes (Mapourida, dots; and Alexandrina, unshaded) when exposed to the parasitic trematode *Microphallus* sp. from the two lakes. After Lively (1989).

interactions with hostile species in the evolution and maintenance of sexual reproduction.

WHAT ARE MALES AND FEMALES?

The majority of organisms exhibit anisogamy, the production of gametes of more than one size, usually producing one very large type of gamete (eggs), and one very small type of gamete (sperm, pollen). The evolution of anisogamy is reviewed by Hoekstra (1987). Individuals which produce small gametes are males, individuals which produce large gametes are females. There are many different sex-determining mechanisms, well reviewed by Bull (1983). In the mammals with which we are most familiar, the presence of a Y chromosome dictates maleness, though the exact sex-determining elements of the Y chromosome remain elusive. The Y chromosome is small and paired to a larger X chromosome (heterogamety), while females have two X chromosomes (homogamety). By contrast, in birds females are the heterogametic sex, while males are homogamous. The diversity does not stop there. In some reptiles, for example, the sex is determined by the temperature at which the eggs are incubated. In some groups males are produced exclusively at low temperatures, in others males are produced at hot temperatures, and still others the production of males is confined to intermediate temperatures. In some species of arthropods, most notably the Hymenoptera, males are produced from unfertilised haploid eggs and females are produced from fertilised diploid eggs (haplodiploidy or arrhenotoky). In some groups (e.g. birds), all members exhibit similar sex-determining mechanisms, while in some close relatives there may be a great diversity of mechanisms (notably in the Coccoidea, or scale insects).

Sequential hermaphroditism

In some species, both sorts of gametes can be produced by the same individual. Such hermaphroditism may be sequential, where the organism switches from producing small gametes to large gametes or vice versa within its lifetime. Sequential hermaphroditism has been predicted to occur wherever one sex gains fertility with increased age or size more rapidly than the other sex (Ghiselin 1969). For example, in the blue-headed wrasse *Thalassoma bifasciatum* the majority of individuals begin their lives as small females but ultimately change into large brightly coloured males (Warner *et al.* 1975). A large male spawns many times per day, while a female has a more conservative spawning rate. There are some individuals which begin life as males but look like drab females. These males either sneak copulations early in life until they grow large enough to assume bright coloration, or enjoy enhanced growth rates (Warner 1984).

In contrast to the relative advantage enjoyed by large male fish, in most organisms bigger females should have an advantage over small females in the production of either more young or larger young (Chapter 6). However, there is likely to be less correlation between size of males and the size or abundance of offspring they sire, because all that males contribute to the zygote is genetic material. Despite problems with the over-zealous application of this principle (Shine 1988), this size–fecundity advantage hypothesis helps explains why males are smaller than females in almost all groups of animals (Greenwood & Wheeler 1985).

Therefore in strong contrast to the bluehead wrasse, pandalid shrimps change from males to females, presumably as the fecundity of males remains constant from year to year, while female fecundity increases with age (Charnov 1979). In one extraordinary example, mating pairs of the polychaete *Ophryotrocha puerilis* switch sex simultaneously, with males becoming females and vice versa (Berglund 1986). The production of large eggs is debilitating, while the production of small sperm is cheap. The worm therefore loses condition as a female but gains condition as a male, and alters its sex when it passes a threshold of body condition.

Simultaneous hermaphroditism

In simultaneous hermaphrodites both maleness and femaleness are expressed at the same time, so there is some probability of self-fertilisation. While simultaneous hermaphroditism in unitary organisms is usually rather simple, the modular structure of plants allows a bewildering variety of mating strategies. Simultaneous hermaphroditism can be achieved through individual flowers having both pollen and ovules, or, as a consequence of their modular construction, a single individual producing both male and female flowers (monoecy). There may also be individuals with male and hermaphrodite flowers (andromonoecy), or female and hermaphrodite flowers (gynomonoecy). There are numerous combinations which represent intermediate states between hermaphroditism and the full separation of the sexes (dioecy), such as shifts in ther relative importance of pollen and seed production, or functional gender (Lloyd 1984b; Lloyd & Bawa 1984). For example, functional gender in monoecious *Pinus sylvestris* can vary from 0.069 (or close to exclusively male) to 0.825 (or predominantly female) (Ross 1990). Some populations contain both female and hermaphrodite individuals (gynodioecy), and there are very occasionally populations containing both males and hermaphrodites (androdioecy; e.g. *Datisca glomerata*, Liston *et al.* 1990).

Once again the ratio of maleness will depend on size–fecundity relationships. For example, monoecious plants which are stressed, produce a greater proportion of male flowers than unstressed plants (Freeman *et al.* 1980). Stress induces maleness in plants capable of changing sex, and improved environmental conditions or increased size promotes femaleness (Heslop-Harrison 1957; Policansky 1981; Bierzychudek 1982; Solomon 1985).

A variety of mechanisms control the extent of self-fertilisation among hermaphroditic plants (Fig. 2.2, Table 2.5), which either show very high or very low levels of self-fertilisation (Fig.7.7; Schemske & Lande 1985). This strong bimodality is presumably a reflection of the transitory costs of selfing. While selfing in a previously outcrossed population will lead to the exposure of deleterious recessives and severe inbreeding depression, a population which survives this temporary flush will be purged of the recessives and is therefore unlikely

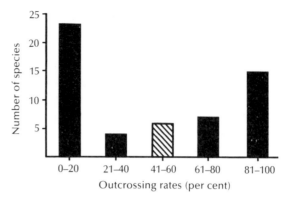

Fig. 7.7 Frequency distribution of mean outcrossing rates for plant species in natural populations. Solid bars are for species with limited interpopulation variation. Hatched bars are for species with marked interpopulation variation. After Schemske & Lande (1985).

to suffer further depression. The circumstances where selfing is favoured are analogous to those where sex is favoured. The reproductive assurance provided by self-fertilisation probably has special importance (Lloyd 1988).

Indeed, the evolution of dioecy from monoecious ancestors has often been attributed to this reduction in selfing (Baker 1959, 1967). Recent analyses have detected a suite of plant characters which are associated with the evolution of dioecy, including the production of fleshy fruits, a woody habit, small, insect- or wind-pollinated flowers, island habitat, and heterostyly, though these correlations have not yet been subject to rigorous analysis which disentangles covariation between traits and phylogenetic influences (Thompson & Brunet 1990). Some of these characters are easily reconciled with an important role for avoidance of selfing. For example, because of their size and growth architecture, woody plants may suffer an increased probability that pollen transfer between flowers on the plant will occur, and a higher probability that the various parts of the plants will be physiologically uncoordinated, so temporal separation of male and female function will be more difficult to maintain (Thompson & Barrett 1981). However, other associations are less easily explained in this way, suggesting that a multifactorial hypothesis for the evolution of dioecy is probably necessary, as is a more comprehensive data base (Bawa 1980, 1982; Givnish 1980, 1982; Willson 1982; Baker 1984; Thompson & Brunet 1990).

SEX RATIO VARIATION

In the previous discussion I was concerned with the way in which gender is expressed in an individual. The remaining component of the study of sex allocation is how an individual allocates resources to male and female offspring. The standard axiom is that all things being equal, it is evolutionarily stable for parents to invest equally in male and female offspring (Box 7.1; Fisher 1930; Shaw & Moehler 1953), and indeed equal sex ratios are very commonly observed. However, an equal sex ratio could also arise in birds and mammals purely as a consequence of heterogametic chromosomal sex determination. This makes it difficult to discern whether the equal sex ratio is an artefact of the sex determining mechanism, or the sex determining mechanism is a way of achieving the optimum sex ratio (Williams 1979; Maynard Smith 1980). In the following discussion I show a variety of reasons why all things are not equal, so that substantial biases in the sex ratio can arise.

Improved condition affects the fitness of one sex more than the other

The size–fecundity advantage hypothesis predicts that mothers capable of producing large offspring should bias sex allocation towards daughters, in order to enjoy the fecundity advantage of those daughters. By contrast, the fecundity of males will vary less with size, so mothers in poor condition should produce daughters. The mermithid nematode *Romanomeris culicivorax* exhibits environmental sex determination, and Petersen *et al.* (1968) showed that sex is influenced by the size of host and the degree of crowding. The size of the nematodes decreases with the number of worms per mosquito larva they infect. However, for any given infection level, the emerging worms are larger if the host is larger. Where emerging worms are larger, they tend to be female (Fig. 7.8).

In contrast to most other animals, in both birds and mammals males are more likely to be the larger sex. Two causes warrant discussion. Greenwood & Wheeler (1985) suggest the heterodox hypothesis that the critical influence is homeothermy. Because heat is teratogenic, and large animals have more trouble dumping heat than small animals, the upper size of females is constrained. The second cause is sexual selection. Increased male size relative to females will commonly result from intrasexual selection for increased male fighting prowess, and hence size, or intersexual selection, with females preferring larger males.

Where there is strong intersexual or intrasexual selection for increased body size, the mating opportunities of small males are likely to be much restricted. By contrast, females are unlikely to have difficulty acquiring a mate, but are unlikely to enjoy benefits of large size to the same extent as heterotherms because their fecundity is limited by the intense parental care necessary before the young achieve independent homeothermy. Under these circumstances variance in male reproductive success may be much greater than variance in female reproductive success. Therefore the advantage enjoyed by a very healthy male may greatly exceed that of a very healthy female, favouring investment in sons by mothers in good condition (Trivers & Willard 1973). This

Box 7.1 Equal investment in males and females is an ESS. After Maynard Smith (1978).

A pair can produce m male offspring and f female offspring, and an increase in m causes some decrease in f (formally, we say (m, f) must lie within a fitness set. We want to determine the evolutionarily stable sex ratio (m^*, f^*).

Consider a random-mating population in which typical pair (genotype $++$) produces m^* sons and f^* daughters. A rare dominant gene M causes females to produce m sons and f daughters (the results would be the same if the gene affected fathers). The frequency of $M+$ females is P and of $M+$ males is p. Because P and p are very low and the population is random-mating, we can ignore the very rare MM individuals (and the P^2 and p^2 in the following equations; see Chapter 2).

Matings			Total offspring			
Female	Male	Frequency	Male		Female	
			$M+$	$++$	$M++$	$++$
$M+$	$++$	P	$\frac{1}{2} mP$	$\frac{1}{2} mP$	$\frac{1}{2} fP$	$\frac{1}{2} fP$
$++$	$M+$	p	$\frac{1}{2} m^*p$	$\frac{1}{2} m^*p$	$\frac{1}{2} f^*p$	$\frac{1}{2} f^*p$
$++$	$++$	$(1 - P - p)$	$-$	$m^*(1 - P - p)$	$-$	$f^*(1 - P - p)$
Total		1		$m^* + P(m - m^*)$		$f^* + P(f - f^*)$

The frequencies of the rare allele in the next generation are P' and p' and are given by

$$P' \approx \frac{1}{2} \frac{m}{m^*} P + \frac{1}{2} p$$

$$p' \approx \frac{1}{2} \frac{f}{f^*} P + \frac{1}{2} p$$

and by addition and manipulation

$$P' + p' \approx P + p + RP$$

$$\text{where } R = \frac{1}{2} \left(\frac{m}{m^*} + \frac{f}{f^*} \right) - 1$$

If $m = m^*$ and $f = f^*$ then $R = 0$, and the frequency of M in the population does not alter (hardly surprising, as M has identical effect to $+$). For (m^*, f^*) to be an ESS, then R must be negative for any values of (m, f) not equal to (m^*, f^*); when $R < 0$ the frequency of M will decrease. That is, $m/m^* + f/f^*$ must be a maximum when $m = m^*$ and $f = f^*$. This occurs when the product of $m^* \times f^*$ is maximal. This need not occur when $m^* = f^*$ if the cost of one sex is greater than the other.

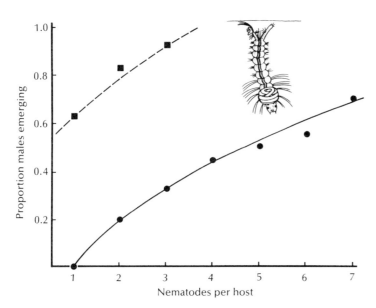

Fig. 7.8 Sex ratio in the nematode *Romanomeris culicivorax*, as a function of infection density and whether the mosquito larva host is large (●) or small (■). If worms are likely to be small, they are likely to be male. After Charnov (1982).

effect has been demonstrated most clearly in red deer. Male reproductive success is much more variable than the reproductive success of females (Clutton-Brock *et al.* 1982). Mothers with high dominance rank produce a greater proportion of sons than mothers with low dominance rank, and maternal dominance exerts a much greater effect on the fitness of sons than on the fitness of daughters (Clutton-Brock *et al.* 1984, 1986; Fig. 7.9).

Where populations are highly structured

One of the restrictive assumptions of the Fisher model is random mating (Box 7.1), which is only possible where complete mixing of the population occurs. In most populations this simply does not occur. Two special forms of population structure have clear influences on sex allocation.

Local mate competition

When a patch of unoccupied habitat is colonised by a pregnant female, her offspring have an excellent opportunity to saturate the patch. Sons can mate with their sisters to produce a generation of grandoffspring for the original mother. Under these circumstances the number of offspring produced by the female potentially depends on the sex ratio, as the rate of growth of the matriline will be slowed by overproduction of males, who compete unnecessarily among themselves to mate with their sisters. This local mate competition is therefore predicted to lead to a female-biased offspring sex ratio (Hamilton 1967). Although these

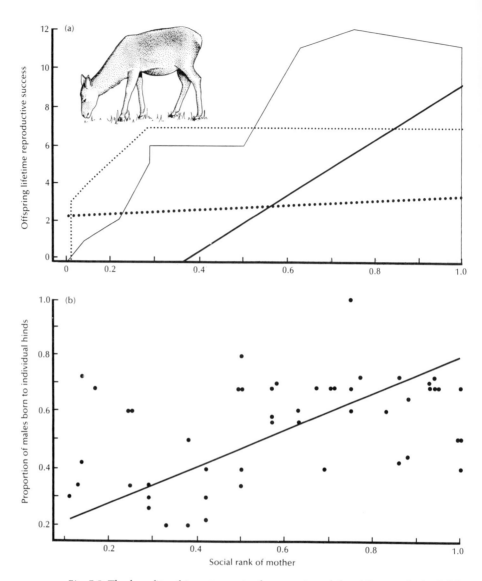

Fig. 7.9 The benefits of investment in the sexes in red deer (*Cervus elaphus*). (a) Lifetime reproductive success (LRS) of sons (continuous lines) and daughters (dotted lines) in relation to the social rank of the mother. The straight lines are the reduced major axes; the faint lines enclose the scatter of points. LRS of males increases more with increase in maternal rank. (b) Percentage of male offspring born to different hinds plotted against the hind's dominance rank. After Clutton-Brock *et al.* (1986).

conditions at first sight seem fanciful, there is abundant evidence for the significance of local mate competition in nature. The two most speciose groups of animals, specialist herbivorous insects and parasitoids, frequently conform to the required population structure, and also often exhibit sex ratios biased in the appropriate direction (Hamilton

1967). Phylogenetic inertia is an inadequate explanation of the pattern. Scolytid beetles which colonise trees by digging into bark have two mating systems. Some females mate before they disperse and their offspring then colonise the tree with several generations of inbreeding. Males may even be flightless. Others mate at the site of colonisation after dispersal. In the former group sex ratios are strongly female-biased: in the latter they approach parity.

One unusual consequence of this pattern is the case of superparasitism. What happens when several females contribute to the founding population? In the parasitoid wasp *Nasonia vitripennis*, the sex ratio is strongly male biased when there is one founder, but quickly converges to parity as the number of founders increases (Werren 1983). The proportion of sons laid by the second female to parasitise a host depends on the number of offspring she produces. Where her clutch is much smaller than that of the original female, she may even produce exclusively sons (Fig. 7.10), no doubt because she is able to exploit the original female's provision of eligible daughters. When she fails to detect prior parasitism and lays an ordinary clutch, her brood is female-biased.

Local resource competition and enhancement

Another factor which may affect the cost of sons and daughters to their parents is any sex bias in the extent to which members of one sex hinder or help their parents. For example, if one sex disperses from

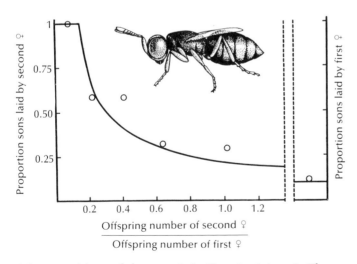

Fig. 7.10 Superparasitism and the sex ratio in *Nasonia vitripennis*. The proportion of sons is plotted in relation to the ratio of the brood size of the second female to lay to the first female to lay. The proportion of sons produced by the first female to lay is also shown. After Charnov (1982).

their natal range while the other sex is philopatric, as is commonly the case in mammals and birds (Greenwood 1980), then any parent–offspring competition will be more prolonged for the philopatric sex. The value of the dispersive sex to their parents may be increased. Such local resource competition could be overcome by investing more heavily in the dispersive sex prior to their dispersal, in order to balance investment in the sexes (Johnson 1986), or overproduction of the dispersive sex. The most convincing evidence for local resource competition comes from primates, where the male bias in the sex ratio at birth is highly correlated with whether female reproductive success declines as the size of female groups increases (Johnson 1988; Fig. 7.11).

The presence of relatives is not always costly. In some social systems philopatric relatives help their parents with reproduction and territory defence (local resource enhancement). In some species this cooperative breeding is sex-biased. Male red-cockaded woodpeckers (*Picoides borealis*) help their parents but females disperse from the group. Sons are overproduced when cooperative groups are small, and this effect declines when cooperative groups are large, probably because the benefit of adding helpers quickly reaches an asymptote (Gowaty & Lennartz 1985).

In summary, there is growing evidence of adaptive manipulation of sex allocation in a variety of species, as predicted by theory. Unfortu-

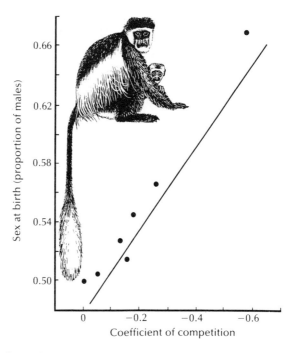

Fig. 7.11 Relation between sex ratio at birth in primate species with female-biased philopatry and an index of competition between females. The level of competition was determined by calculating the effect of increasing female group size on average reproductive success. Points are averages for genera. After Johnson (1988).

nately the success of theory has not been matched by research on
the physiology of how adaptive manipulation may come about. With
haplodiploidy, manipulating the sex ratio is comparatively easy; the
female need simply manipulate whether or not an egg is to be fertilised.
In species with chromosomal sex determination (including all birds
and mammals) the problems are much greater, and present a rewarding
area for future research.

SEXUAL SELECTION AND MATING SYSTEMS

The way in which males and females associate varies greatly within
and between species. Some individuals form lifelong pairbonds to a
single partner while others release their gametes into the ocean or
wind without any possible hope of distinguishing which conspecifics
will be fertilised by those gametes (if indeed there are any). Some
males contribute nothing to parental care while others take responsibility for everything but laying the eggs. A simple classification of
mating systems based on the formation of pairbonds, even if these
pairbonds are temporary, is given in Table 7.2.

Differences between mating systems have enormous consequences,
and are a particularly ripe area for study because there is tremendous
variation within taxonomic groups, and within species, meaning that
both comparative analysis and direct experiment are feasible. For
example, the hominids (humans, chimpanzees, gorillas and orang-
utans) are phylogenetically so extremely similar that determining
which species are most closely related has challenged even the most
sophisticated of genetic techniques (see Caccone & Powell 1989 for
some recent discussion). However, none of the species have remotely
similar mating patterns (Harcourt *et al.* 1981). Chimpanzees live in
large groups in which the opportunities for copulation are enormous,
as sexually receptive females will copulate with most of the males in
the group. By contrast, and belying their 'King Kong' image, gorillas
may have a copulatory frequency of less than once per year. There are
a variety of morphological correlations of this diversity. Chimpanzees

Table 7.2 Mating systems classified according to the formation of pairbonds
between individuals. Thus a bird may be called monogamous even if extra-pair
copulations occur.

		Male pairs with	
		1 female	> 1 female
Female pairs with	1 male	Monogamy	Polygyny
	> 1 male	Polyandry	Polygyandry
No pairing (usually many mates) — promiscuity			

have very large penises and conspicuous testes, while the same organs on the gorilla are tiny, incidentally increasing the susceptibility of captive gorillas to sterilisation when they contract mumps.

Sperm competition

As the chimpanzee example suggests, males do not only compete directly among themselves for access to mates, but their sperm may also compete for access to the egg, or pollen may compete for access to the ovaries. There is very great morphological diversity in both insemination behaviour, and the morphology of the sperm themselves. In a truly bizarre example, the clerid beetle *Divales bipustulatus* has sperm which are about 10 mm long, longer than the animal itself (Mazzini 1976). The acrosome and nucleus are only about 16 microns long. The functional significance of this incredibly long tail is not known.

Females may manipulate the ability of males to supply sperm in a variety of ways. Many females possess complex sperm storage organs (spermathecae) in which sperm assortment is possible. In some cases the sperm and ejaculatory fluid can be transferred to the gut and used as a source of nutrients. Counteradaptation on the part of the male is therefore expected. For example, the sperm scrapers on the penises of damselflies have developed to remove the sperm of other males, and traumatic insemination in bedbugs to counter female assortment of sperm (see Figs 1.7 and 4.7).

Differences in the opportunity for selection on the sexes

Paternal care is less common than maternal care. There are at least three reasons why this should be the case. As we have seen, sperm competition means that males may be uncertain about the paternity of their offspring. Second, because by definition females produce eggs and males provide sperm, females are more likely to be in close contact with offspring at the time of egg-laying or hatching. Good evidence for these effects comes from the distribution of paternal care in bony fishes. Fish with internal fertilisation only rarely show paternal care, but it is reasonably common in species with external fertilisation, as the male can guard the eggs against predators (Gross & Shine 1981).

Last, sperm are 'cheaper' to produce, so males invest a far smaller proportion of their lifetime reproductive effort in each copulation, and presumably face a trade-off between using paternal care to enhance the prospects of survival of a brood that they have sired, or investing in acquiring additional mates. Paternal care is sometimes essential to survival of the brood, enforcing monogamy. For example, in some dense seabird colonies the risk of intraspecific or interspecific nest predation is so high that it is unlikely that an unattended clutch would ever be successful. The time required to incubate the eggs and fledge chicks exceeds the capacity of one parent, so paternal care is obligate.

However, in many cases, females bear the responsibility of rearing the young alone. This has led to the proposition that selection on males should favour traits which enhance the acquisition of mates, while selection on females should favour traits which help them raise young, such as selection of habitat which affords resources during the period of parental investment (Darwin 1871; Bateman 1948; Trivers 1972).

The strange case of the dunnock

The dunnock, *Prunella modularis*, is the quintessential little brown bird, and has been used as a symbol of propriety and chastity (Davies 1985). However, in an extraordinary analysis by Davies and his co-workers of a small population in the Cambridge Botanical Gardens, a bizarre and complicated social organisation has been revealed. First, some males pair with more than one female (either on their own, or in polygynandrous association with another male). Some males are monogamous, some share a single female with another male (polyandry), while some unfortunates remain unpaired, despite gaining a territory (Davies & Lundberg 1984). This ranking is on an approximate scale of the ability of males to monopolise matings.

The more successful males defend larger territories, thus encompassing the ranges of more females (Fig. 7.12). Female range size decrease as the amount of good quality feeding habitat in their range increases (Fig. 7.12), supporting the argument that the predominant motivation for males is to increase their access to mates, while the predominant influence on females is the need to provide high quality food resources to their offspring. Indeed, experimental addition of food caused a reduction in female range, causing a shift in the modal mating system away from polyandry and monogamy towards polygyny (Davies & Lundberg 1984).

However, females are not just passively responding to the availability of food. Females derive benefit from copulating with more than one male because chicks are then fed by both males, and survive better when two males attend the nest (Davies 1986; Davies & Houston 1986). Indeed, if two males are in attendance one is behaviourally dominant to the other, and this dominant male will actively try to drive away the subordinate. By contrast, the female deliberately attempts to evade the attentions of the dominant male and to consort with the subordinate male (Davies 1985; Davies & Houston 1986). The contribution of the subordinate male to feeding is directly proportional to the amount of exclusive access to the female he has enjoyed (Fig. 7.13), presumably as this enhances his chances of shared paternity. Males do not apparently recognise their 'true' young, but are forced to rely on indirect measures to guess at their prospect of siring young

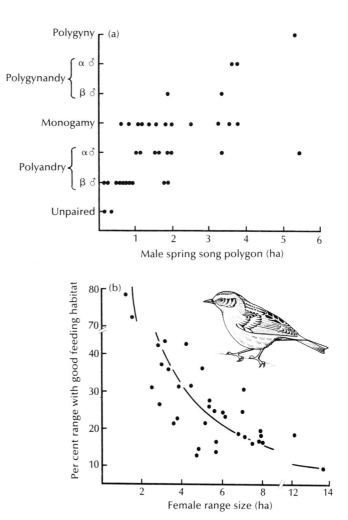

Fig. 7.12 The determinants of range size in dunnocks, *Prunella modularis*: (a) the number of mates monopolised or shared by a male increases as the polygon surrounding the points at which it sings increases; (b) female range size declines as the percentage of good quality feeding habitat increases. After Davies & Lundberg (1984).

(Burke *et al.* 1989). Copulation is preceded by a complicated display in which the male pecks at the females cloaca, from which she extrudes a drop of sperm, perhaps to increase the probability that males will believe that their own sperm has precedence, so that providing some help is worthwhile (Davies 1983).

INTERSEXUAL SELECTION AND FEMALE CHOICE

Two quite different sorts of selection could conceivably enhance the ability of a male to acquire mates. Some traits enhance the competitive ability relative to other males, such as the ability to sustain courtship,

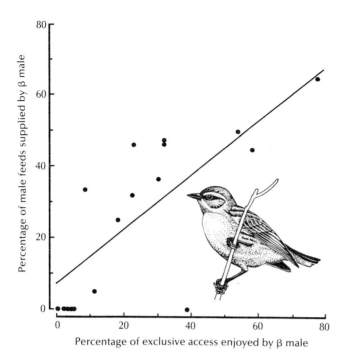

Fig. 7.13 The proportion of feeds supplied to the nestlings in relation to the exclusive access to the female enjoyed by the β male dunnock, *Prunella modularis*, living in polyandrous or polygynandrous groups. After Burke *et al.* (1989).

to fight against other males for access to mates or the resources those mates require, or to provide more conspicuous signals likely to be noticed by females. Selection of this sort is called intrasexual selection, and is presumably very common, leading to traits as diverse as the antlers of deer and the frequency of the croak in some frogs.

Because females usually have no difficulty in attracting a mate, yet many males do not mate at all, females are more likely to have the opportunity to choose their mate than are males. The consequences of female selection for particular male genotypes is called intersexual selection (Darwin 1871). In contrast to intrasexual selection, the role of female mate choice in influencing male characters is highly controversial, and forms the basis for the rest of this Chapter.

Before proceeding to that discussion, it is worth noting that sexual selection is usually distinguished from natural selection because of the possibility that sexually selected traits may have viability costs. For example, devoting energy to the production of antlers may reduce the ability to invest in maintenance, and singing vigorously or displaying brilliant plumage may increase the risk of predation, so singing and coloration are sometimes greatly reduced outside the mating season. Indeed, the size of the nuclei that control bird song is also reduced, suggesting that the maintenance of additional neural tissue to support mate attraction is also energetically costly (Nottebohm 1981).

Darwin (1871) was the first to ascribe evolutionary importance to female choice, and based his enthusiasm on observation of courtship in many species. It is not difficult to repeat Darwin's observation in a farmyard or in the wild. Males appear to show limited fussiness, to the point that their lack of discrimination can be exploited by other species or biologists. Some orchids resemble female bees and are pollinated when the male attempts to copulate with the flower, sperm used in the artificial insemination of cattle is collected using a false cow bearing only passing resemblance to the original, and some beetles will attempt to copulate with beer bottles.

By contrast, females will often successfully resist (or ignore) the courtship attempts of males, and several females will mate with a particular male even though other males remain unmated. Females could base such discrimination between males on a variety of features. The most easily understood are immediate benefits to herself or her offspring. For example, a male may be chosen because he provides some resource she needs. The duration of copulation and prospect of effective sperm transfer in the scorpion-fly *Hylobittacus apicalis* is dependent on the size of the nuptial gift presented by the male (usually the carcass of some choice insect such as a blowfly; Thornhill 1980). These food gifts provide the female with a direct nutritional reward but in species with paternal care may also indicate the ability of the male to deliver food to her offspring (e.g. terns; Nisbet 1973).

Of greater interest from an evolutionary viewpoint is female choice based on genetic criteria. Females may choose between the genotypes of their mates for a variety of reasons. Short-term genetic benefits are most easily understood. For example, incest avoidance is a form of mate choice based on the genetic quality of partner, but the genetic benefits are confined to the heterotic combinations of the offspring, and have no long-term effects. We would also anticipate selection against mating with males of the wrong species, but again the choice is directed against temporary effects caused by hybrid infertility (but see Ryan 1990).

Greater difficulty surrounds the proposition that females may discriminate between mates according to the quality of their genes; and in mating with males with 'good genes' they may enhance the quality of their descendants. The problems with this hypothesis concern the likelihood of heritable variation for fitness. In a population at equilibrium we expect heritability for fitness to approach the low values that can be maintained by mutation pressure. While there may be heritable variation in components of fitness, females choosing their mates on the basis of any one component may end up with offspring which perform inadequately in other components of fitness.

We can summarise these results by saying that most theoretical arguments and some empirical research cast doubts on the ability of females to derive long-term genetic benefits from mate choice. Unfortunately, this confidence that mate choice should not work is shaken by observational data. One of the mating systems in which female choice is most conspicuous, and where male adornment is often highly developed, is called lek promiscuity. By definition, lekking involves male aggregation at mating arenas, which females visit for the purpose of copulation. The females depart from the arenas, or leks, and rear their young alone. Females derive nothing from the male but sperm, yet one or a few males at each lek typically achieve most of the copulations (e.g. Lill 1974; Bradbury 1977). While patterns of male–male interaction might explain some of these results, much of the asymmetry in mating success derives from direct female preference (Bradbury *et al.* 1986).

In some populations a genetic basis for female choice has been demonstrated, though it is not immediately obvious why that choice arises. It is certainly curious that in two well-studied systems where a genetic basis for preference has been demonstrated, selective and non-selective genotypes coexist in the population (Majerus 1986; Engelhard *et al.* 1989). Engelhard *et al.* suggest that the two genotypes can coexist in seaweed flies (*Coelopa frigida*) because mate choice is only adaptive at high density. At low density, reduced propensity to mate may mean loss of mating opportunities altogether.

Although these results are of great interest, it is difficult to see how polymorphic behaviour of this sort could drive the evolution of the bizarre and exaggerated male ornament seen in many animals. Darwin (1871) was most puzzled by bizarre plumage (Fig. 1.3), inexplicable in terms of intrasexual or conventional natural selection. Choice seems important, but what benefits could females be deriving from this choice, and what are the consequences of persistent choice? Controversy surrounds two hypotheses which attempt to explain both female choice and development of dimorphism. Both are focused directly on the problem of the heritability of fitness; one seeks a solution assuming no heritability; the second seeks an environmental agent which would maintain heritability.

Fisher's runaway hypothesis

The first hypothesis we owe to Fisher (1930), who aimed to explain the paradox he had first identified: at equilibrium there should be no heritable benefit for fitness. Fisher pointed out that both female preference for and male expression of a trait will be under independent polygenic control. However, there will be an almost inevitable tendency for genes for female preference and male trait to become associated, or to be in linkage disequilibrium. The reason is clear. A male with a long tail probably had a father with a long tail and a mother with a preference for a long tail. The same is true of the male's sisters. The son will therefore pass on to his offspring genes for *both* preference and expression, as will his sisters. Because female choice will greatly influence male fitness, there will be pressure on males to conform to female selectivity. Nonetheless there are times where average male expression and average female preference are not the same. We could call this difference choice discrepancy. The female widowbirds I described in Chapter 4 prefer longer tails than those exhibited by the males in the population (Andersson 1982).

What are the consequences of linkage disequilibrium and choice discrepancy? Fisher (1930) realised, and Lande (1981, 1982b) subsequently demonstrated theoretically, that this will very much depend on the extent to which female and male offspring tend to covary, plus the strength of natural selection on the trait and the extent and variance of female preference. The outcome of Lande's analyses has been represented graphically by Arnold (1983) (Fig. 7.14). First, where the slope of the genetic covariance is shallow, the population will come to lie along a line-of-equilibrium, with all points along the line having equivalent fitness. Therefore the population is free to move up and down the line as genetic drift affects the genes controlling female

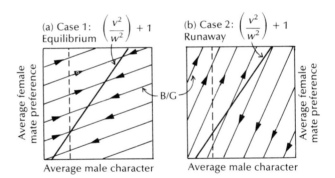

Fig. 7.14 The evolution of average female mate preference and the degree to which male's express a trait in Lande's (1981) model. There are five important variables: First, there is a tail size which is optimised by natural selection. Next, consider the progeny from a particular male parent mated to a large sample of females. We can calculate the extent of genetic covariance between male character and female preference, B (determined by the extent of disequilibrium), and the genetic variance in male tail size, G (variance is required for evolution to occur). The genetic regression of daughters' mate preference on sons' tail size is B/G, and determines the evolutionary trajectories. The other two values determining the outcome of sexual selection are the degree to which female preference is stereotyped w, and the extent of natural selection on the trait v. The slope of the equilibrium line is given by $(v^2/w^2) + 1$: (a), when B/G is shallow relative to the equilibrium line the trajectories converge; (b) when the slope is steeper than the regression line the runaway effect occurs, and may produce exaggeration or diminution of the trait. The end point is determined in both cases by the initial discrepancy, and in the latter case by when natural selection ultimately restrains the development of the trait. After Arnold (1983).

fitness. The initial point at which the population reaches equilibrium will depend on the extent of choice discrepancy in the first place. Second, if the slope of the covariance is very steep, there is potential for explosive divergence. This is called the runaway effect. Only minor perturbations from equilibrium are necessary to provoke massive escalating departures away from the line-of-equilibrium, which will then become exaggerated until the male display becomes so expensive that it is restrained by natural selection.

The theoretical feasibility of the Fisher hypothesis has been demonstrated using theoretical quantitative genetic models on several occasions (Lande 1981; Seger 1985; Kirkpatrick 1987), most recently for monogamous species (Kirkpatrick *et al.* 1990). It produces one conclusion apparently not forseen by Fisher (1930). There is no initial necessity for the female preference to be adaptive at all. Female whim is sufficient to drive male evolution. Despite a century of intense scepticism, this is very much the result anticipated by Darwin (1871).

There are a number of other ways that the runaway effect could become established. For example, it has been suggested that many male traits represent a type of sensory exploitation of pre-existing female biases (Ryan 1990; Ryan *et al.* 1990; Ryan & Rand 1990). If a female has a search image for red berries, then a male with a red breast might be chosen more often, founding the conditions necessary for the runaway effect to proceed.

The condition-dependent handicap hypothesis

Zahavi (1975, 1977) suggested some time ago that extravagant male traits such as a peacock's tail may be a reliable indicator of the prowess of the male, as the female could observe the male's ability to withstand a handicap, and thus his underlying vigour. In its original form, this model has attracted considerable criticism, principally because the female's offspring might inherit the handicap but not the vigour. More recently, Hamilton & Zuk (1982) have argued that a special form of this 'handicap principle' might indeed be the driving force of development of bizarre traits. They point out that heritability of fitness may persist because populations are rarely at equilibrium. One source preventing attainment of equilibrium will be the persistent evolution of enemy species (the Red Queen effect), particularly those species with lifespans which are short relative to the lifespan of the species of interest, such as some parasites. The ability to resist disease is almost always subject to some genetic variation (Chapter 5, Chapter 10), and offspring would benefit from disease resistance. Females can therefore be expected to scrutinise males for their ability to resist disease, and select traits which display resistance 'honestly' (Kodric-Brown & Brown 1984). Full development of elaborate traits such as complicated prolonged song and epigamic adornment is unlikely to be possible in individuals suffering from disease (we cannot sing like Caruso in the shower if we have influenza), providing strong intersexual selection for males that display their disease resistance with elaborate display.

Once again, quantitative genetics theory has been used to examine the range of conditions under which this hypothesis might operate. Despite some initial negative results (e.g. Kirkpatrick 1986), it does appear the mechanism is feasible (Andersson 1986; Pomiankowski 1987a, 1987b; Charlesworth 1988; Tomlinson 1988; Grafen 1990a, 1990b; Michod & Hasson 1990), though under more restrictive conditions than the Fisher hypothesis, no doubt because there are a greater number of explicit predictions and assumptions inherent in the model.

Can the hypotheses be tested empirically?

Given that the theoretical plausibility of the two models has been confirmed, it would be rewarding to know whether one or both frequently contribute to the evolution of sexual dimorphism. There is growing evidence that male expression of a trait and female preferences do evolve in parallel, a requirement of both models. In the most direct demonstration, Houde & Endler (1990) have shown that female guppies differ in the extent to which they prefer orange coloration on the male. The degree of female preference based on orange is strongly correlated between populations with the amount of orange exhibited by the male, strongly supporting parallel evolution.

The obvious way to distinguish the hypotheses would be examination of the genetic parameters and assumptions in the models. Unfortunately measurement of most of the large number of parameters in the models is currently beyond our ability (see Andersson 1987 for an assessment of the number of features requiring measurement), and the vagueness of our understanding of quantitative variation in seemingly more tractable systems like morphological traits does not inspire confidence that a solution is just around the corner (see Chapter 2). Instead, we have to rely on indirect tests of the models.

The same features which limit the theoretical plausibility of the condition-dependent handicap hypothesis facilitate testing of its predictive capacity relative to the Fisher model. Because the assumptions in the Fisher model are fewer, so are its unique predictions. Tests of the condition-dependent models have focused on two predictions, one direct and one indirect. First, at an intraspecific level, the hypothesis directly predicts that females should mate preferentially or longer with males with low levels of disease. A correlation of this sort has been observed in many organisms, including some of the groups in which elaborate male epigamic traits are common, such as insects (Zuk 1987, 1988), fish (Kennedy *et al.* 1987; Milinski & Bakker 1990) and birds (Borgia & Collis 1989; Møller 1990). Unfortunately there are several alternative explanations for this correlation (Borgia & Collis 1989; Howard & Minchella 1990). Males with low parasite loads may be able to perform courtship displays more vigorously than diseased males, and therefore more quickly pass the threshold stimulus required for receptivity in the female. Females also have another motive for discrimination, avoidance of disease themselves. Reliable tests distinguishing these alternatives have not been performed. Borgia & Collis (1989) have recently shown that while female satin bowerbirds (*Ptilinorhychus violaceus*) discriminate between males on the basis of their infection with the louse *Myrsidea ptilonorhychi*, the data are more consistent with direct female avoidance of parasites than with a good genes effect, as there was no obvious negative correlation between parasite load and either condition or plumage quality.

Second, at an interspecific level, Hamilton and Zuk (1982) claimed that epigamic adornment would be most pronounced in those species with high parasite levels or diversity, as in these species display of resistance would provide a consistent selective advantage. The appropriate correlation was detected for songbirds from both North America (Hamilton & Zuk 1982) and Europe (Read 1987), though there are substantial statistical and methodological problems which hinder the unambiguous interpretation of these data (Cox 1989; Read & Harvey 1989b), and some negative data (e.g. Scott & Clutton-Brock 1990). Other interpretations of the correlation are also possible. For example, if the strength of sexual selection for whatever reason is correlated with polygynous and promiscuous mating systems, and venereal transmission of parasites is common when multiple matings occur, the correlation could arise for reasons unrelated to the parasites *per se*. It is possible to approach this problem by partitioning effects statistically, but some uncertainty must surround any interpretation (Read 1988).

A largely unaddressed problem with the hypothesis is the difficulty of interpreting measures of parasite load. Neither parasite abundance nor species diversity need be a reliable indicator of the genetic aspect of host resistance. Parasites vary dramatically in their effects on host fitness in complex ways (Chapter 5). For example, in concomitant immunity, a mammal may be unable to rid itself of a particular parasite, but may enjoy protection against other parasites of the same species and even other species. The famous ability of cowpox to reduce infection by smallpox is an effect of this sort, prompting one of the greatest revolutions in medical treatment. Many other parasites have limited effect unless the immune system is compromised by some other agent. Parasite loads may therefore reflect nutritional or social factors rather than genetically based resistance.

Blood parasites like malaria do have exactly the rapid population cycles which would favour the operation of the Hamilton & Zuk model. The effects of malaria on human fitness are certainly significant, as many children are killed by their first exposure to the disease. However, an adult who has recovered from infection as a child is thereafter unlikely to die from the disease. They may or may not become ill at regular or irregular intervals, and there is only a weak correlation between the extent of visible suffering at a phenotypic level and the number of parasites visible in a blood smear. Ideally, the extent of genetically based resistance to the parasite should be clearly discernible at the time of mate choice, though this may well only be true of diseases which have their effect predominantly in adulthood. While venereal diseases are of this sort, they are also most likely to generate selection for active female avoidance of diseased males in order to minimise the risk of selection.

The 'good genes' element of the hypothesis has received far less scrutiny, though this is without doubt its most crucial element. Partridge (1980) allowed *Drosophila melanogaster* females to mate under two conditions: in a small vial with only two males, or in a larger cage with two hundred males. She then transferred the mated females to laying jars, and carried out the tedious task of collecting first instar larvae as eggs hatched. The 'choice' larvae and 'no-choice' larvae were then allowed to compete in separate tiny vials against a control strain which was easily distinguishable when they emerged. The choice larvae were better competitors against the control strain, supporting the idea that females can improve the fitness of their young by mate choice. The source of the benefit has never been identified, and the possible role of heterotic effects could not be distinguished from the long-term genetic benefits envisaged in the 'good genes' model.

More recently, von Schantz *et al.* (1989) have reported the intriguing result that the ability of male pheasants, *Phasianus colchicus* (Fig. 7.15) to attract females depends strongly on the length of their spurs (and not, incidentally on their bizarre plumage; but see Hillgarth 1990). This advantage can be augmented by experimental elongation of the spurs. Dominance and other intrasexual interactions between males does not appear to depend on spur length. Although males contribute no parental care, female reproductive success increases with the spur length of the male, suggesting that both males and females derive advantage from the male trait. Unfortunately these data cannot be simply related to the Hamilton & Zuk model because the heritability of spur length and its relation to parasite resistance has not yet been determined.

In summary, despite exciting data from pheasants, there is no species in which all the predictions of the condition-dependent model have been verified, and some reason to be cautious about all of the evidence cited in favour of the hypothesis. This, of course, is no reason for acceptance of Fisher's runaway model, and a unique test of this model has thus far proved elusive. One novel prediction of the viability-independent model is that there should be a degree of unpredictability of the direction in which the runaway effect occurs. Dawkins (1986), following Borgia (1979) and Darwin (1874) asks if whim can lead to selection for long tails, why not short tails? Certainly explosive reduction of ornaments is predicted by theory as frequently as exaggeration of the same ornaments (Fig. 7.14). Tail-shortening is not predicted by the condition-dependent model, as the failure to develop elaborate plumage is a dishonest signal, and may be indicative of inadequate condition as much as an ability to fly in spite of the handicap. Male

Fig. 7.15 Epigamic tail-shortening in golden-headed cisticolas (*Cisticola exilis*) (a) and spur-length in pheasants (*Phasianus colchicus*) (b) provide support for the 'good genes' and Fisher runaway model respectively, but definitive empirical tests have not proved possible.

golden-headed cisticolas (*Cisticola exilis*) moult from an eclipse plumage with a tail of about 50 mm to an epigamic plumage of about 35 mm (Fig. 7.15). Females suffer some tail-shortening at the time of the nuptial moult, but the moult affects males much more. Could this be driven by female whim? Continued empirical attempts to discriminate between these hypotheses is one of the most exciting and active areas of research in evolutionary ecology; at the moment it is certainly conceivable that both effects are important.

Mate choice in plants?

Darwin (1871) was drawn to conclude that female choice was a significant force in evolution by his observations of animal behaviour, particularly the elaborate courtship of birds, in which female discrimination seemed self-evident. As he commented (1874): 'The males display their charms with elaborate care and to the best effect; and this is done in the presence of the females ... To suppose that the females do not appreciate the beauty of the males, is to admit that their splendid decorations, all their pomp and display, are useless; and this is incredible.' However, as we have seen, experimental scrutiny of this supposedly obvious behaviour has proved very awkward, and at least in birds, has yet to deal with such intractable problems as the variation in attractiveness of an animal throughout its lifespan, and the relation of such variation to the chief hypotheses attributing evolutionary importance to sexual selection.

Surprisingly, some of the most exciting and elegant experiments on sexual selection are currently emerging from a group in which female fussiness is anything but obvious, the flowering plants. I will conclude my discussion of sex and its consequences with a brief review of the rapidly developing research on the extraordinary sex lives of radishes.

The female flowers of plants are well-designed to collect the pollen they require for fertilisation. Indeed, the stigma will often receive pollen from many males. There is therefore potential at the time when this interaction takes place for both intra- and intersexual selection (Willson & Burley 1983). Competition between pollen can be intense. Indeed, in wild radish, *Raphanus sativus*, each maternal plant typically has between six to eight mates, and the seeds of individual fruits are usually sired by one to four (mean = 2.1) mates (Ellstrand & Marshall 1986). Paternity in this case can be resolved through electrophoretic analysis of allelic diversity. One obvious component of success in pollen will be the rate of pollen-tube growth and the success of individual grains in fertilising ovules. There is some evidence that both variation in growth-rate and the ability to fertilise ovules has both a genetic basis, presumably leading to intrasexual selection (Mulcahy 1974; Johnson & Mulcahy 1978; Stephenson & Bertin 1983), and an environmental basis depending very much on the condition of the male plant (Young & Stanton 1990).

There are several additional factors which could influence the success of the pollen (Queller 1987; Marshall 1988). First, the degree of genetic complementation between parents may be important (e.g. Waser *et al.* 1987). The female plant may reject or retard the growth of pollen that is genetically too similar to itself. This is a case of avoidance of inbreeding. There may also be direct competition between embryos for maternal resources. Of greater interest is any direct female intervention in the fate of the pollen, or in the fate of the embryos. Queller (1987) argues that it is only in the former case that the choice is strictly analogous to choice based on heritable additive genetic variance in male genotypes, as part of the selection between embryos could rely on epistatic or dominance variance arising in the embryo or endosperm because of interactions between male and female genomes.

However, there is substantial evidence that selective seed abortion may also be a way of distinguishing between male genotypes. Discrimination of this sort among radishes takes the form of position-dependent seed abortion and resource allocation within fruits (Marshall & Ellstrand 1988). Stressed plants in particular select for pollen donors that sire seeds in the basal and middle part of the fruit, rather than the stylar part (Fig. 7.16). Pollen reaching these areas are likely to produce big seeds and fruit that are likely to mature, but must grow past fertilisable stylar ovules in the process.

Substantial evidence suggests that male success is not just a consequence of competition between pollen (Marshall 1988). First, there

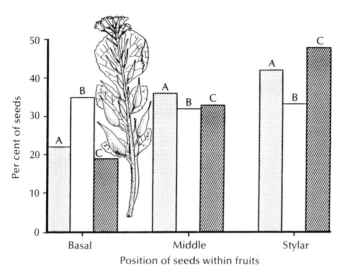

Fig. 7.16 The percentage of seeds sired by three pollen donors A, B, C in the basal, middle and stylar thirds of multiply-sired fruits of wild radishes. The stylar third has more seeds because of the method of assignment to the three classes. Basal seeds are likely to be larger and more likely to survive in stressed plants. After Marshall & Ellstrand (1989).

are strong interactions between maternal genotype and paternal genotype, supporting a role for mate choice or genetic complementation (Marshall 1988). Second, multiply-sired fruits are allocated more resources than single-sired fruits, consistent with mate choice (Marshall & Ellstrand 1986; Marshall 1988). Last, maternal condition has a strong effect on both variables, in a pattern suggesting that when mothers are in good condition and there is no nutrient limitation they invest in all seeds, but when they are stressed so that all seeds cannot be filled to normal weights, maternal fitness is enhanced by aborting the genetically poorest seeds or providing them with fewer resources.

The heavy investment in multiply-sired fruits could have two causes; a high quality male is more likely to have fertilised ovules (though of course so is a low quality male), or there may be advantages in producing diverse progeny. Advantages to diversity include the possibility of producing a rare genotype favoured by frequency-dependent selection (the Red Queen hypothesis), producing an individual which is more likely to exploit the heterogeneous local environment, and offspring which are unlikely to compete for the same vacancy in the local environment (the Tangled Bank hypothesis) (Karron & Marshall 1990). These results are therefore relevant to the key issues in both the theory of intersexual selection and the evolution of sexual reproduction. When the number of potential fathers is increased in experimental cross-pollinations, the number of fathers per fruit levels off, suggesting that there is some upper limit to the benefits of multiple paternity (Marshall & Ellstrand 1989). Two mates per fruit seems to be an optimum number. Further, having multiple mates does not appear to increase the mean dry weight of seedlings over plants fathered by a single male, but it does sharply increase the variance in seedling weights (Karron & Marshall 1990). There is thus little evidence that reduced competition among offspring is the factor favouring increasing the number of mates. The signficance of the other hypotheses awaits testing, but the elegance with which experiments can be applied to plants suggests that they will make a great contribution to unlocking the uncertainty surrounding sex and its consequences.

SUMMARY

The evolution of sex has proved particularly troubling to evolutionary theory, because it appears to depend on group selection, yet imposes formidable short-term costs. There is little doubt that sex has several beneficial functions, including the facilitation of DNA repair. Sex assists the long-term persistence of eukaryote lineages by preventing the accumulation of unacceptable loads of deleterious mutations (Muller's ratchet). Because sexual populations are freed from the ratchet, the total genome size may increase. Mutation load may also be important in the short-term, as the efficiency of selection against the continuous input of deleterious mutations is greatly enhanced by sex. Segregation further allows favourable mutations to become homozygous. These functions contribute to an explanation of the taxonomic distribution of sex and genome size, but say little about the relative importance of sexual and parthenogenetic forms in different habitats. Abundant evidence from both plants and animals suggests that sex is most prevalent in old, stable and complex communities. There are two reciprocal influences on this correlation. Parthenogenesis removes the risk that reproduction will not be possible when temporarily favourable conditions arise, and would be expected early in succession and in extreme habitats where density is very low. However, sex is favoured where strong gene/environment correlations mean that local success cannot necessarily be translated in time and space. The most important influence of this sort appears to be the frequency-dependent selection imposed by parasites and other natural enemies, which ensures that a successful genotype in one generation will usually succumb to parasite infection in later generations.

Parents derive strong benefits from manipulating their allocation of resources to male or female reproduction, in the case of hermaphrodites, or male and female offspring, where the sexes are separate. Theory predicts that investment in male and female offspring should be balanced. In randomly mating populations this means that male and female offspring should be produced in equal numbers, unless they impose different costs on their parents. However, several circumstances also predict strongly biased sex ratios. Population structuring means that parents will often benefit from investment in the sex which costs less. Because the sexes often show different fitness gains from increased condition or investment, there should be strong selection for facultative response to parental condition. Sex ratio is a clear example of a trait with strong fitness consequences which cannot be approximated by lifetime reproductive success. Ideally, the number of grandoffspring should be assessed.

Sexual selection arises because traits which enhance mating success may be favoured even if they decrease other aspects of viability. Although intrasexual selection, or competition within a sex for mates, is common, it appears not to explain some of the extreme cases of sexual dimorphism. Females are more likely to invest heavily in their offspring and be selective about their mates than are males. Female choice for particular male characters (intersexual selection) could explain the evolution of bizarre adornment but it is unclear what might be the criterion of choice. Models based on choice of 'good genes' suggest that heritability of male fitness is maintained by coevolutionary races with hostile species, particularly parasites. By contrast, models suggest that the effect of linkage disequilibrium between male trait and female preference is adequate to drive the runaway evolution of bizarre adornment without any fitness heritability. Despite the elegance of the theory, empirical tests have not been forthcoming. Plants are elegant experimental models for investigating sexual selection.

FURTHER READING

Sex has been a popular topic for monographic reviews, perhaps supporting the notion that we write most about the things we least understand. The evolution of sex is discussed by Bell (1982), Michod & Levin (1988) and Stearns (1987), though the emphasis is heavily zoological. Sex determination has been treated by Bull (1983) and sex allocation by Charnov (1982). Bateson (1983) and Bradbury & Andersson (1987) are excellent discussions of the controversies surrounding the study of mate choice and sexual selection. Mating systems have been reviewed for birds by Oring (1982), for mammals by Clutton-Brock (1989), and for insects by Thornhill & Alcock (1983). Willson & Burley (1983) is a heterodox but stimulating attempt to drag plants into the framework found useful by zoologists.

TOPICS FOR DISCUSSION

1 One organism often lives completely within the other in obligate symbioses. Discuss the predictions of the Tangled Bank and Red Queen model for the distribution of sex in endosymbionts and ectosymbionts. This topic is discussed by Law & Lewis (1983).

2 Postulate hypotheses other than mate choice by females to explain the tendency of a small number of males to garner most of the matings on leks.

3 Females occasionally prefer males of a congeneric species over males of their own species (Ryan 1990). Discuss the significance of this result for the Fisher runaway hypothesis of sexual selection.

Chapter 8
The Ecological Context
of Speciation

In the remainder of this book I will be interested primarily in the changes in ecological communities through history, and the evolutionary processes which shape those changes. Evolutionary changes are ultimately a consequence of three processes. The first process is anagenesis, change within a lineage or species (Fig. 8.1). The second process is cladogenesis, or the branching of lineages during phylogeny (Fig. 8.1). The third process is extinction, the elimination of a lineage. Ultimately the diversity of a lineage is dictated by the balance between cladogenesis and extinction. In this Chapter, I consider the causes and consequences of the simplest form of cladogenesis, the origin of new species.

WHAT IS A SPECIES?

Defining a species is a useful starting point for any discussion of speciation, but is less easy than recognising one. To a taxonomist, a species is usually the name given to a particular specimen. Despite intense debate about the efficacy of different taxonomic and systematic methods, there is usually concordance between different methods in defining what is and what is not a species. Indeed, several studies have demonstrated excellent agreement between folk knowledge of local faunas and floras and the conclusions of taxonomists (e.g. Patton *et al.* 1982). There are two common sources of error. First, where members of species are polymorphic, different morphs are often incorrectly assigned specific status. The literature contains several cases where

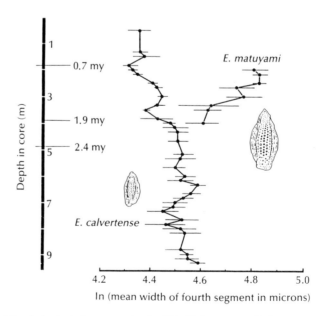

Fig. 8.1 Morphological divergence in the Plio-Pleistocene radiolarian genus *Eucyrtidium* is caused by gradual anagenesis which postdates cladogenesis. The sequence represents approximately 3.5 million years, and was taken from a core at 39°N latitude in the northern Pacific Ocean. After Gingerich [1985].

males and females have initially been called different species! Second, the problem of sibling species. Genetic differences and morphological differences are often uncorrelated, so great differences in evolutionary history are not recorded by morphological diversification (Wake *et al.* 1983).

The biological species concept describes a species as a group of interbreeding or potentially interbreeding populations incapable of breeding with other such populations. At the level of taxonomic practice, this is virtually never tested. The biological species concept has its own problems. First, it is irrelevant to parthenogenetic organisms where interbreeding does not take place. Second, it is awkward when we try to add the temporal scale so important in evolution. In the case of anagenesis, the fossil record may reveal very distinct forms which nonetheless intergrade gradually into each other (Fig. 3.11). The limits to interbreeding have generated substantial confusion. Are the characteristics which enforce species identity designed to ensure interbreeding does not take place, or do they arise as a result of selection on the mating behaviour of the populations, more or less independently of any selective pressures imposed by other species?

Whether we focus on limits to breeding, or the aspects of mating behaviour which maintain the integrity of populations, there is a general belief that species are real entities, and that the formation of new species (speciation) as a consequence of natural selection is hindered by this integrity. Two discrete problems need to be overcome, gene flow and coadaptation of genetic information.

Gene flow

First, gene flow might erode any local adaptation which would allow differentiation into discrete morphs to arise. For example, two morphologically distinct forms of gulls occur in England and exhibit all the characteristics of good species. Yet close analysis reveals that each population is capable of interbreeding with an adjacent population to the east or west (Fig. 8.2). A series of contiguous subspecies ring the entire Arctic Circle. At the extremes of the distribution morphological differentiation is clearly apparent, yet genes may move from one to the other, albeit by a circuitous route. Such ring species are an extreme example of two very common patterns of distribution. Clines are geographic gradients in either a measurable character (e.g. height) or the frequency of a gene, genotype, or phenotype. A Rassenkreis is a group of subspecies connected by clines.

The adaptive landscape

Second, Wright's metaphor of the adaptive landscape suggests populations are not free to move through adaptive space in any case. Selection is incapable of shifting a population from one peak to another where epistatic interactions cause valleys in the adaptive landscape. In the

Fig. 8.2 Overlap in races of gulls of the *Larus argentatus* group. In North America one variant has developed as a separate species *L. glaucoides*, dashed lines. Elsewhere the various races form a more-or-less complete circumpolar ring. Where the ring overlaps in north-west Europe there is virtually no hybridisation between the European end of the chain, (a) the lesser-black-backed gull (*L. fuscus*) and (b) the American end of the chain, the herring gull (*L. argentatus*) as a consequence of habitat separation and ethological isolation. After Mayr (1963, 1970).

following discussions I will outline current models for speciation and how they attempt to overcome these problems.

ALLOPATRIC SPECIATION

The first hypothesis for speciation attributes importance to the disruption of gene flow. Two populations which are separated by a geographic barrier will diverge as a consequence of drift or selection caused by the separate environmental pressures prevailing on either side of the barrier. Such separation is called allopatry, in contrast to overlapping populations, which are said to be sympatric, and contiguous populations which are called parapatric (Fig. 8.3). Populations diverging in allopatry may ultimately become incapable of interbreeding because

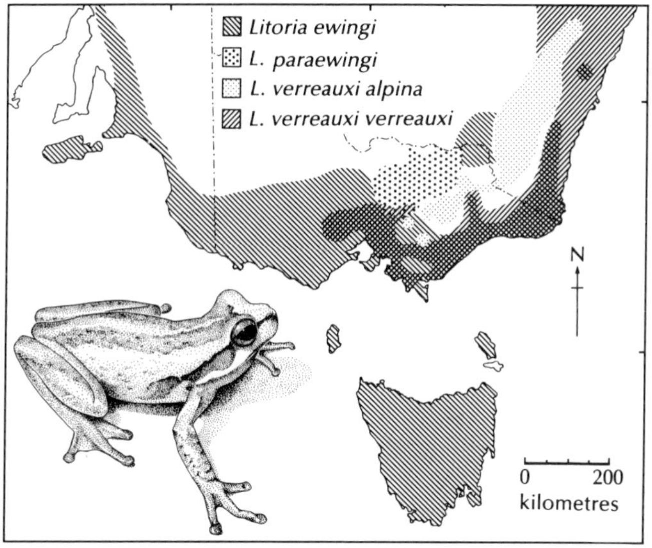

Fig. 8.3 Patterns of distribution in the frog genus *Litoria* in south-eastern Australia. Note the extensive sympatry between *L. ewingi* and *L. verreauxi* in the south eastern part of the mainland; the parapatric distribution of *L. ewingi* and *L. paraewingi* and the various populations of *L. ewingi* isolated in allopatry from their conspecifics on islands, or in disjunct mainland populations. After Littlejohn & Watson (1985).

isolating barriers arise as a pleiotropic consequence of the divergence. This hypothesis is entirely plausible and assumes no special genetic mechanisms, ecological conditions or population structure. Scepticism about its general importance arises from its failure to deal with the problem of the adaptive landscape. If species are well-integrated genetic entities, the effects of drift and selection in large populations may be quite small.

Ecological and genetical conditions promoting divergence

One way of overcoming this difficulty is to look for ecological conditions which accelerate divergence. An obvious way to facilitate divergence would be the breaking of a cline. If one of the intermediate populations in the gull Rassenkreis depicted in Fig. 8.2 was broken, any retardation of divergence caused by gene flow could be overcome. A number of alternative methods have been suggested that allow acceleration of divergence, and normally rely on aspects of the genetic architecture or genetic correlations present within populations.

Molecular drive

The genetic processes that cause concerted evolution might feasibly accelerate divergence above that expected by drift (Dover 1982; Krieber & Rose 1986), though it is currently impossible to test this hypothesis.

Sexual selection

Darwin (1871) observed that many closely related species differ most in the sexual characters of males (e.g. Figs. 1.3, 8.4). These characters are therefore among the most rapidly evolving and taxonomically important morphological traits (Lande 1987). The models of intersexual selection described in Chapter 8 suggest two ways that sexual selection could exaggerate the rate of allopatric divergence. First, one evolutionary consequence of the line of equilibria is a plateau on the adaptive landscape. Because there are several possible outcomes with approximately equal fitness, the direction of selection and drift are less determinate, and changes in traits important to reproduction are easily accomplished. Second, if conditions favouring runaway sexual selection emerge in only one population, that population can diverge extremely rapidly. Under these circumstances the propensity to speciate is determined by sexual behaviour within populations. Speciation may take place even with a modest amount of gene flow (Lande 1981, 1982b), though it will be facilitated by allopatric isolation.

Mutualism

Analogous arguments have been advanced to explain the extraordinary speciose radiations of some pairs of mutualists, e.g. the figs and fig wasps described in Chapter 1. Mutualisms may evolve to the extent that matching the characteristics of its partner can be an important selective force on a species. Under these circumstances reciprocal selective forces create lines of equilibria for the phenotypes of the mutualists (Kiester *et al.* 1984). Diversification can then occur easily, as allopatrically isolated populations are free to drift along the line of equilibrium. The prospect of divergence will be a simple function of the time since isolation. There are two additional interesting outcomes from these quantitative genetic models. First, the rate of drift is a function of the harmonic mean of the effective sizes of the two species, so where one

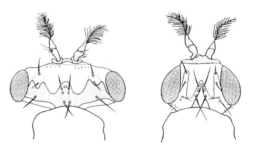

♂ *Drosophila heteroneura*   ♂ *Drosophila silvestris*

Fig. 8.4 Morphological differentiation between the closely related Hawaiian fruitflies, *Drosophila heteroneura* and *D. silvestris*. As is very often the case, divergence in male characters is much more pronounced than divergence in female traits. After Maynard Smith (1989).

species has an effective size much smaller than its partner, this small size will disproportionately affect the harmonic mean. Therefore the potential for diversification of both partners by drift will be higher than might be assumed by examining the coevolving species with a large effective population size. The second outcome is that where sexual selection in one species is combined with coevolution, runaway sexual selection in one species can cause rapid coevolution, and greatly enhance the prospect of speciation of both partners.

Kiester *et al.* (1984) consider a number of radiations where these effects might be important. For example, the relationship between figs and fig wasps involve several attributes that may increase the intensity of coevolution. The pollinators are both highly specialised and totally dependent on figs for resources, including both food and oviposition sites. The extreme host-plant specialisation also provides the opportunity for flowering times of host trees to drift out of synchrony with the rest of the population, which both restricts the effective population size of local populations and provides a basis for reproductive isolation.

Other coevolutionary interactions

These very intense mutualisms are not the only coevolutionary interactions that have been implicated in speciation. Eichler (1948) has suggested that good concordance between the phylogenies of parasites and the phylogenies of their hosts occurs with sufficient frequency to warrant inscribing this relationship as a fundamental principle of evolution (Fahrenholz's rule). Subsequent phylogenetic analysis has given strong support to this rule on some occasions (Hafner & Nadler 1988), but only patchy support in some other lineages (Mitter & Brooks 1983). As a corollary, Szidat (1956) has proposed that the more primitive the hosts, the more primitive the parasite. Again there are clear exceptions (Hart 1988).

The possibility that speciation occurs in concert in parasites and hosts does not necessarily mean that coevolutionary interactions accelerate the process of divergence. Thompson (1989) defines two processes whereby acceleration could occur. Diversifying coevolution is a process where the coevolutionary interaction causes one or both species in the interaction to speciate. If speciation affects only one partner the process is sometimes called mixed-process coevolution (Thompson 1987). Many interspecific interactions have different forms in different environments. Interactions that are antagonistic in one location may be mutualistic in another. For example, mycorrhizal fungi can benefit plants in infertile soils by increasing the surface area available for capture of nutrients, but in fertile soils they can harm their hosts (Bowen 1980). Indeed, Thompson (1982) points out that many mutualisms may evolve from initially hostile interactions. Because alleles that increase the probability of interaction will be favoured in environments where the relationship is mutualistic, and selected against when the relationship is hostile, rapid divergence should be possible even in the presence of moderate gene flow (Thompson 1987). Rapid reciprocal diversification under these conditions is most likely where one species lives as a symbiont within a host and also influences the transfer of gametes between the hosts. For example, in some plant–pollinator interactions the pollinators have larvae which feed on the same plants that the adults pollinate. The mutualism probably arose from seed parasitism. Once again, figs and fig wasps are the classic example. However, similar interactions are found between yuccas and yucca moths and globeflowers (*Trollius* spp.) and the *Chiastocheta* spp. that pollinate them (Thompson 1989).

Escape-and-radiation coevolution is a process originally invoked by Ehrlich & Raven (1964) in a classic paper on the diversification of flowering plants and the butterflies that feed on them, but it can be generalised to any interaction between parasites and their hosts. Unlike most coevolutionary models, the close association between the species is broken for parts of their evolutionary history. Thompson (1989) points out that the model has five steps. First, hosts develop novel defensive mechanisms such as plant secondary compounds. Second, these defences reduce the acceptability of the host, and are strongly favoured by selection. Third, the protected hosts undergo evolutionary radiation into a new adaptive zone in which they are free of their former parasites or herbivores. Fourth, a novel mutation appears in the parasites that enables them to exploit the new host radiation. Last, these parasites or herbivores radiate, generating a new clade. The critical difference from diversifying coevolution lies in the time of speciation. In escape-and-radiation coevolution, the speciation in hosts occurs when the interaction is broken. By contrast, speciation through diversifying coevolution is caused directly by the interaction. Test of the hypotheses is hindered by our inadequate knowledge of the phylogeny of both hosts and parasites (e.g. compare Berenbaum 1983 with Miller 1987), and no unequivocal demonstration of this effect is available (Thompson 1989).

Reinforcement

What happens when populations which have diverged to a certain extent in allopatry come into contact and hybridise? Because hybrids between dissimilar populations often show reduced fitness, it has been suggested that any tendency to avoid mating with members of the other population will be selected, reinforcing the distinction between the two species (Dobzhansky 1937). This reinforcement serves only to preserve the purity of locally adapted genomes, and will accelerate the process of allopatric speciation. Reinforcement is thus a component of speciation. Selection for divergence in mating tendency could also occur when reproductive isolation, and hence speciation, had already taken place. Assortative mating would reduce wastage and may allow previously parapatric forms to live in sympatry. Butlin (1987) suggests this should be called reproductive character displacement, and should not be used as evidence for speciation by reinforcement.

In considering the hypothesis of reinforcement it is useful to distinguish two components of reproductive isolation (Table 8.1). Prezygotic isolation is the prevention of gene flow due to characters which have their effects before the formation of the zygote, such as assortative mating or pollen−stigma incompatability. Postzygotic isolation is the prevention of gene flow due to reduced fitness of the hybrids. Reinforcement can be a mechanism whereby prezygotic isolating mechanisms are selected to reduce the costs of postzygotic isolation.

Most plant and animal species are isolated from their close relatives by some sort of prezygotic barriers to gene flow, and this has contributed to the enthusiasm for the hypothesis of reinforcement. Nonetheless, the idea has been subjected to intense criticism, particularly by Paterson (1978, 1982). First, he suggests that the genetics of reinforcement are analogous to the single-allele case of heterozygote inferiority. Because such polymorphisms are unstable, it is unlikely that the two types could coexist for long enough to drive selection for reproductive isolation to completion. Second, the intensity of selection will diminish as reinforcement proceeds, lessening the prospect of final isolation. Third, it is difficult to see why the prezygotic mating mechanisms causing reproductive isolation should spread into the area of allopatry, as changes to the mating system should be deleterious outside the zone of recontact. There is a substantial theoretical and empirical basis for these claims (Templeton 1981; Spencer *et al.* 1986; Butlin 1987; Chandler & Gromko 1989; Sanderson 1989), suggesting that reinforcement should never be assumed in interpreting speciation mechanisms.

Table 8.1 A classification of isolating mechanisms. Modified from Mayr (1963).

Prezygotic isolating mechanisms
1 Mechanisms that prevent interspecific crosses (premating isolating mechanisms)
   (a) Potential mates or their gametes do not meet (geographical isolation)
   (b) Potential mates have overlapping ranges but reproduction takes place at different times (temporal isolation)
   (c) Potential mates meet but do not mate (ethological isolation)
   (d) Copulation may be attempted but there is no transfer of sperm (mechanical isolation)
2 Mechanisms that prevent fertilisation
   (a) Sperm transfer takes place but the egg is not fertilised (gametic mortality)
   (b) Pollen reaches the style but does not grow; gametes meet but there is no fertilisation (genetic incompatibility)

Postzygotic isolating mechanisms
3 Mechanisms that prevent the development of interspecific hybrids
   (a) Eggs are fertilised but the zygote dies (zygotic mortality)
   (b) Zygote produces an F_1 hybrid of reduced viability (hybrid inviability)
   (c) F_1 hybrid zygote is fully viable but partially or completely sterile, or produces deficient F_2 (hybrid sterility)

The most striking evidence that reinforcement can be important comes from *Drosophila*, where Coyne & Orr (1989) were able to assemble data on 119 pairs of closely related species with known genetic distances (and hence divergence times), mating discrimination, strength of hybrid sterility and inviability, and geographic ranges. Mating discrimination and sterility or inviability of hybrids increase gradually with time. Among allopatric species, mating discrimination and postzygotic isolation evolve at similar rates, but among sympatric species, strong mating discrimination appears well before severe sterility or inviability. Prezygotic isolating mechanisms are much stronger in sympatric pairs (Table 8.2), and species status appears to be attained twice as quickly in sympatric pairs as in allopatric pairs.

The formation of species
in small isolated populations

The next set of allopatric models take as their underlying assumption the view that species are well-integrated entities that are very hard to change radically. Epistatic interactions in the fitness effects of different loci lead to coadapted gene complexes that should prove resilient to change. The underlying model is Wright's notion of the adaptive landscape, where adaptive peaks are separated by valleys of low fitness which cannot ordinarily be crossed. Occasionally small populations are founded in isolation from their ancestral population. The founder event and recovery from it can affect the fate of the small population in a number of ways. The most likely result of a founder events will be no effect at all, or an immediate return to the adaptive peak from which the population has emerged. Because drift is exaggerated in small populations, and the selective environment is homogeneous because of the low spatial heterogeneity in small areas, there is the possibility of disruption of the coadapted gene complexes. Therefore, the second most likely outcome is extinction of the founder population. However,

Table 8.2 Degree of isolation between pairs of sympatric *Drosophila* spp. Pairs were only included if electrophoretically similar. Prezygotic isolation was measured from mate-choice experiments.

$$\text{Pre-zygotic isolation} = 1 - \frac{\text{frequency of heterospecific mating}}{\text{frequency of homospecific mating}}$$

Postzygotic isolation was calculated by dividing the number of sexes that were infertile or inviable in both reciprocal matings by four. The difference is not significant for allopatric pairs but is significant for sympatric pairs. Data from Coyne & Orr (1989).

	Allopatric pairs	Sympatric pairs
Pre-zygotic isolation	0.21	0.63
Post-zygotic isolation	0.35	0.34

great interest has focused on the populations which undergo disruption yet recover over time.

Genetic revolutions

Three models have been suggested to describe the genetic outcome under these circumstances. In the most influential of all models of speciation, Mayr (1954, 1963, 1982b) argued that the founder event would be characterised by a great increase in the level of homozygosity, followed by a persistent loss of additive genetic variance because the population remains small. Selective forces change as a consequence of the increased homozygosity, triggering a genetic revolution. If populations survive the revolution they will be free of epistatic constraints and able to move to a new adaptive peak, where a new coadapted gene complex is acquired. The critical element of the model is the loss of genetic variance, and the suggestion that selection is maximal at, and driven by, the time of greatest homozygosity (Carson & Templeton 1984). This argument was motivated largely by empirical evidence. Mayr had noted that while mainland populations were often homogeneous, island populations often show strong differentiation. However, there are a number of theoretical problems with Mayr's model. First, the postulated loss of additive variance is unlikely to occur unless the population remains small for long periods, in which case extinction is much more probable than a peak shift (Lewontin 1965b; Nei *et al.* 1975; Lande 1980; Barton & Charlesworth 1984). Second, reduction in additive variance reduces the probability of peak shifts, because both drift and selection require genetic variance (Rouhani & Barton 1987). Third, Mayr suggests that founders come primarily from peripheral demes, which Templeton (1981) points out would already be characterised by inbreeding and high homozygosity. It is therefore unlikely that the genetic revolutions envisaged by Mayr play an influential role in evolution (Barton & Charlesworth 1984; Carson & Templeton 1984).

Founder-flush

By contrast, Carson (1982) places greater emphasis on the break up of coadapted gene complexes caused by drift. The disorganisation is accentuated as the population establishes itself in its new environment. In absolute contrast to the genetic revolution hypothesis, the founding event and subsequent period of population increase are thought to be a time of reduced selection pressure, and little additive genetic variance is thought to be lost. Additive variance may even increase as a result of the production of normally deleterious variants by recombination. The reduction of selection pressure is thought to result from the ecological release accompanying colonisation of a new habitat unpopulated by conspecifics and competitors, and perhaps even parasites (Freeland 1983). Selection re-emerges when the carrying capacity of the new habitat is approached, and the resultant coadapted gene complexes may occupy a new adaptive peak. Once again, Carson's arguments are undoubtedly motivated by his extensive empirical experience with the dramatic radiation of many hundred species of *Drosophila* in Hawaii, where founder events seems to be extremely important. Indeed, the dramatic radiations characteristic of the Hawaiian fauna have taken place against a pattern of emergence and loss of entire islands through volcanic activity, as well as the fracturing of habitat and generation of new successional regimes caused by lava flows (Simon 1987). Many of the important cladogenic events in the Hawaiian *Drosophila* precede the emergence of any of the current islands (Beverley & Wilson 1985).

Genetic transilience

Templeton (1980) suggests a modification of this hypothesis that requires that the ancestral population is relatively outcrossed and polymorphic for coadapted gene complexes centred around major loci. Inbreeding and drift can alter the frequencies of the major loci, and may fix certain alleles. This can cause a drastic change in the fitness weighting of the pleiotropic effects associated with the major genes, and intense selection may result, provided that there are large amounts of genetic variance at numerous modifier loci. Once again a peak shift may result.

Recent empirical and theoretical work suggests an additional source of genetic variance upon which selection could act during founder-induced speciation. Founder events can easily be simulated in the laboratory by passing breeding laboratory colonies through bottlenecks. Bryant *et al.* (1986) report the heterodox result that additive variance for several morphological traits of houseflies actually increased after bottlenecks, particularly those of intermediate size (4 to 16 mating pairs). The increase took place despite evidence for inbreeding depression. Robertson (1952) suggested that increased variance could result from a chance increase in the frequency of rare recessive alleles at loci influencing a particular trait, and therefore increase the variance for that trait. Goodnight (1987, 1988) points out that epistatic variance can also be converted to additive variance by a founder event. Production of additive variance in founder events also overcomes the traditional expectation that traits closely related to fitness will typically show low variance, and hence have only a limited capacity to respond to selection (Chapter 2; Carson 1990).

Sexual selection and the founder models

Sexual selection may work in an unusual way during a founder event. Kaneshiro (1976, 1980) has suggested that the founding of new colonies will be accompanied by a loss of courtship elements as a consequence of drift. There should then be strong selection against intense discrimination between mates by females. The loss of discrimination is a consequence of the cost of choice when the range of partners is not great. Ancestral females will discriminate against males from the derived population, but derived females will not. This hypothesis rests on the largely untested assumption of significant genetic polymorphism for courtship elements (Chapter 7).

Evaluating the founder models

Criticisms of the founder-flush and genetic transilience models focus on the generality of their assumptions. The conditions required for founder-induced speciation include both ecological factors such as the frequency with which founder events occur and the selective milieu in relation to density changes, and assumptions concerning the genetic organisation of the species involved (Carson & Templeton 1984).

It is certainly true that the selective milieu differs in response to a changed demographic environment. For example, increased fecundity is strongly favoured in expanding populations while increased viability may be favoured in stable populations (Chapter 5). However, the relative intensity of selection and extent of fitness differences in different demographic environments has not been reliably assessed (Barton & Charlesworth 1984).

We clearly need to know more about the frequency of founder events in continental biotas (Endler 1977). Hawaii is an extraordinary natural laboratory for evolutionary ecologists (Simon 1987), yet we need to beware of incorporating its special features in attempts to derive general models. Similar problems arise with the analysis of hybrid zones, where much of the best data come from areas of post-glacial contact (Hewitt 1988). Is this because hybrid zones are often formed as a consequence of secondary contact of populations that have diverged in allopatry, or because North American and European biologists often work close to postglacial environments? Even among the Hawaiian *Drosophila* there appear to be some lineages whose diversification is better explained by subdivision of large populations than by founder events (De Salle & Templeton 1988).

Although it is generally conceded that epistatic interactions are common, their relative importance in different organisms needs to be addressed. Comparative data are not really adequate to make predictions about those groups in which transilient modes of speciation should be most prevalent. In modular organisms like most plants, epistatic interactions may be of less significance than in unitary organisms, because developmental integration is less pronounced (Gottlieb 1984; Buss 1987). For example, genetic control of morphological differentiation in plants falls into two classes. Differences in structure, shape, orientation, and presence versus absence are frequently discrete and determined by as few as one or two genes (Fig. 8.5). By contrast, differences in dimensions, weight and number usually reflect polygenic control. Morphological differentiation in critical reproductive functions may therefore arise far more easily in modular organisms (Gottlieb 1984).

One situation in which founder events are almost certainly crucial

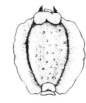

Coreopsis saxicola   *Coreopsis grandiflora*

Fig. 8.5 Discrete morphological differences in plants such as these alternative forms of the seeds in *Coreopsis* are often controlled by one or a few loci, in contrast to the complex polygenic control of morphological differentiation in animals with unitary development. After Gottlieb (1984).

is the fixation of unusual isolates that would otherwise be incapable of interbreeding. For example, many species of plants are derived via polyploidy, the multiplication of the haploid set (Grant 1981). Reproduction by polyploid variants is obviously assisted in populations capable of selfing (Chapter 7). Variants with unusual chromosomal arrangements are also likely to be fixed in small populations. This rapid fixation is also more likely in reproductively subdivided populations, where mating is non-random. The rapid rate of morphological change, chromosomal fixation and speciation in the highly social mammals has been attributed to an effect of this sort (Larson *et al.* 1984).

PARAPATRIC SPECIATION

Some authors have argued forcefully that allopatric models are adequate to account for virtually all speciation events (Mayr 1963). This enthusiasm stems both from vigorous advocacy, from biogeographic evidence, and from the belief that gene flow and epistatic interactions are insurmountable difficulties in contiguous populations. The biogeographic evidence is straightforward. Many island populations differ abruptly from their mainland antecedents, often to the extent that they can be described as distinct species. It is frequently possible to establish the precise time at which islands were formed, and thus set a minimum period for divergence to have arisen. On the mainland, populations broken up by clear obstacles to dispersal likewise can show obvious differentiation.

Unfortunately these patterns may have led to the unconscious exclusion of alternative modes of speciation. Endler (1977) points out that although most terrestrial species are found on continents, most discussions of genetic diversity are primarily concerned with islands. The extent of genetic divergence through clines can be very great. A variety of evidence suggests that gene flow may often be greatly limited by the restricted dispersal evident in many natural populations (Slatkin 1985b). In the best studied Rassenkreis, Wake & Yanev (1986) showed that genetic discontinuities between subpopulations of *Ensatina eschscholtzii* were much more abrupt than indicated by the comparatively gradual morphological intergradation, and that many of the incipient stages necessary for clinal divergence into new species were present within the Rassenkreis. In this study and in many other cases it is simply not possible to tell whether a zone of intergradation represents secondary contact between populations that have diverged in allopatry, or a local steepening of a cline (Endler 1977).

Evidence that alternative phenotypes can be maintained within a single population was presented in Chapter 5, and may influence our understanding of speciation in two different ways. Phenotypic diversity can be interpreted as an incipient stage in sympatric speciation via disruptive selection. Alternatively, West-Eberhard (1986, 1989) argues that if a population that contains the capacity for developmental conversion suddenly finds itself in conditions that favour only one of two or more alternatives, then the population may evolve rapidly because the genome is released from the constraints imposed by the necessity to accommodate multiple alternatives. The initial failure to express more than one phenotype may occur without genetic change, via the consistent environmental induction of only one of the facultative forms. A possible example of the fixation of alternative phenotypes is provided by social parasitism in Hymenoptera. Some species are capable of independent reproduction and parasiting the reproductive labours of other species. Some species have become obligate parasites of social species.

Divergence along clines is the only explanation for some conspicuous genetic differentiation. A famous example comes from the establishment of the grasses *Anthoxanthum odoratum* and *Agrostis tenuis* on the polluted slag heaps surrounding mines (Bradshaw & McNeilly 1981). Seedlings growing on the slag heaps must be tolerant to the high concentrations of heavy metals in the soils. Heavy metal tolerance must involve some cost, as tolerant phenotypes perform poorly when transplanted back into the surrounding grassland from which they must originally have colonised. Populations of *Agrostis tenuis* growing on the slag from copper mines show considerable tolerance to copper (Fig. 8.6a). In the field their progeny are much less tolerant than their parents (Fig. 8.6b). This is because local adaptation is retarded by gene flow from nearby non-tolerant populations, as progeny from adults reared in isolation inherit tolerance (Fig. 8.6c). At the boundary between copper-tolerant forms and intolerant pasture forms there is marked divergence in flowering time, possibly to reduce the costs of hybridisation (Fig. 8.6d). This suggests that reinforcement may be contributing to divergence in the direction of parapatric speciation (Caisse & Antonovics 1978; but see Butlin 1987).

SYMPATRIC SPECIATION

Some groups contain such an extraordinary profusion of species that it is difficult to believe that any of the preceding models are adequate to

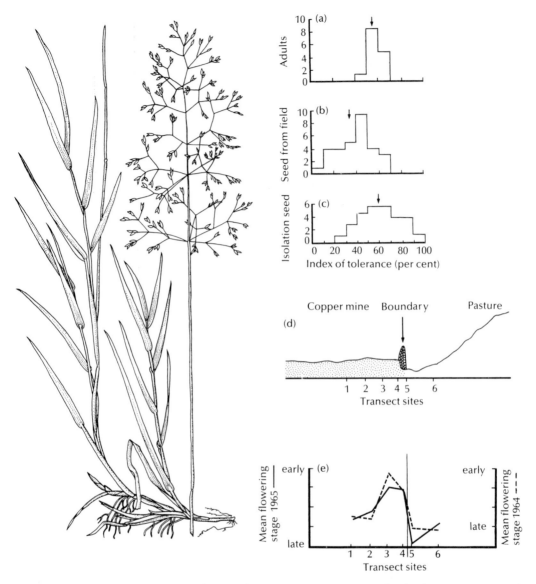

Fig. 8.6 Populations of *Agrostis tenuis* growing on the slag from copper mines show considerable tolerance to copper (a). In the field their progeny are much less tolerant than their parents (b). This is because local adaptation is retarded by gene flow from nearby non-tolerant populations, as progeny from adults reared in isolation inherit tolerance (c). At the boundary between copper-tolerant forms and intolerant pasture forms (d) there is marked divergence in flowering time, possibly to reduce the costs of hybridisation (e). After Bradshaw & McNeilly (1981).

account for all the necessary speciation events. The most notable cases are specialist parasitic and phytophagous insects (Chapter 1; Strong *et al.* 1984; Butlin 1987). Sympatric speciation, where daughter species arise within the range of their antecedents, would encounter fewer difficulties in explaining this pattern, as allopatric isolation or clinal variation are unnecessary to the speciation event.

The most enthusiastic advocate of the importance of sympatric speciation is Bush (1966, 1974, 1975), who was motivated by his research on the evolution of fruit flies. *Rhagoletis pomonella* is a tephritid fly endemic to eastern North America, where its native hosts are hawthorns (*Crataegus* spp.) (Fig. 8.7). During the nineteenth century its host range expanded to include apples, and still more recently it has been found to infest cherries, roses, dogwood and pears. All hosts belong to the family Rosaceae. Flies originating from apples accept apples for oviposition attempts much more than flies originating from hawthorn; though hawthorn fruit are accepted nearly equally by apple and hawthorn origin females, presumably because hawthorn was the original host (Prokopy *et al.* 1988). Flies in this experiment had no prior host experience, and larval experience does not influence fruit acceptance by adults, so these behavioural differences presumably have a genetic basis. This behavioural specialisation is not reflected in increased survival of maggots on the host from which the fly originated. Evolution of behavioural specialisation therefore appears to precede any difference in survival, recalling the comparable results discussed in Chapter 5.

There is electrophoretic evidence of significant differences in allele frequencies between sympatric populations of flies that utilise hawthorn or apple (Feder *et al.* 1988, 1990a, 1990b; McPheron *et al.* 1988). Reproductive isolation is exacerbated by differences in fruiting phenology and development times. Apples are suitable for oviposition earlier in the season than hawthorns. Even more important, apple flies take about 46 days to develop under standard conditions in the laboratory, while hawthorn flies take about 72 days to develop (Smith 1988). Because this seasonal asynchrony simultaneously restricts the availability of potential hosts and mates, adaptation of development time to the respective host species provides a parsimonious explanation for initial barriers to gene flow. It therefore appears likely that many of the steps necessary to accomplish speciation have resulted from a host shift which occurred in the last 200 years.

These results are exciting because they represent a case where we can observe a rapid diversification *in situ*. Generalisation of the sympatric model to the large number of specialist phytophages and parasitoids awaits appropriate genetic analysis, and is a major challenge for the next generation of evolutionary ecologists.

IS SPECIATION THE PIVOT OF MORPHOLOGICAL EVOLUTION?

The 'genetic revolution' model for speciation and its corollary of species homeostasis have been used by Eldredge and Gould as a component of their hypothesis of punctuated equilibrium (Eldredge & Gould 1972; Gould & Eldredge 1977), the notion that most morphological change is compressed into the events at the time of speciation and recovery from speciation. The evolutionary history of organisms

Fig. 8.7 Rhagoletis pomonella.

thereafter is characterised by morphological stasis, with little anagenetic change. Gould (1983) offers the operational definition that the period of stasis will usually represent 99 per cent of the evolutionary history of the organism. They contrast punctuated equilibrium with phyletic gradualism, the hypothesis that anagenesis is the major contributor to phenotypic change. Debate over the distinction between punctuated equilibrium and phyletic gradualism needs to focus carefully on what is and what is not controversial. Stasis is an easily established fact which can be observed directly in the fossil record (e.g. Eldredge 1985; Cheetham 1986; Stanley & Yang 1987; Fig. 8.8), and inferred from a combination of biogeographic and genetical data about living species, including 'living fossils' such as the horseshoe crabs (Fig. 8.9).

Evidence that morphological change is compressed to the time of cladogenesis, or can result from anagenesis should also be detectable in the fossil record. There is certainly evidence of abrupt changes in some lineages, and of gradual changes in others, and some lineages where both occur (e.g. Wei & Kennett 1988). However, assessing their relative importance is hazardous, and greatly confounded by the obvious limitations of the fossil record. Measurement of evolutionary rates depends directly on the sampling interval (Gingerich 1983, 1985). Where sampling is infrequent, and there are occasional reversals in the direction of evolution, morphological trends will be smoothed.

By contrast, the causal importance of speciation in the punctuations is not directly measurable, as we usually cannot detect biological species in the fossil record (but see Jackson & Cheetham 1990). Any attempts to do so are hindered by the spectre of circularity. If species are defined by morphological discontinuities, then speciation will appear to be correlated with the appearance of morphological novelty. We know that speciation can occur without any morphological change, and have reason to believe that morphological novelty can arise rapidly without speciation occurring (Palmer 1985; Newman *et al.* 1985).

The problems inherent in interpreting the fossil record are well illustrated by an example. Fossils typically record a sequence of events at one locality. Does the sudden appearance of one species in the record indicate its near instantaneous origin in a founder event elsewhere, or does it represent the slow infiltration of a widespread species poorly recorded at other sites? Bell *et al.* (1985) observed anagenesis of a variety of morphological characters in a fossil sequence of the Miocene stickleback *Gasterosteus doryssus*. Several characters such as pelvic structure and dorsal spine number change abruptly at one point in the sequence, yet quickly approach their previous condition (Fig. 8.10). Others characters show very little change at the same point in the sequence. How can we interpret these results? Bell *et al.* suggest that the original population went extinct, was replaced by a differentiated population, which presumably then converged rapidly on the locally selected optimum.

The final criticism which can be directed at the punctuated equilibrium model is its emphasis on the founder event model of speciation.

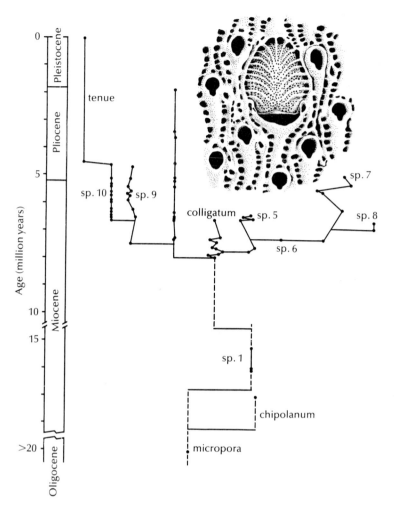

Fig. 8.8 Evidence for morphological stasis and apparent punctuation in a fossil sequence of bryozoans. The degree of horizontal divergence is a depiction of morphological separation using multivariate analysis to summarise several intercorrelated morphological traits. After Cheetham (1986).

Because Mayr's original model attracts little favour from geneticists, it is a shaky basis on which to build a general theory of evolutionary change.

STASIS AND SPECIES HOMOGENEITY

Regardless of whether speciation is the important pivot argued by Eldredge & Gould (1972), or whether stasis dominates 99 per cent of the phenotypic history of species, the apparent resistance of phenotypes to what often appears to be dramatic ecological change demands explanation and analysis. Indeed, Gould (1983), comments: 'Of the two claims of punctuated equilibrium — geologically rapid origins and

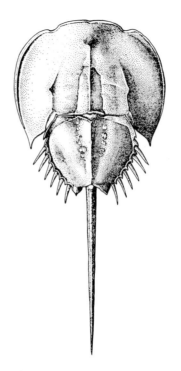

Fig. 8.9 The horseshoe crab *Tachypleus tridentatus*, a living fossil. After
Bonaventura *et al.* (1982).

subsequent stasis—the first has received most attention, but Eldredge
and I have repeatedly emphasised that we regard the second as more
important. We have, and not facetiously, taken as our motto: stasis is
data.'

Autopoiesis and canalisation

The simplest explanation of stasis suggests that organisms evolve
resistance to change, or undergo canalisation. Wake *et al.* (1983) use
the metaphor of the living system as a balloon, impinged upon by the
environment as countless blunt probes. This autopoietic system com-
pensates environmental and genetic changes, and persists by evolving
minimally. Provided the compensation mechanisms are well developed,
the system can remain stable in the face of massive environmental
challenges, and in the face of genetic perturbations such as speciation.
Wake *et al.* (1983) support their argument with the salamander genus
Plethodon, illustrated in the Chapter vignette, which shows negligible
morphological change over 60 million years despite frequent speciation,
and enormous changes to the communities in which its constituent
species live. These salamanders show great feeding flexibility, as a
consequence of a rather general innate prey stimulus pattern which
can be focused greatly in the first few months of life as a consequence
of early experience. As a result of coupling of behavioural plasticity

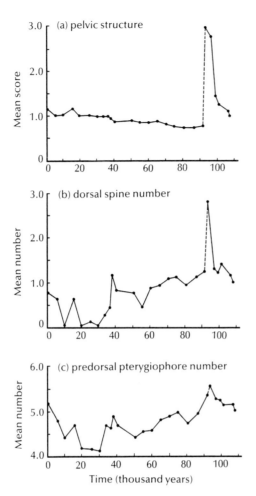

Fig. 8.10 Difficulties in interpretation of morphological change in the Miocene stickleback *Gasterosteus doryssus*. Several characters change abruptly, but then approach their previous condition. Others change slowly or not at all. After Bell *et al.* (1985).

with a generalised tongue and tooth morphology, they can respond to changing environmental conditions without changes to morphology. In contrast, the bolitoglossine salamanders have specialised tongues, and have rather limited prey recognition. The morphological specialisation enhances capture of small compact prey, but has constrained dynamic interaction with the environment, and has channelled further evolution towards specialisation (Wake & Larson 1987). In turn, the vast bolitoglossine radiation has been sorted ecologically into distinct niches, largely through modification of the locomotor apparatus (hands, feet and tails).

Stabilising selection

The most common Darwinian explanation for stasis suggests that the

uniformity of phenotypes through time arises from stabilising selection
(e.g. Stebbins & Ayala 1981; Charlesworth *et al.* 1982). Unusual or
deviant phenotypes are purged in each generation, unless environmental
changes create circumstances favouring the deviants, promoting
anagenetic change or fixation of the deviant forms.

Some paleontologists are intensely sceptical of this explanation,
pointing to the absence of anagenetic change in lineages which apparently undergo enormous changes in the selective environment in
which they live (Gould 1983). Instead, they emphasise the importance
of constraints to evolution driven by selection.

However, the difference between epigenetic constraints and the re-
strictions imposed by genetic interactions is not always clear (Chapter
2; Charlesworth 1984b; Cheverud 1984). The distinction between
developmental constraint and stabilising selection becomes blurred if
stabilising selection acts predominantly to purge developmental ab-
normalities (Cheverud 1984; Williamson 1987). It is clear that many
traits of most populations respond to artificial selection, yet the response
is often accompanied by developmental instability and reduced fitness.

Brown (1984) points out that the frequency of stabilising selection
may be most important in those populations likely to be sampled by
the fossil record. The common positive relation between distribution
and abundance of species (Chapter 5), means that widespread species
will occur over a range of habitats and be most abundant in the centre
of their ranges. Stabilising selection will predominate at the centre of
the range, and sheer numerical abundance means that the abundant
central populations are most likely to be sampled in the fossil
record. By contrast, rapid directional selection will be confined to
isolated peripheral populations with low abundance. In order to become
sufficiently abundant to achieve representation in the fossil record,
newly formed species would have to benefit from environmental change
or undergo an adaptive morphological transition in order to compete
with widespread abundant species.

Gene flow as a constraint
and creative force

Clinal variation is a problem for both the idea of a species resistant to
selection, and for interpretation of the fossil record. Variation along
clines is extremely common, supporting the idea that selection can
maintain some morphological differentiation in the face of gene flow,
and that populations are capable of responding to selection. What then
is recorded by a fossil sequence at a given locality? In the case of global
clines, morphological change may reflect the movement of locally
adapted variants in response to climatic change rather than the appearance of morphological novelty (Koch 1986; Wei & Kennett 1988).

Despite evidence that gene flow can be overcome by selection in
certain circumstances, gene flow will normally be a retarding force
(Slatkin 1985b, 1987). A correlation between speciation and morpho-

logical transition could therefore arise if speciation could facilitate morphological change not by disrupting homeostasis or accelerating the response to selection, but by disrupting gene flow along clines, freeing isolated populations to move in morphological space (Grant 1977; Futuyma 1987).

SUMMARY

It is now clear that several modes of speciation have played a critical role in evolution, but it is not yet possible to determine the relative importance of each mode. Species may diverge in allopatry, form in parapatry, or arise comparatively rapidly in sympatry. Sexual selection and coevolution may have special roles in facilitating rapid divergence. The species is a fundamental unit of biological organisation, and some lineages appears resistant to morphological change, despite continued genetic divergence. Some authors have argued that speciation represents a time of liberation from these constraints, and that the time and process of speciation is intimately associated with the origin of morphological novelty. We cannot test this hypothesis because speciation cannot reliably be determined in the fossil record.

FURTHER READING

Endler (1977) is a readable account and lucid definition of patterns of geographic variation and their relationship to the theory of speciation. Otte & Endler (1989) contains a series of interesting papers discussing species concepts, the modes of speciation, and the consequences of speciation. Barton & Charlesworth (1984) and Carson & Templeton (1984) represent interesting alternative statements of the importance of founder events in evolution. The hypothesis of punctuated equilibrium has prompted a heated correspondence over two decades. Levinton (1988) is a scathing attack, and Gould (1983, 1985a) and Gould & Eldredge (1986) represent defences and elaborations of the hypothesis.

TOPICS FOR DISCUSSION

1 Re-read the description of figs and fig wasps in Chapter 1, and attempt to assess the relative importance of the various modes of speciation for this species.

2 As part of a discussion of stasis in *Plethodon* salamanders, Wake *et al.* (1983) comment that 'fit adaptations are those that persist through time'. Discuss the problems with this point of view.

3 The hypothesis of punctuated equilibrium was originally based on a model of speciation which some population geneticists now believe is unlikely to be very important in evolution. Discuss how comfortably the alternative founder models of speciation fit with the idea of punctuated equilibrium.

Chapter 9
Why are There
So Many Species?

Speciation is obviously the ultimate key to the diversity of life, but an understanding of the mechanisms of speciation does not answer why some taxa are so diverse, while some 'boom and bust' and others persist over vast time scales without either radiating or suffering extinction. This chapter cannot hope to answer those questions completely, but will hopefully draw attention to what May (1986) has described as the greatest intellectual challenge facing biologists in the next few decades, attempting to answer why there are so many species. The urgency of this task is heightened not just by our curiosity about the unresolved problems posed by nature, but by growing evidence that after a long history of increasing diversity of life, human insult to the environment has started to reduce species diversity by causing wholesale extinctions.

The first part of this Chapter tackles the question of just how many species exist on earth, and how that number has changed through time. This focuses attention on whether species diversity has attained equilibrium, or whether our current perspective represents a snapshot of a continually diversifying biota. Put another way, has nature filled the vacuum it purportedly abhors? I then examine the three critical questions about the development and maintenance of diversity. First, why do some communities and regions support more species than others? For example, why do tropical communities often contain more species than their temperate analogues? Second, what are the factors which maintain species diversity within particular communities and regions? For example, are the processes that allow coexistence of several hundred species of wasp in a tropical forest similar to those which allow coexistence of a smaller community in temperate habitats. Last, why are some taxa much more speciose than others? Why do so many phyla contain only a few species, while the Arthropoda contain many millions (Fig. 1.6)?

HOW MANY SPECIES?

There are currently fewer that two million species of organisms which have been formally described by taxonomists. The tally changes daily as taxonomists revise each other's work, and describe species completely new to science. The rate of description of new species varies with taxonomic group, and with geographic area. At one end of the taxonomic scale, four completely new species of primate (two lemurs and two monkeys) have been discovered since the late 1970s. At the other end of the spectrum, a taxonomist sent a small sample of mites from a tropical area which had not previously been sampled would be surprised if there were not several new species in the sample. The intriguing and unanswered question remains, how many species are there still to be described?

The most dramatic upward revision of the tally results from the application of a new technique, the use of knockdown insecticides to sample canopy-dwelling arthropods (Chapter 1). Four principal results emerge. First, it is clear that the canopy of tropical rainforest contains a vast fauna which has not previously been adequately sampled. Erwin (1982) estimates the canopy to be twice as rich in species as ground-dwelling arthropods. The first 3099 individuals sampled by Erwin (1988) in a rainforest in Peru contained 1093 species! In Bornean rainforests, 1455 chalcid wasps included 739 species, of which 437 were represented by only a single individual, and only eight species had more than ten individuals in the sample (Stork 1988). In neither of these studies was there any evidence that the number of species was reaching an asymptote as sampling intensity increased. Second, there are enormous differences between the faunas sampled at forest plots separated by only short distances. At Manaus in the Amazon basin, Erwin (1983) showed that of 1080 species analysed, there was only a 1 per cent overlap of species in the four forest types sampled. Third, many beetles appear to specialise on individual trees, and parasitic wasps specialise on only a few hosts. Based on these very preliminary data, Erwin (1982) calculated a diversity of about 30 million species of tropical arthropods alone, suggesting we have previously underestimated the number of species of arthropod by an order of magnitude. Erwin (1988) is the first to concede the uncertainty of his estimates, but recent studies which refine his analyses tend to support or even elevate the figures he has proposed (Stork 1988).

Indeed, Stork (1988) sounds the following caution: 'If it has taken 230 years since Linnaeus to describe one million species of insects what chance do we have of describing the other 9 (or 79!) million ...? Given that the British Museum (Natural History) fills a six-floor building with representatives of about 0.5 million species of insects, where would we house representatives of the rest, plus the associated literature?' Or as May (1988) comments: 'Amazingly, there is as yet no centralized computer index of these recorded species. It says a lot about intellectual fashions, and about our values, that we have a computerized catalog entry, along with many details, for each of the seven million books in the Library of Congress but no such catalog for the living species we share our world with. Such a catalog, with appropriately coded information about the habitat, geographical distribution, and characteristic abundance of the species in question (no matter how rough or impressionistic), would cost orders of magnitude less money than sequencing the human genome; I do not believe such a project is orders of magnitude less important.'

Knowing approximately how many species exist on earth is only the first step. A far more difficult question is why there are so many. It has been estimated that humans now consume or appropriate more than a quarter of all the organic material assimilated by plants, probably a greater proportion than has ever been achieved by a single species (Vitousek *et al.* 1986). So why not four gluttonous animals like *Homo sapiens*, instead of one *H. sapiens* and millions of others.

WHY ARE SOME COMMUNITIES
MORE SPECIES RICH THAN OTHERS?

Regardless of the uncertainty over full inventories of species for virtually all communities, it is clear that there are enormous differences in species diversity between areas, and in some cases these differences are repeatable and indicative of clear patterns. Indeed, despite the presumed complexity of community organisation, the patterns are usually much more obvious than explanation of those patterns. In the following discussion, I outline some of the obvious correlates of diversity and introduce the various hypotheses which attempt to explain that diversity.

Types of diversity

Following Whittaker (1972), it has proved useful to separate measures of diversity into three categories:
α-diversity: or the number of species within a community or patch of habitat.
β-diversity: or the way that species change from one habitat to another.
γ-diversity: the total number of species in a region, or a combination of α- and β-diversities.

Species diversity increases with area

Habitats that are common support more species than habitats that are rare. For example, the number of species on an island within an archipelago is very often closely related to the size of the island, usually as a power function that can be linearised with a double logarithmic plot (Fig. 9.1). Island area can be interpreted liberally. Geographically widespread species of plants have a more diverse fauna of insects feeding from them than similar species with limited distribution, and large patches of remnant native vegetation in the middle of the western Australian wheatbelt are likely to support a more diverse fauna of vertebrates than small patches.

There are three hypotheses commonly applied to this pattern (Strong *et al.* 1984). First, large areas include more microhabitats, and therefore even if α-diversity is the same in different microhabitats, the effects of habitat specialisation will lead to β-diversity, in turn increasing γ-diversity. The number of species of birds found within forests is very strongly correlated with the structural diversity of those forests (Recher 1969). The increase in microhabitat diversity may occur in subtle ways. For example, structurally complex trees may support more insects than shrubs. Architecturally complex members of a single genus may support many more phytophagous insects than congeneric species which are comparatively simple (Fig. 9.2). However, there is good evidence that habitat specialisation is more pronounced in species-rich communities (Brown & Maurer 1989; Stevens 1989; Pagel *et al.* 1990), so

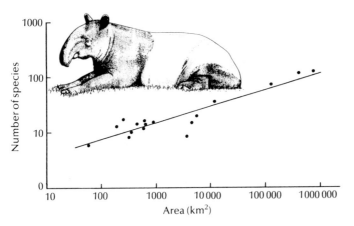

Fig. 9.1 The relation between island size and diversity of mammals (excluding bats) for the Sunda Islands between Malaysia and Wallace's line. Note the logarithmic axes. Modified from Wilcox (1980).

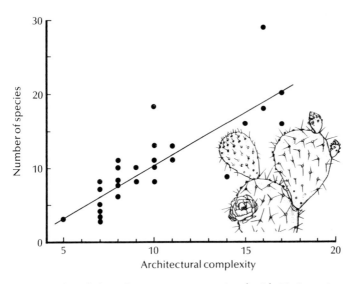

Fig. 9.2 The number of phytophagous insects associated with 28 *Opuntia* spp. in North and South America in relation to the 'architectural complexity' of each species of cactus. Complexity increases with plant size, branching complexity, development of woody stems and the surface morphology of the cladodes. After Moran (1980) and Strong *et al.* (1984).

the increase in diversity cannot merely be a consequence of adding microhabitats.

Second, large areas of habitat are encountered more often than smaller areas. This is a particularly compelling argument for increased diversity of phytophagous insects on widespread host plants. The ability of a plant to colonise a host often depends on the ability to counter the chemical and structural defences of that plant (Chapters 5 and 8). One important source explaining deviation from the species-area curve is the degree of taxonomic isolation of the host (Fig. 9.3). Plants with few close relatives have fewer species than would be predicted from the species–area relation, suggesting that species richness depends in part on sources of potential colonists. Widespread plants recruit more of these partly preadapted hosts than is likely for rare species that are encountered infrequently. The importance of the difficulty of colonisation and adaptation to hosts in determining species richness is illustrated by Kennedy & Southwood (1984), who showed that although the number of species of phytophagous insect using a particular tree in Britain is determined largely by the area that is occupied by that tree, there is a strong secondary effect from the time that the plant has been grown in Britain. Coope (1987) has examined the change in British beetle faunas throughout the late Quaternary, a period of history characterised by abrupt climatic changes associated with the cycle of glaciation. Wholesale and often very sudden extinction reflect rises and falls in the regional temperature. By contrast, the arrival of colonising species is much more gradual, as it depends not just on suitability of the environment but the routes and rates of spread of species from their different refuges. Community composition in these cases is determined not by local environmental conditions but by regional history, and much of the Quaternary may have been spent in non-equilibrium conditions.

Third, the Equilibrium Theory of Biogeography suggests that species richness is determined by a balance between colonisation and extinction (MacArthur & Wilson 1967). Although most representations of this hypothesis have stressed the distance between the source and new patch as the principal determinant of colonisation probability (e.g. Fig. 9.4), colonisation rates will also be determined by the emigration behaviour and size of the source population, and also the size of the

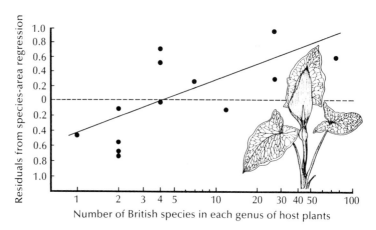

Fig. 9.3 The effect of taxonomic isolation of the host plant on the species richness of phytophagous insects associated with monocotyledons. The residuals from a relation between number of phytophagous insects using a plant and the range of that plant are plotted against the intrageneric diversity of the plant. Species-poor genera have fewer associated insects than species-rich genera. After Strong *et al.* (1984).

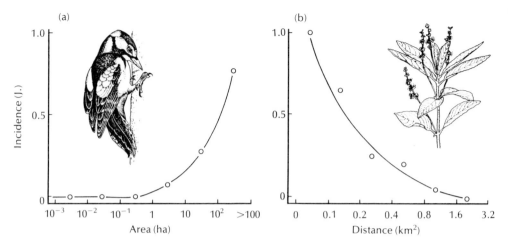

Fig. 9.4 Incidence (the proportion of habitat fragments in a given class containing the species) of: (a) the great-spotted woodpecker, *Picoides major* in relation to area of woodland; (b) dog's mercury, *Mercurialis perennis*, in relation to the distance to the nearest forest containing the species. After Wilcove *et al.* (1986).

patch to be colonised, for reasons discussed in the previous paragraph. Extinction rates are assumed to increase as population size dwindles, and population size to increase with island area. It is certainly true that persistence of some species depends on patches of a critical size (Fig. 9.4).

According to this view, species richness should reach equilibrium through time. One of the best neontological data sets comes from the Krakatau Islands, from which life was obliterated by the famous volcanic eruption of 1883. In their classic elaboration of the equilibrium theory of island biogeography, MacArthur & Wilson (1967) used data to 1933 to conclude that equilibrium species diversity of landbirds had been attained about 30 years after the eruption, even though vegetation diversity continued to increase. More surprisingly, they estimated that extinction was the fate of 2 to 5 per cent of the bird species per annum. More recent surveys in 1951 and 1984–1986 suggest that species diversity of all groups (except rats, usually thought to be superb colonisers) has continued to increase, and that extinction rates were overestimated by an order of magnitude (Thornton *et al.* 1988). Because primary forest trees have not yet become established on the Krakataus, there is no reason to believe that further colonisation should not be possible, particularly as early seral stages are consistently replenished by periodic, though less dramatic irruptions on Anak Krakatau. The centenary of the original irruptions has clearly been insufficient to achieve equilibrium on Krakatau.

In practice, distinguishing between these three hypotheses can be extremely difficult, and there is no reason to believe that all three effects do not contribute to this almost ubiquitous relationship.

A strong negative relation between latitude and diversity has been documented on many occasions from both plants and animals, and from marine and terrestrial species. Most reported exceptions to the generality of this phenomenon focus on taxonomic groups which are replaced by more speciose counterparts at equatorial latitudes (Stevens 1989).

An area effect?

Two hypotheses are directly related to those invoked to explain the species−area relation. Terborgh (1973) has suggested that the average area available for colonisation in the tropics is much greater, allowing a much higher equilibrium diversity. Not only is there more area in the tropics for any number of degrees of latitude, but the areas at a given distance north and south are the equator are contiguous at the tropics, yet separated by ever greater distances as latitude increases. The increase in area is correlated with an increase in habitat complexity. It is not difficult to believe that a tropical rainforest contains more microhabitats than the tundra. Coupled with the increase in micro-habitat availability, the degree of habitat specialisation permissible in tropical species appears to be greater than that for temperate species (Stevens 1989), so more species can be accommodated within the habitat gradients that do occur.

Time for colonisation?

The second argument we have already encountered suggests that the tropics are older and more stable, and hence have been exposed to the processes of speciation (and particularly specialisation) and colonisation by recruitment for much longer than temperate areas, particularly those recovering from glaciation. There are two potential limitations to the diversity of biotas, one set by the rate of colonisation and one set by the rate of speciation. In a review of the effect of human introductions of plants and animals, Simberloff (1981) found that in 678 out of 854 cases there was no discernible effect on the resident community, a clear indication that the species occupy a vacant habitat which could have been exploited by an earlier successful coloniser.

The hypothesis that species are continuing to proliferate can be illustrated with some examples from parasites. Teleost fish carry a much more diverse burden of monogenean gill parasites in the Pacific Ocean than in the Atlantic Ocean, though the response to environmental features of the two oceans such as water temperature is approximately the same, suggesting that niches remain unfilled in the Atlantic (Fig. 9.5). Nematodes are one of the most prominent groups of endoparasites of terrestrial and freshwater plants and animals, yet only parasitise the marine fauna patchily, despite the diversity of free-living marine nema-

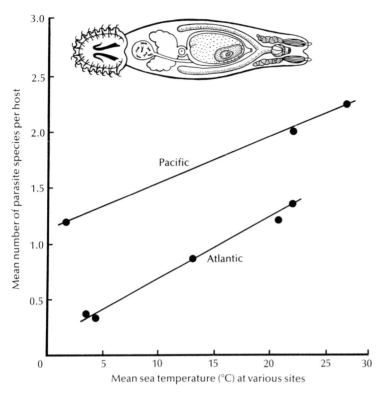

Fig. 9.5 Number of monogenean gill parasites of fish in relation to sea temperature for the Pacific and Atlantic oceans. Although diversity appears to be related to the same proximate cause (increased temperature) as in the Atlantic, Pacific fish bear an increased parasite burden. After Rohde (1982).

todes (Nicholas 1983). Curiously, most of the marine parasitic species have closer taxonomic affinities with terrestrial taxa. This suggests that marine nematodes evolved the terrestrial habit, terrestrial nematodes the parasitic habit, and now parasitic nematodes are secondarily evolving the marine habit, slowly exploiting vast untapped opportunity.

Despite such evidence of empty niches, there is also evidence that the relation between latitude and diversity is a very ancient one. The fossil record shows that there has been a clear latitudinal gradient in species diversity of fossilised planktonic foraminifera for eighty million years (Fig. 9.6). Although angiosperms first became prominent at low latitudes and have since greatly expanded their latitudinal range, there has been a clear latitudinal gradient in species diversity for over 100 million years, again negating a hypothesis of recent occupation of northerly latitudes (Crane & Lidgard 1989). This suggests either that niche filling is extraordinarily slow, or that the trend is indicative of patterns which are to all extents and purposes at equilibrium, and not solely a result of slow exploitation of habitats which have been perturbed at some time in the past.

Further, there is some evidence that while major environmental

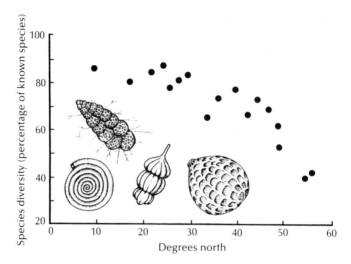

Fig. 9.6 Diversity of late Cretaceous planktonic foraminifera in the northern hemisphere in relation to latitude. Diversity is expressed as the percentage of the total known species which occur at any latitude. After Stehli *et al.* (1969).

catastrophes such as glaciation could lead to local impoverishment of biotas, moderate catastrophes can lead to enrichment through speciation in the genetically isolated refugia (Robinson & Gibbs Russell 1982; Grubb 1987; but see Connor 1986 for a note of caution about this hypothesis).

The species—energy hypothesis

Several authors have pointed out that the gradients described above could be a consequence of the increased energy available in large areas relative to small areas, and in the tropics relative to temperate environments (Brown 1981; Wright 1983). High amounts of energy sustain larger populations and increase resilience to extinction. Because of strong correlations between all the relevant variables, and the obvious difficulties in experimental analysis of latitudinal gradients, it is not easy to assign unambiguously a role to weather, except in comparative small scale tests (e.g. Brown 1975). Turner *et al.* (1988) provided an ingenious test by dividing the small insectivorous birds of Great Britain into three groups (year-round residents, summer visitors and winter visitors). Species diversity in 75 quadrats was then regressed against a variety of climatic variables. The diversity of summer visitors was correlated with summer temperature but not winter temperature, the diversity of winter visitors with winter temperature only, and the diversity of year-round residents with both summer and winter temperatures. The climatic variables explained more of the variance than latitude alone. Similar arguments can be developed for the British Lepidoptera (Turner *et al.* 1987) and North American plants (Currie & Paquin 1987).

However, it is unlikely that the energy hypothesis can easily explain all geographic patterns. For example, there are pronounced longitudinal gradients in species diversity of North American mammals, with diversity highest in areas with complex topography (Brown & Maurer 1989; Pagel *et al.* 1991). It is not easy to see how this pattern could be simply related to the availability of energy. In addition, at least one study of plants in a large geographical regions suggests that species diversity is not obviously related to energy availability (Richerson & Lum 1980).

Of course, there are no particular reasons to believe that several factors do not contribute to latitudinal and other geographic gradients in species diversity, and despite the consistency of the pattern, there is no reason to assume that this is indicative of a single cause.

Exceptions to community convergence

In spite of these comparatively clear patterns we are still a long way from predicting the diversity of any particular community. Ricklefs (1987) contrasts the mangrove vegetation of the New World tropics and West Africa where there are usually no more than three or four species within a region, each occupying a distinctive zone, with the mangrove forests of Malaysia where there are 17 principal and 23 subsidiary species organised into five loosely organised zones.

PROCESSES WITHIN COMMUNITIES AND REGIONS?

Given that there are obvious differences between communities, we can ask what are the processes which influence diversity within communities? Most attention has focused on biological interactions and their capacity to influence the niche of coexisting species.

Competition

The role of competition in regulation of species diversity has already been introduced obliquely in Chapter 5, and in the discussion of geographic gradients. In the presence of interspecific competition there is an upper limit to the number of individuals from the competing species which can coexist on a given resource. Where the resource is diverse there are more ways in which it can be partitioned, and hence more species can be accommodated (Fig. 5.4). Communities should move to an equilibrium species diversity whether they are coevolution-structured or invasion-structured. Of course this claim has an element of circularity. If we are to refine it to a useful predictive statement we need to know how similar species can be in order for them to persist without the exclusion of a particular species from the community by its competitors (Hutchinson 1959).

Assessing the presence or magnitude of these limits to similarity is by no mean easy. For example, how do we determine whether two coexisting species could be more similar? One escape from circularity is to look for the regular partitioning of several species along a gradient of resource use. For example, body size might reflect prey use and therefore we would expect an even spread of prey size. Early enthusiasm for the widespread occurrence of this pattern was inspired by a highly influential paper by Hutchinson (1959). However, it is now recognised that detection of the regular partitioning of species will usually depend on the development of a set of questionable assumptions about what is an appropriate null model against which the supposed pattern can be compared (Harvey *et al.* 1983; Colwell & Winkler 1984; Harvey & Ralls 1985).

Nonetheless, the predicted pattern does emerge unambiguously from sufficiently disparate taxa as raptors (Schoener 1984), finches (Schluter 1988b, 1988c) and North American desert rodents (Bowers & Brown 1982; J.H. Brown 1987) to suggest that it has widespread importance. In a very compelling example, Dayan and his co-workers (Dayan *et al.* 1989, 1990) have measured canine diameter for mustelids and felids in North America and Israel. First, there is both striking sexual dimorphism and great variation in the size of individual species at different sites. Yet when males and females are treated as separate 'morphospecies', the ratio of sizes of adjacent morphospecies tends to be very evenly spaced (Fig. 9.7). The same very even spacing is absent or less clear when other body traits are used (see also Ralls & Harvey 1985). Canines are used to seize and kill live prey, and so are likely to be closely associated with the ability to kill certain prey types and hence with partitioning of food resources. Other measures of body size are presumably influenced by confounding factors.

Coupled with evidence that the degree of habitat specialisation is greatest in productive, species-rich communities (Chapter 5), it seems certain that resource partitioning within groups of species with similar requirements contributes significantly to determination of species diversity within communities.

Predation and disturbance

The resource partitioning model assumes that maximum species diversity will be obtained when full competitive resource partitioning has occurred. An alternative argument suggests that species will be most diverse when competition is prevented, as monopolisation of resources by one species is prevented, and any species which succeeds in establishing itself will be able to flourish. In a famous experiment, Paine (1966) removed the starfish *Pisaster* from a rocky shore. *Pisaster* was a predator of several herbivorous, planktivorous and carnivorous invertebrates, including chitons, limpets, bivalves, snails, and barnacles (Fig. 1.5). The ultimate consequence of the removal was dominance by one or two species in the removal plots. The 'top predator' depressed densities of all species and prevented competitive displacement, allowing coexistence. These results are of considerable interest as they

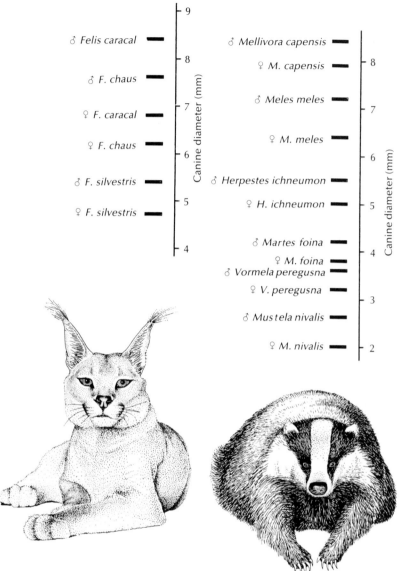

Fig. 9.7 The mean diameter of the upper canines of each sex of three species of *Felis* (left) and five species of mustelids and one species of viverrid (right) in the Middle East. The ratios of adjacent canines are more even than would be expected by chance alone. After Pimm & Gittleman (1990).

suggest that the dominant contributor to the regulation of diversity will depend on trophic position, so groups such as phytophagous insects in which competition is uncommon will be affected quite differently to carnivores.

Disturbance may act in a similar way to predation (Petraitis *et al.* 1989). Reefs protected from cyclones have fewer coral species than reefs recovering from a cyclone. Species diversity rises after the disturbance as more and more species settle in a plot, but then decline as one dominant species overgrows the other species (Fig. 9.8; Connell 1978). Indeed, it has been suggested that one of the causes of the high diversity of trees in tropical rainforests is the diversity of disturbance conditions created by treefalls (Connell 1978; Hubbell & Foster 1986). The opening of the canopy and subsequent availability of light on the forest floor caused by these falls represent the only opportunity for establishment of new seedlings. In certain circumstances a single tree species can come to dominate a rainforest, and not all of these can be explained in terms of synchronous recruitment after a catastrophe or unusual soil conditions (Hart 1990).

The relative importance of
predation and competition

Comparative reviews suggest that competition is common in natural communities (Connell 1983; Schoener 1983a, 1985). However, there is considerable evidence that it is much more important in some communities and types of organisms than others. The first important theoretical contribution to this argument was the suggestion that the relative importance of predation and competition alternate within a food web (Hairston *et al.* 1960; Slobodkin *et al.* 1967). Thus animals in the highest position in a food web (large carnivores) will compete, while herbivores will be largely contained by these carnivores, so they will not compete, allowing the producers to reach high densities and hence compete.

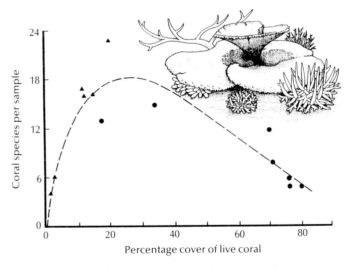

Fig. 9.8 Species diversity of corals in the subtidal outer reef slopes at Heron Island in relation to the amount of live coral in a plot. ▲, data from northern slopes heavily damaged by a cyclone three to four months previously; ●, southern sites protected from the cyclone. After Connell (1978).

Despite neglect of this hypothesis, considerable evidence has accumulated that competition is less prominent among some herbivorous groups than the animals that eat them. For example, competition is rarely detected in studies of herbivorous insects (Lawton & Strong 1981; Strong et al. 1984; but see Southwood 1987). Several important modifications and caveats to the original model have been suggested. Schoener (1989) argued that seven major trophic roles dominate the composition of terrestrial food webs in a set of seventeen webs compiled by Briand & Cohen (1987). These are:

1   Producers, the basis of all food webs.
2   Small phytophagous herbivores, the highly diverse arthropods.
3   Large phytophagous herbivores, usually mammals ranging in size from a vole to a giraffe.
4   Small carnivores, largely arthropods such as spiders, predatory insects and parasitoids.
5   Medium carnivores, usually vertebrates, that feed primarily upon small carnivores and herbivores. These include passerine birds, lizards and shrews.
6   Large carnivores, which feed mainly on medium carnivores or large herbivores. Examples would include eagles, canids and felids, and some snakes.
7   Medium omnivores, which eat both plant and animal matter. There are many examples, including many birds and mammals.

Schoener (1989) is the first to concede the deficiencies in this analysis. First, a bias towards data sets from temperate mesic communities probably underestimates the importance of frugivores and termites in the tropics, and granivores in the deserts. Second, parasites and decomposers are usually not included in food web compilations. While the latter omission can be justified because decomposers do not obviously influence the populations of the species with which they interact (though they may compete with consumers, (Janzen 1977), the absence of parasites almost certainly reflects grossly inadequate attention from field ecologists. We have already seen that parasites can influence the outcome of interactions between competitors, and can limit the restricted niche of organisms (Chapter 5). There are several reviews which lend support to the widespread importance of these interactions, and in particular suggest that the presence or absence of parasites can profoundly influence the prospects of successful colonisation by a host species (Freeland 1983; Dobson & Hudson 1986; Price et al. 1986, 1988).

Nonetheless, it is easy to see that if a particular element (say large carnivores) is absent from a web, predictions of the relative importance of competition and predation can easily be reversed for all lower links in the web. Schoener (1989) provides a careful review of the expectations for more complex webs, and Spiller & Schoener (1990) show experimentally that the effect of removing a top predator depends on the degree of omnivory and the number of links in the food chain.

It has been argued that regular switches in the relative importance of competition and predation along the length of a food web are less likely to be relevant in marine ecosystems because omnivory is very significant, to the extent that all higher elements in a food web may feed on the lower levels. Thus the importance of predation relative to competition will increase the lower a population is in the food web (Menge & Sutherland 1976, 1987; Menge, *et al.* 1986).

In mobile organisms, competition for resources recurs throughout the lifetime of an organism, and any individual may interact and compete with many other individuals. But what of an organism which settles in one place and then never moves, like a plant or some sessile marine invertebrates? First, the species of plant may be less relevant than its size, for a large established plant can prevent any seedling from gaining access to light and soil nutrients (Goldberg & Werner 1983). Second, the number of neighbours with which an individual interacts may be limited. Success will depend on being the first to establish in an unoccupied site, such as a gap in the canopy or a site on a rocky shore where a storm has moved a boulder, exposing uninhabited rock. The outcome of contests for establishment can have deterministic components. For example, dispersal of offspring at the correct time of year can enhance their prospects of success (or reduce their risk of failure). However, the outcome will also be influenced by many stochastic components, so that sessile organisms compete in a lottery to have a reproductive propagule in the right place at the right time. In this context 'sessile' can be interpreted liberally. Indeed, the hypothesis of lotteries for space was originally invoked to explain the coexistence of fish living on coral reefs (Sale 1977, 1982).

The ability of such lotteries to allow coexistence among competitively 'equivalent' species has been examined theoretically on a number of occasions, and all analyses support the premise that species coexistence could occur without resource partitioning (Huston 1979; Warner & Chesson 1985; Chesson & Case 1986; Hubbell & Foster 1986; Fagerström 1988; Pacala 1988; Chesson & Huntly 1989).

WHY ARE THERE SO MANY SPECIES OF BEETLES?

Latitudinal gradients in species diversity occur in most taxa. However, the elevation of the curves differ radically between taxa. On a graph of the diversity of beetles (an order of insects) against latitude, the equivalent data points for Carnivora (an order of mammals) would scarcely be visible. Therefore, it is important to focus on the properties that cause these enormous differences between taxonomic groups. The number of species in any taxon is a result of the balance of three factors. The rate of origin of new species through speciation, which was considered in Chapter 8, the rate of loss of species through local or complete extinction, and the time that those processes have occurred. For example, Coleoptera were the first of the phytophagous insect orders to emerge in the fossil record. Could their numerical predominance reflect a greater period of diversification?

EXTINCTION

Extinction appears to be the fate of most organisms (Simpson 1952). The palaeontological and neontological records suggest that extinction may usefully be divided into three classes: 1, Mass extinctions, where the usual extinction rate is briefly elevated over a wide geographic and taxonomic range; 2, Background extinctions, or the disappearance of taxa through time; and 3, Anthropogenic extinctions, or the large class of direct and indirect effects of human activities which eliminate species. Consideration of this last source of extinctions will deferred until Chapter 10, though two points require mention. First, while 'modern' extinctions can often be directly attributed to human action, past extinctions events such as the extinction of the megafauna at the end of the Pleistocene are more difficult to attribute unambiguously (e.g. Martin & Klein 1984; Markgraf 1985; Barnosky 1986). Second, there is growing evidence that modern anthropogenic extinctions could approach the magnitude of the worst mass extinctions, creating one of the major political and ethical problems for our generation (Wilson 1988).

Mass extinctions

Mass extinctions have recently attracted much attention and controversy. This renewed interest stems from several recent developments. First, it has been argued these extinctions randomise evolution by eliminating whole clades and interrupting any of the conventional notions of progress (Benton 1987). If this is true then much of the pattern observed in nature may be governed by these extinction events, and not by the process of adaptation. Gould (1989b) provides an entertaining account of the attempts to force the bizarre fauna of the Burgess Shale (Fig. 3.8) into the straightjackets of contemporary taxonomy, and speculates how different life on earth might have been if a different set of phyla had emerged from this extraordinary Cambrian explosion of diversity. As Conway Morris (1989) concludes:

> What if the Cambrian explosion was to be rerun? At a distance the metazoan world would probably seem little different; even the most bizarre of Burgess Shale animals pursue recognizable modes of life, and therefore the occupants of the ecological theatre should play the same roles. But on close inspection the players themselves might be unfamiliar. Today, barring a mass extinction, the predictability of replacement is high. But at the outset of diversification the range of morphologies, combined with the majority of species failing to leave descendants, means that the processes of contingent diversification might produce a biota worthy of the finest science fiction.

Second, it has become clear that the important extinction at the Cretaceous/Tertiary (K/T) boundary is associated with high rates of deposition of otherwise rare elements, particularly iridium (Fig. 9.9). This indicates a direct causal link, as iridium is usually only present in large quantities in meteorites, or in unusual volcanic eruptions.

Third, it has been argued that extinctions of marine metazoa are periodic (Raup & Sepkoski 1984, 1986; Raup & Boyajian 1988), and are therefore likely to have a common cause. We can examine these arguments by looking at two of the mass extinctions in some detail.

The Cretaceous/Tertiary boundary

The mass extinction which took place 65 million years ago was sufficiently dramatic to have achieved public fame as the time of demise of the dinosaurs. The effect on some parts of the marine ecosystem were even more sensational, notably causing the loss over a very short period of calcareous and siliceous nanoplankton and a presumed cascading effects on the food chain. Land plants were much less affected (though see Johnson *et al.* 1989), as were eutherian mammals, snakes, and deep sea benthic organisms. Recent excitement has been generated by the discovery of a very high concentration of iridium precisely at the boundary. Iridium is extremely rare in crustal deposits. The first attempt to explain the anomaly hypothesised a huge asteroid impact which produced a world-wide dust cloud and freezing conditions, inhibiting photosynthesis and causing a collapse of the marine food chain (Alvarez *et al.* 1980, 1984). This hypothesis generated great excitement, and interesting parallels between this ancient event and the likely consequences of a nuclear conflagration. The Caribbean has

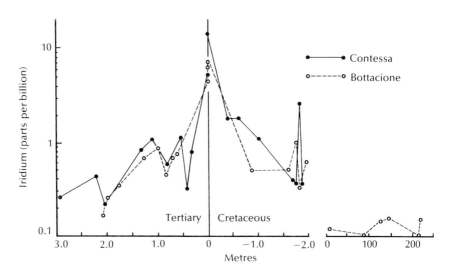

Fig. 9.9 Concentration of iridium in two sections of the Gubbio Shale which span the Cretaceous (K)/Tertiary (T) boundary. Control samples reveal only trace levels of iridium. After Hallam (1987).

recently emerged as the favoured site for a massive impact (Hildebrand & Boynton 1990; Kent *et al.* 1990).

Critics of the asteroid hypothesis argue that many of the groups that met their ultimate demise at the K/T boundary were in decline prior to the extinction event, though there does appears to have been a *coup de grâce* at the K/T boundary (e.g. among the ammonites depicted in Fig. 9.10). The selectivity of the extinctions is also not easily reconciled with an instantaneous event. In response to these criticisms Hut *et al.* (1987) suggest that multiple cometary impacts extending over long geological periods may be necessary to explain the stepwise periods of decline. However, comets have a much lower iridium level than the iron meteorites originally thought to be a candidate for impact, and other K/T boundary events are not explained satisfactorily by this hypothesis (Hallam 1987).

The alternative scenario suggests a volcanic origin for the iridium, as the aerosols from some volcanic eruptions from the deep mantle show enormous iridium enrichment. Increased vulcanism is compatible with the rapid fluctuations in sea level which characterise (and frustrate analysis of) the K/T transition.

The importance of a global winter has been weakened by the discovery of diverse late Cretaceous dinosaur faunas at extreme and highly seasonal latitudes in both the Alaskan North Slope and in Australia, which at that time was in a much more southerly position (Brouwers *et al.* 1987; Rich *et al.* 1988), suggesting that dinosaurs would be quite tolerant to low temperatures. Indeed an equally plausible scenario has attributed the breadth of the extinction to global warming (Rampino & Volk 1988).

These arguments can be summarised by saying that something unusual probably happened at the K/T boundary which contributed to the final demise of the dinosaurs and many other organisms, but left some other taxa relatively unaffected. We remain unsure whether extraterrestrial agents need to be invoked. What is clear is that the world was changed forever by the extinction event. The demise of the

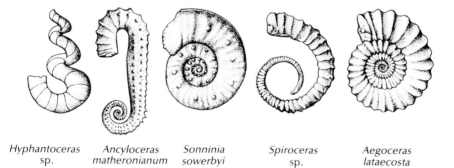

| *Hyphantoceras* sp. | *Ancyloceras matheronianum* | *Sonninia sowerbyi* | *Spiroceras* sp. | *Aegoceras lataecosta* |

Fig. 9.10 The history of the ammonites is one of boom-and-bust. The two species on the left are from the Cretaceous; the three on the right from the Jurassic. After Tasch (1973); Clarkson (1979); British Museum of Natural History (1983).

dinosaurs as the dominant terrestrial group was followed by the dramatic emergence of the mammals. At the same time, the diversification of the angiosperms was gathering pace (Fig. 3.12), giving rise to the world we know today.

The Permian/Triassic boundary

Some estimates suggest that as many as 96 per cent of all marine metazoan species met their demise at the end of the Permian Period 253 million years ago (Sepkoski 1989). This extinction terminated the Palaeozoic Era and is the most significant event in the history of animals since their emergence several hundred million years earlier. Attempts to find an iridium anomaly at the Permian/Triassic boundary have not been successful (Clark *et al.* 1986). The most widely argued scenario for the Permian extinction suggests the importance of the fusion of the continents into a single supercontinent (Pangea) at about that time. This had the consequence of reducing inshore habitat and merging previously separated faunas, and stopping sea-floor spreading and causing a major regression in sea levels (Newell 1952; Fig. 9.11). A further regression ensued as Pangea drifted north, increasing the latitudinal gradient in temperatures, and increasing seasonality and climatic instability. The number of recognisable marine provinces declined. As we have seen, the species diversity in an area will increase exponentially with habitat area, but different species should persist in different isolates. This collapse in habitat area and heterogeneity would contribute to the extinction.

It has proved extremely difficult to assess whether the demise or survival of species was correlated with some particular trait. Erwin (1989) suggests that families and genera that were species-rich and had a broad geographical distribution and environmental tolerance were more likely to persist. This pattern would be expected if the extinction of species was random.

Are mass extinctions periodic?

Strong periodicity in mass extinctions is a highly provocative claim, as it is suggestive of an extrinsic influence on the history of life, which while physically deterministic, acts a randomising agent on biological material. As Gould (1985b) points out, the pattern that exists may be seeded in a way that is so random and blind as to demand the 'greatest revision of cosmology (at least for our little corner of the heavens) since Galileo'. Indeed, a variety of corollary hypotheses have been generated, including the existence of a hitherto undetected solar companion star called Nemesis, which periodically perturbs the Oort comet cloud, creating a shower of comets (Raup 1986). The critics of the claim of periodicity focus on two problems, both fundamental: the quality of the data, which is fragmentary and bedevilled by artefacts caused by inconsistent taxonomic practice (Patterson & Smith 1987;

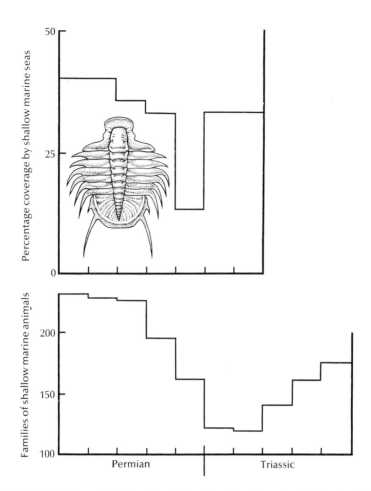

Fig. 9.11 Correlation of the diversity of families of marine animals living in shallow water with the reduction in the area covered by shallow marine seas during the Permian crisis. After Schopf (1974).

Smith & Patterson 1988), and the quality of the statistical modelling (Quinn 1987; Stigler & Wagner 1987; Quinn & Signor 1989). I share their scepticism.

Background extinctions

Our knowledge of the proximate causes of background extinctions is surprisingly weak. We have observed many extinctions, but almost all of these are direct consequences of our own actions (Chapter 10). In any case, the proximate cause of extinction in a very small population, although of profound interest in population management, is less relevant to a discussion of long-term influences of species diversity than an analysis of what characters of species might predispose them to extinction.

It is obvious that the risk of extinction is greatest for very small, geographically restricted populations, as both demographic and environmental stochasticity will be most pronounced under these circumstances (Chapter 10). It is also reasonably obvious that fluctuating populations will be more susceptible to extinction than stable populations. Jablonski (1986a, 1986b) has been the strongest advocate of the view that susceptibility to extinction is very different in mass and background extinctions. Once again, interpretation of data from the fossil record, from which we deduce most extinctions, is beset by methodological difficulties (Raup 1987 is a good summary). For example, among marine invertebrates the extinction rate within communities appears to be higher offshore than in shallow environments closer to the shore. This is perplexing because nearshore environments should fluctuate more than offshore environments. Sepkoski (1987) has recently argued that this pattern obscures two influences. First, genera within individual taxonomic classes which are broadly distributed, such as trilobites, have their highest extinction rates close to shore. Second, genera in classes with low characteristic extinction rates, such as bivalves and gastropods, are concentrated in nearshore environments, while groups with higher extinction rates, such as articulate brachiopods and anthozoans, occurred in outer shelf environments during the Palaeozoic.

Life cycles in marine invertebrates

Larval development in marine invertebrates is extremely diverse, but a simple classification by Jablonski & Lutz (1983) based on dispersal capacity distinguishes planktotrophic larvae, which spend a proportion of their time swimming freely and rely on other planktonic organisms for nutrition, from non-planktotrophic larvae, which either do not become free-swimming or remain in the plankton for only a short time. They are nourished by the yolk from which they develop. Many higher taxa of invertebrates show both forms of development, though in certain cases the vast majority of members of a particular higher taxon exhibit only one form of development (e.g. all modern articulate brachiopods and most cheilostome bryozoans have nonplanktotrophic larvae). Planktotrophic development appears to develop in tropical communities, where habitats are patchily distributed, and also becomes more common with increasing depth across the continental shelf (Jablonski & Lutz 1983).

These differences could influence probabilities of speciation and extinction in a variety of ways. First, the potentially massive dispersal of planktotrophic larvae is likely to retard genetic differentiation, reducing speciation rates, yet increase geographic range and perhaps total population size, reducing the influence of both demographic or environmental stochasticity and hence extinction rates (but see Chatterton & Speyer 1989). Effective colonisation also enhances the probability that areas subject to local extinction will be rapidly repopulated. Large geographical ranges and lower extinction rates have been documented for planktotrophic species (Hansen 1980; Jablonski & Lutz

Table 9.1 Effect of larval ecology on taxonomic survivorship in late Cretaceous molluscs. Data from Jablonski & Lutz (1983) and Jablonski (1986b).

	Planktotrophs	Non-planktotrophs
Background extinction	(*n* = 50)	(*n* = 50)
Range (km)	1600	250
Median survival (million years)	6	2
Extinction rate (spp. per lineage million years)	0.17	0.34
Mass extinction	(*n* = 28)	(*n* = 21)
Percentage survival of genera at the Cretaceous/Tertiary boundary	39	39

1983; Table 9.1). By contrast, differences in larval ecology appear to exert less effect during mass extinctions (Valentine & Jablonski 1986; Table 9.1), again emphasising the randomising effect of these extinctions.

The Red Queen and the constancy of extinctions

One of the puzzling results that emerges from the fossil record is a slow decline through time in the extinction rate of families of marine invertebrates (Flessa & Jablonski 1985; see also Fig. 3.7). This could indicate a gradual increase in resilience to extinction as a consequence of adaptation. However, a plausible alternative explanation suggests that although the rate of loss of species has remained approximately constant, the number of species assigned to each family has increased through time. As all the members of a family must become extinct, the rate of family extinction will decline (Flessa & Jablonski 1985).

In a related observation, Van Valen (1973a) suggested that within a particular taxonomic group the probability of extinction of a genus or family is independent of its prior existence. It was this observation that prompted Van Valen to propose the Red Queen hypothesis that we have encountered in various guises throughout this text. Because hostile species are evolving all the time, there will be continual pressure for adaptation just to maintain a level of adaptedness. Some species will fail to adapt, and extinction will ensue at an approximately constant rate. Several preliminary attempts have been made to generate predictive models of the patterns which we would expect in the fossil record if the Red Queen is a more important contributor to background extinction rates than is change to the physical environment (Stenseth & Maynard Smith 1984; Rosenzweig *et al.* 1987), and some attempts have been made to apply the models (Hoffman & Kitchell 1984).

The alternative side of the balance sheet to the rate of extinction is the rate of diversification. While there appears to have been a general increase in the number of species through time, there appear to have been a number of explosions of diversity which warrant special mention.

Adaptive radiations

Throughout the history of life there are conspicuous explosions and diversification of some taxa to fill a variety of roles. These adaptive radiations are often presumed to occur in response either to ecological opportunity or to the acquisition of some key adaptive trait, or both (Simpson 1953). Some authors have recently attacked the idea of progressive improvement and key adaptations. However, as we saw in Chapter 3, acquisition of improved vascular architecture and reproductive efficiency of reproductive organs assisted the dramatic radiation of the angiosperms (Fig. 3.12). This in turn has created extraordinary ecological opportunity for the diversification of the insects, particularly pollinators and herbivores (Niklas 1986). In turn, these insects may have influenced speciation in angiosperms (Chapter 8), proved such efficient pollinators as to transform reproductive performance of plants, and also provided a platform for diversification of parasitoids, another group of exceptional diversity.

Ecological opportunity can arise in a variety of ways. For example, mass extinctions are often followed by an explosive radiation of a previously much less diverse clade. The rapid origin of the contemporary mammalian orders after the mass extinction at the end of the Cretaceous is a classic example. Island faunas often exhibit radiations of taxa that would otherwise be unlikely to fill such a diversity of roles. For example, the terrestrial fauna of Christmas Island is dominated by crabs, despite the physiological difficulties these organisms experience living on land. The adaptive radiation of island avifaunas proved very influential to both Darwin and Wallace in their development of the Principle of Natural Selection, and the analysis of groups such as Darwin's finches on the Galápagos remain among the most important evolutionary models (Lack 1947b; Grant 1986).

In a sophisticated comparative extension of Darwin's analysis, Schluter (1988b, 1988c) has shown that there are fundamental differences between the radiations of finches on islands and continents (Table 9.2). In the Galápagos, granivorous finches exploit many types of seeds and show enormous diversity of morphology, while mainland radiations do not exploit all the available seeds (as these are consumed by other granivorous taxa), yet show strong resource partitioning between finch species.

One of the most extraordinary adaptive radiations are the proliferation of species of cichlid fishes found in some of the rift lakes of Africa. For example, Lake Victoria contains 200 species although it has

Table 9.2 Characteristics of adaptive radiations in the finch communities of Kenya and the Galápagos. Asterisks denote features that may differ generally between continents with complex faunas and isolated archipelagos where greater diversification is possible. After Schluter (1988c).

Features	Galápagos	Kenya
Entire community		
Diversity of beak sizes*	High	Low
Diversity of seeds consumed*	High	Low
Utilisation of seed species	Complete	Incomplete
Individual finches		
Separation by habitat*	Infrequent	Frequent
Separation by microhabitat*	Moderate	Frequent
Separation by seed species of similar size*	Infrequent	Frequent
Separation by seed size	Frequent	Infrequent
Diet specialisation	Low	High
Morphology–diet relations	Strong	Weak
Sensitivity to cover*	No	Yes

only been in existence for about one million years. Genetic and morphological evidence support the view that the species have radiated within the rift lake, so the diversity is not a consequence of repeated colonisation and recolonisation (Greenwood 1981; Meyer *et al.* 1990). Some of their morphological specialisations are extraordinary, and certainly evolved in response to the habits of other cichlids. Some species feed on the young brooded in the mouths of other cichlids! This is a fascinating counterexample to the *Plethodon* salamanders described in Chapter 8, and it would be of great interest to know if there is some aspect of the genetic architecture which predisposes these fish to speciate.

The bottom-heaviness of clades

An alternative way to look at the phenomenon of radiation is to seek patterns of diversity within individual clades through time. Gould *et al.* (1987) suggest that we can distinguish three evolutionary patterns of diversity for a clade. First, bottom-heavy diversity where the generation of diversity is concentrated early in the history of the clade, then eroded by extinctions (Fig. 9.12). Second, symmetrical clades which wax and wane through time, but reach their height of diversity in the middle of their history. Third, top-heavy clades, which increase steadily in diversity (angiosperms are an excellent example), but may disappear suddenly, perhaps because of mass extinction. Analysis of fossil data suggests that the bottom-heavy pattern is most frequent, an important result as the 'pull of the Recent' (Raup 1978), or the likelihood that our comparatively complete knowledge of the extant biota will cause over-representation of modern forms, suggests exactly the opposite effect

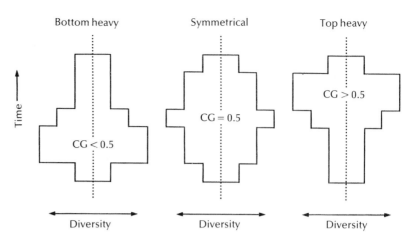

Fig. 9.12 Patterns of diversity in clades through time. In bottom-heavy clades radiation is concentrated into the early history of the clade. Although the 'pull of the Recent' predicts that top-heavy clades should predominate, the bottom-heavy pattern is most commonly detected. CG, centre of gravity of the observed distribution. After Gould *et al.* (1987).

should predominate. The pattern of bottom-heaviness occurs across taxonomic levels and taxa, and is retained in groups such as Miocene horses where taxonomic resolution is extremely fine (MacFadden & Hulbert 1988), again suggesting that this result is unlikely to be an artefact (though see the simulation modelling by Kitchell & MacLeod 1988, which suggests that the null model used by Gould *et al.* is questionable).

If innovations are concentrated at the bottom of clades, it becomes important to ask whether the ecological vacuum created by a mass extinction itself can become a stimulus for speciation and cladogenesis. Rosenzweig (1978) has suggested that such patterns are common, and represent a type of sympatric speciation. An important component of the model is the expectation that a poorly adapted genotype may be able to exploit unoccupied habitat, and once established, may rapidly adapt to the pressures of the new environment. Wilson & Turelli (1986) use a variety of models to confirm and extend this expectation.

A case study: the diversification of the Metazoa

Some of the difficulties that arise in interpreting the causes of radiations are illustrated in the two great explosions of the Metazoa into ecological vacuums, the first of which took place at the start of the Cambrian as multicellular organisms first started to diversify on a grand scale; the second of which took place in the Mesozoic during the recovery from the massive Permian extinction (Fig. 3.7). Although they spanned a similar period and led to an equivalent increase in diversity, the two radiations are fundamentally different in character. While the early Cambrian explosion was characterised by the development of many novel body plans or *Baupläne*, higher taxa did not originate as a result of the Mesozoic explosion. There are two broad arguments as to why the radiations differed so fundamentally, one emphasises ecological opportunity; the second an increase in genetic and cellular organisation which precluded fundamental change thereafter.

Strong support for an ecological basis for the differences in the nature of the radiations comes from careful analysis of the ecological roles of the organisms which founded each radiation. Bambach (1985) shows that while the number of species involved may not have differed greatly, most roles were 'filled' by the survivors of the Permian extinction, so overall ecological opportunity was lower. According to this view the rapid Palaeozoic expansion represents a relaxation of selection in the face of ecological opportunity, but as ecological space became saturated only the best-adapted phenotypes would persist. Key innovations also contributed directly to the radiation (McMenamin 1988). Vermeij (1973a, 1973b) has suggested that individual taxa tend to develop increase in morphological diversity through time, allowing them to fill adaptive zones previously inaccessible to them (but see Strathmann 1978).

By contrast, the genetic hypothesis suggests that because the *Baupläne* of the early Metazoa were less canalised, mutations could easily generate new variants. As greater genetic integration and canalisation occurred, major shifts were rendered impossible.

Buss (1987) makes the challenging suggestion that this reduction in the diversity and absence of innovation in *Baupläne* through time may represent the result of ancient conflict between the interests of the individual and the interests of the cell. He argues that development should not be interpreted as a cooperative enterprise but reflects ancient interactions between cell lineages in their quest for increased replication. The interests of cells in simple modular organisms are best served by replication, but this is inimical to the interests of a multicellular organism, as evidenced by the grim effects of cancer. The evolution of development required some sort of control over cell lineages which abandon somatic function for unhindered replication. Isolation of germ-line cells and embryological development has allowed resolution of the conflict between replication and cooperation, but may have increasingly restricted the potential for innovation.

Buss (1988) further argues that metazoan phyla with comparatively late or variable germ line determination are more species rich, though his conclusions are greatly weakened if we reject his claim that the Nematoda, with early germ-line determination, contains only about 10 000 species (cf. Barnes 1987; May 1988; Fig. 1.6). In a related argument, Gould (1977) argues that empirical evidence suggests that within any clade, paedomorphic forms (see Chapter 3) are more likely to found great radiations because more developmental options are available at this early stage of development (but see McNamara 1986, 1988).

In summary, it appears that the origin of the metazoan phyla conforms to the bottom-heavy diversity pattern of Fig 9.12, with most of the phyla originating early in the Palaeozoic. The remnant phyla are a small subset of the original, but were able to replenish biological diversity after the Permian extinction.

There are more small species

Perhaps the simplest correlate of diversification among animals is size. At virtually any taxonomic level, there tend to be more with intermediate sizes than large or very small sizes, with a very strong skew to the smaller organisms (e.g. Van Valen 1973b; May 1978, 1988; Morse *et al.* 1988; Brown & Maurer 1989; Fig. 9.13). Indeed, May (1978) suggests that most insect diversity can be accounted for by this factor alone. The importance of small size stems principally from the fact that the environment can be divided more finely between small species than is possible between large species (Southwood 1978; Janzen 1987).

The predominance of small species has important evolutionary consequences. If mass extinctions do operate more or less randomly, the numerical predominance of small species would suggest that small species are most likely to found lineages and be the starting point for adaptive radiations. This predominance will be exacerbated by the ability of small species to use spores or some other diapausing stage to insulate themselves from temporarily bad conditions. Indeed, it has been known for some time that lineages have a tendency to be founded by small organisms, and to increase in size through evolutionary time (Cope's law of phyletic size increase). A possible contributor to this pattern is the ability of large organisms to monopolise a disproportionate share of the total resources available (Chapter 5; Brown & Maurer

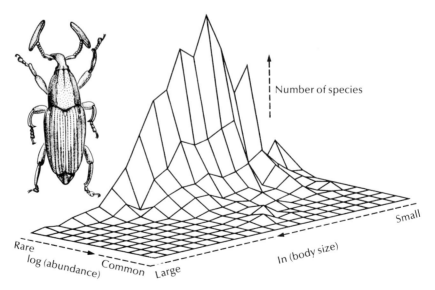

Fig. 9.13 The relation between species diversity and log (abundance) and log (body size) in the beetles of Borneo. Species richness is greatest for species of intermediate body size. After Morse *et al.* (1988). The beetle (*Cercidocerus securifer*) is drawn after Reitter (1960).

1986), suggesting that one aspect of the alternation between mass and background extinctions will be the relative importance of small and large organisms.

Small species also have a higher rate of extinction. Following Diamond (1969), turnover per unit time (from t_2 to t_1) can be expressed as:

$$\frac{\text{extinctions of spp. already present} + \text{no. of new spp. immigrating}}{\text{no. of spp. at } t_1 + \text{no. of spp. at } t_2}$$

The limited values available in the literature suggest a strong relation with generation time (Schoener 1983b; Fig. 9.14). Small species have shorter generation time than large species, which may increase their susceptibility to environmental stochasticity as they are less likely to survive a temporary period which does not permit breeding, and, if they are more specialised, are more likely to lose all of their habitat. A related argument suggests that small species may require a larger population size to insulate them from extinction. For example, introductions of large animals are more successful than introductions of small animals, in taxa from nematodes to mammals (Lawton & Brown 1986).

SUMMARY

We do not currently know how many species there are on earth, even within an order of magnitude, let alone understand the processes that lead to that number. The acceleration of extinction rates caused by human habitat destruction and the inadequacy of our understanding of the former question points to the importance of taxonomists and systematists in answering fundamental biological problems. Understanding why there are so many species is arguably the greatest intellectual challenge facing humanity in the decades ahead. A good starting point for understanding local diversity is analysis of the conspicuous

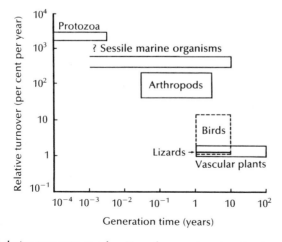

Fig. 9.14 Relative turnover as a function of generation time in six types of organisms. After Schoener (1983b).

between-habitat differences in diversity, such as the tendency of diversity to decline with latitude, and to increase with area of habitat. Current analysis suggests that a pluralistic approach will be most helpful, instead of assigning overwhelming importance to single-factors. Biological interactions are of great importance in limiting local diversity. Predation, parasitism and competition are all of great importance, and we are beginning to understand the circumstances in which each has greatest effect. Global species diversity is a consequence of the balance between rates of speciation and rates of extinction. The history of life has been perturbed by a series of mass extinctions, the cause of which remains very controversial. These mass extinctions reset the competitive environment and may have profound effects on the subsequent diversification. Diversification accelerates in response to ecological opportunity, yet the founders of radiations are likely to be determined by features such as body size and developmental plan.

FURTHER READING

Animal data are emphasised in Raup and Jablonski's (1986) compilation of thoughtful and comprehensive articles on patterns in the fossil record and their causes. Brown & Gibson (1983) is an excellent treatment of the principles of biogeography. White (1986) is a beautiful and inspiring description of the history of plant life. Wilson (1988) is a chilling series of articles on the problems faced as a result of our continued insult to the environment.

TOPICS FOR DISCUSSION

1 What does evolutionary 'success' mean? See Wilson (1987) for one definition.
2 Contrast the belief that the radiation of clades on islands is a consequence of exploitation of ecological opportunity driven by selection and adaptation, and the view that it reflects the high frequency of founder events and genetic revolutions driven by drift (see also Chapter 8 and Coyne 1990).
3 Janzen (1985) comments 'I've grown up on a diet of Darwin, Wallace and their cultural offspring. But I'm faced with the reality of the biology of 100 km^2 of lowland dry forest (Santa Rosa National Park) in northwestern Costa Rica. Some of my background does not prepare me for this: almost all of the ecology I see around me could quite easily come to be with virtually no evolution having occurred at Santa Rosa.' Discuss his subsequent argument in support of this view.

Chapter 10
Applied Evolutionary Ecology

Much of evolutionary ecology is 'pure science', driven by curiosity and the desire to understand our natural world. Occasionally such research is portrayed by politicians and captains of industry as a waste of public resources, and irrelevant to the imperatives that drive the economic agenda of the modern world. I hope that reading the preceding nine chapters has convinced you otherwise, and that you agree that the benefits gained from the pursuit of knowledge are so great and long-lasting that they cannot be evaluated by the short-term currencies of political and economic expediency. Indeed, one of the clear lessons of evolutionary biology is to beware anything evaluated purely in terms of its current utility.

Nonetheless, this final chapter is devoted to the application of some of the principles developed in this text. The practical applications that have arisen in the past from ecology and genetics are sufficiently numerous to fill several texts on their own. Here I will concentrate on just three topics, which I have chosen because each is an area of growing or new importance, and each shows clearly how information developed as a consequence of pure research often allows us to anticipate and manage new problems before and when they arise. First, I discuss recent attempts to develop general principles for the management of rare and endangered species. Second, I discuss the problems that might potentially arise from the forthcoming new wave of biological introductions which will stem from the power that recombinant DNA technology gives us to manipulate or 'engineer' the genetic constitution of organisms. Third, I comment on the genetic problems that arise from the exclusive use of new productive strains or organisms which have been produced using new technologies or conventional breeding programmes.

THE MANAGEMENT OF RARE AND ENDANGERED SPECIES

What is rarity?

It is a truism to say that rarity increases the probability of extinction. It is less easy to define exactly what rarity is. Rabinowitz and her co-workers have proposed that patterns of abundance can be arrayed along three axes, in a classification that recalls the discussion presented in Chapter 5 (Rabinowitz 1981; Rabinowitz et al. 1986). First, how broad is the geographical range, from widespread to locally endemic? Second, at any one site does the species occupy a large range of habitats, or is it a specialist? Third, is it common in some parts of its range, or does it occur at low numbers throughout its range? Schoener (1987) also recognised the third dichotomy, calling the former state diffusive rarity and the latter state suffusive rarity. We can therefore recognise eight possible patterns of abundance. While species that are broadly distributed, have a broad habitat range and are abundant in places throughout their range cannot be called rare under any circumstances, Rabinowitz *et al.* argue that the remaining seven combinations represent forms of rarity. Our knowledge of natural distributions is generally inadequate to allow these forms to be distinguished, so their relative frequency cannot usually be assessed. Among the members of the British flora which have been described in the ongoing and ambitious publishing project *Biological Flora of the British Isles*, published as a supplement to the *Journal of Ecology*, Rabinowitz *et al.* (1986) were able to obtain agreement from appropriate experts on the status of 160 species (Table 10.1). The number of species which have low local population sizes throughout their range are very low, and one category is absent.

Although unusually well-known, the British flora is unfortunately not likely to be a reliable model for interpreting natural distributions. Most of the British Isles have had (or suffered) long association with human activity, and the extinctions that appear inevitably to ensue from that activity have probably already eliminated some of the species in the bottom right-hand corner of Table 10.1. Further, the flora is not particularly diverse, and we have no way of anticipating how the results could be generalised to animals.

Perhaps the most dramatic demonstration of the rapidity with which humans can influence distributions comes from the mammals of the Australian deserts. Often portrayed as one of the world's great untouched wildernesses, we might guess that this would be a good example of an unperturbed fauna. However, discoveries of bone deposits and the sketchy reports of early settlers suggested that many mammals had declined in abundance quite recently. In order to assess the magnitude of the decline, Burbidge *et al.* (1988) undertook to survey the local aboriginal knowledge of the desert fauna, by showing tribal elders throughout the northern desert, museum specimens of animals known to have occurred in the region. The results are both fascinating and frightening. Not only has species diversity plunged, but the majority of extinctions have occurred in the last thirty years. Most tribes remembered even the very rare mammals, to the extent of repeatedly pointing out the absence of one species from the set of specimens. Burbidge *et al.* (1988) concluded in hindsight that this was probably *Lagorchestes asomatus*, a hare-wallaby known to science from a single incomplete

Table 10.1 Classification of the types of distribution in 160 British plant species, according to their geographic distribution, local habitat breadth, and whether they ever attain high local population sizes. After Rabinowitz *et al.* (1986).

	Geographical distribution			
	Wide		Narrow	
Local habitat breadth	Broad	Restricted	Broad	Restricted
Local population sizes				
Somewhere large	58	71	6	14
Everywhere small	2	6	0	3

specimen, in stark contrast to its distribution according to aboriginal memory (Fig. 10.1). What is worse, the cause of the decline is not clear, and probably results from the influence of several interacting factors, including grazing by cattle and rabbits, predation by foxes, and changing fire regimes (Morton 1990). Their relative importance will probably never be known. Clearly, any attempts to compile natural diversity and abundance patterns for the Australian mammal fauna is likely to encounter grave difficulties.

There are, however, theoretical predictions which support the distribution of rarity described in Table 10.1. In Chapter 5, I suggested that individuals are usually more abundant at the centre than at the edge of their range. A corollary of this observation is that there is usually, though not always, some correlation between local abundance and regional distribution. Hanski (1982) has modelled the consequences for the probability of extinction, and predicts that local biotas will comprise two sorts of species: core species which are abundant throughout the fauna and resilient to extinction, and satellite species which are subject to repeated local extinctions (see also Caswell 1978). While core species dominate the trophic organisation and biological interactions of communities, satellite species will often be the focus of conservation efforts. A possible example comes from spider communities on small islands in the Bahamas. These communities are very unstable. Average turnover rates are not very different from other arthropod communities averaging more than 30 per cent per year (Schoener & Spiller 1987; cf.

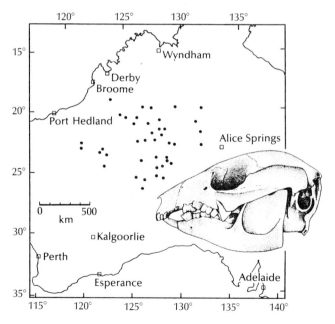

Fig. 10.1 The recent distribution of *Lagorchestes asomatus* revealed by the recollections of aborigines living in the northern Australian desert. The species is known to science by a single specimen, yet it appears to have been common only thirty years ago. After Burbidge *et al.* (1988).

Fig. 9.6). Yet some species are highly resistant to extinction, while others go extinct and then recolonise repeatedly (Fig. 10.2), supporting the belief that even in unstable communities there will be a mix of stable and fugitive species. Although these and other data support Hanski's models, it has not yet been possible to test rigorously all the components of his theory (Gotelli & Simberloff 1987; Gaston & Lawton 1989).

The conservation implications of these data are clear. If we wish to maintain a community in its current state it is inadequate to consider one patch of habitat. Unless satellite species have adjacent areas from which they can recolonise then very rapid declines in diversity could occur. Conservation biology therefore needs to focus on regional processes as well as isolated patches of habitat, as we might have anticipated from the discussion of species diversity in Chapter 9. The importance of regional effects is dramatically illustrated by data on the population dynamics of breeding birds in North America. While densities of resident birds and short-distance migrants have been stable or increased since 1978, there has been a sharp decline in breeding densities of migrants to the Neotropics. Because the decline is most pronounced in species that overwinter in forest, it seems that the destruction of tropical rainforest is having a direct impact on the avian community of the United States and Canada (Robbins *et al.* 1989).

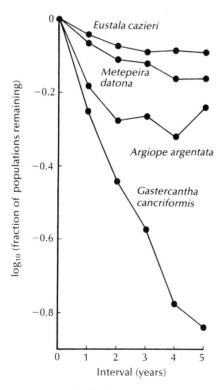

Fig. 10.2 Population persistence for the four commonest orb-spider species on 108 islands in the central Bahamas. After Schoener & Spiller (1987).

Second, we will usually lack adequate knowledge to understand the conservation requirements of all species in a community, particularly invertebrates. This has occasionally led to the suggestion that management strategies should concentrate on a few representative species which might indicate the health of the community, rather than focusing to excess on management of species thought to be endangered (e.g. Gilbert 1980; Terborgh 1986). If the species which dominate interactions are different from those with a high probability of extinction, then this diversity should obviously be taken into account whenever the principle of focusing on keystone species is employed. Interactions can be extremely complex. Consider the effect of the introduction of a mammalian virus to southern England on the large blue butterfly, *Maculina arion* (Ratcliffe 1979). The myxoma virus was introduced to control rabbits and had spectacular success. Grazing and browsing by rabbits had the effect of promoting the growth of certain grasses, which grow from a protected basal meristem. A species of ant (*Myrmica sabuleti*) built its nest in the roots of the grasses, and these ants also tended the caterpillars of the large blue butterfly (Fig. 10.3). Once browsing pressure declined, the grasses were overgrown by woody shrubs, the ants declined and the butterfly went extinct. A diversity of management tactics is certainly desirable.

PROXIMATE CAUSES OF EXTINCTION IN VERY SMALL POPULATIONS

Although extinction appears to be the eventual fate of most species, and that more than 99 per cent of all the species that have ever lived are now extinct (Simpson 1952), in Chapter 9, I pointed out that virtually

Fig. 10.3 The players in an unlikely chain of events leading to the extinction of the large blue butterfly, which followed the introduction of the myxoma virus to England.

all the extinctions about which we have any information have been caused by our own actions. The management implications are profound. If we accept that maintenance of species diversity is an important goal of conservation biology, problems fall into two sorts. First, is it possible to arrest the decline of a species threatened by human activity? Such objectives can often only be realised by stopping the human insult. Second, can we elevate population size and configuration to a size where extinction due to random events is minimised (the Minimum Viable Population, or MVP)? Even for human-induced extinctions, our knowledge of what delivers the final blow to declining populations is surprisingly inadequate (Simberloff 1986).

There are two sources of demographic variation which can drive a population to extinction (Simberloff 1986). The first is demographic stochasticity. Individuals differ in their age-specific probabilities of survival and reproduction, or vital rates. The variance of the vital rates is inversely proportional to population size (Lande 1988b), so there is some probability that all individuals will be doing the wrong thing at the wrong time. As Goodman (1987) puts it, populations can become extinct because of 'bad luck'. For example, a population of sexually reproducing organisms comprised entirely of males is functionally extinct. The probability that a population reduced to four individuals will consist entirely of one sex (assuming random sampling of a population where the sex ratio was 0.5) can be calculated from the binomial theorem as $2 (0.5)^4 = 0.125$. Because survival of the sexes under stress is anything but random (Clutton-Brock *et al.* 1985), the real probability may be much higher.

The second cause is environmental stochasticity which affects all individuals in the population in much the same way. For example, severe drought could precipitate extinction of a small mesic-adapted population. In an extreme example, an environmental catastrophe such as a volcanic eruption could exterminate a species living on the side of a volcano regardless of its prior reproductive performance and demographic stochasticity. Although long-term data are far too scarce, it appears that most populations fluctuate to some extent. In an analysis of some of the best long-term data sets, Pimm & Redfearn (1988) found evidence that the variance in population size increases as the duration of sampling increases (Fig. 10.4), suggesting that most populations follow long-term trends. Such variation in densities inevitably predisposes small populations to extinction (Richter-Dyn & Goel 1972; Goodman 1987). Indeed, Goodman (1987) was drawn to conclude pessimistically that for populations which are sensitive to environmental fluctuations, the minimum reserve size to achieve a reasonably long persistence time is extremely large, and that in the absence of immense reserves, interventionist management aimed at reducing the variance in population growth rate may represent the only possible management strategy.

The rate of environmental change has varied dramatically. For example, the transitions from glacial to interglacial appears to be quite

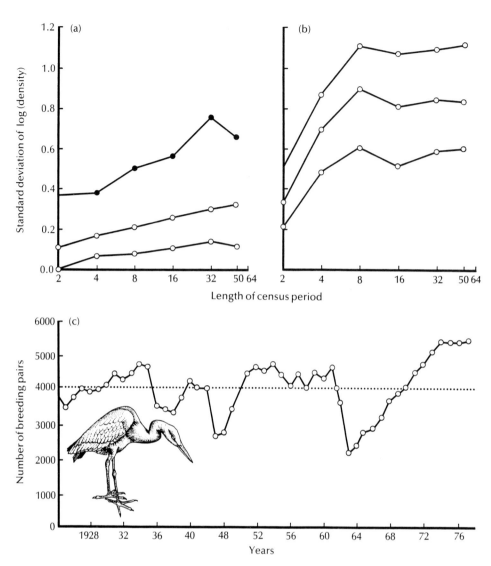

Fig. 10.4 Long-term trends in populations. Population variability (standard deviation of log density) in relation to the number of years over which the censuses were calculated for: (a) 22 British birds and mammals; and (b) four British insect species. The top line is the maximum value for any species, the middle line the mean, and the bottom line the minimum. After Pimm & Redfearn (1988). (c) The number of breeding pairs of grey herons *Ardea cinerea* in England and Wales from 1928–1977. After Lawton (1988).

rapid, and driven by deterministic processes (see Bartlein & Prentice 1989 for a recent review). Unfortunately we are trying to understand the impact of change at a time when the chemical and physical characteristics of our environment are changing rapidly at our own hands (see Dobson *et al.* 1989 and Caldwell *et al.* 1989 for recent reviews).

Pimm *et al.* (1988) have made a number of interesting additional predictions about extinction risk in small populations (Fig. 10.5). They have also developed a set of predictions that relate directly to the body

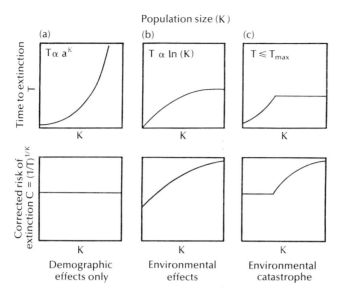

Fig. 10.5 Times to extinction (T) and corrected risk of extinction $(C = (1/T)^{1/K})$ for models of three different sources of extinction: (a) extinctions resulting from demographic accidents in an unvarying environment, T increases as a power of maximum population size K, so that C is independent of K; (b) extinctions where there are demogarphic effects and moderate environmental variation, so T increases as the logarithm of K; (c) for environmental catastrophes that strike after a time T_{max} no population survives beyond T_{max}. After Pimm *et al.* (1988).

size of the organisms concerned. First, populations of long-lived individuals should have a lower risk of extinction than short-lived individuals, as the long-lived individuals have a much greater chance of surviving a period in which reproduction is not possible (Chapter 9). Second, populations with a low intrinsic rate of increase should be more susceptible to extinction than populations which can grow rapidly because they remain at risk of demographic accidents for longer after a density reduction, perhaps caused by environmental events.

However, these predictions are confounded by the well-known allometric effects on both life-span and rate of reproduction. Because large animals will have both longer life-span and slower rates of increase than small animals, there is no simple theoretical basis for predicting which effect should predominate. In an attempt to resolve this question empirically, Pimm *et al.* (1988) used long-term censuses of the number of breeding birds on British offshore islands. Extinctions on these islands are probably determined largely by the interaction of environmental and demographic stochasticity and the small population size common on islands. The relation between risk of extinction and breeding density differed for large- and small-bodied species, so that below about seven pairs (Fig. 10.6), small species have a greater risk of extinction. As Lawton (1989) points out, the data were not adequate to predict what would happen at breeding densities greater than seven pairs.

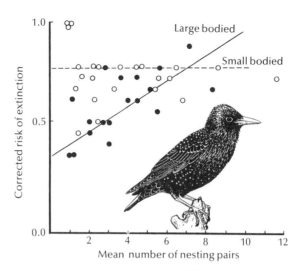

Fig. 10.6 Corrected risk of extinction $(C = (1/T)^{1/K})$ plotted against mean
population size (number of nesting pairs) for birds resident on British offshore
islands. ●, large-bodied species; ○, small-bodied species. After Pimm *et al.* (1988).

The Allee effect

Some populations exhibit social dysfunction at low density, and will
often fail to mate successfully even if given the opportunity to do so
(Allee *et al.* 1949). This Allee effect almost certainly contributed to the
ultimate demise of the passenger pigeon (Fig. 10.7), which until very
recently was the most abundant bird in North America, occurring in
flocks of extraordinarily large size, with population estimates in the
vicinity of 10^9, perhaps half of all the land birds on the continent.
Clearing of mast-bearing trees and trapping reduced the number of
birds dramatically, yet it was the failure to mate among remaining
populations that appears to have caused extinction. In this case it
is likely that the degree of social stimulation required to prompt
reproduction was no longer present.

IS GENETIC DIVERSITY IMPORTANT?

A major research effort has been devoted to analysing the genetic
consequences of very small population size and the need to take
genetic considerations into account in designing management strategies.
Genetic problems include inbreeding depression and the maintenance
of genetic variability within populations. Of course, in one sense the
inevitable death of all organisms means that we are seeking to conserve
genes, not organisms.

Indeed, the interest in application of genetic principles in management has become so widespread that recent reviews have been forced
to emphasise that demographic effects will often be of greater importance (Lande 1988b; Simberloff 1988). Of course, demography and genetics are closely intertwined, and it may be unwise to think of them
separately. Even the Allee effect raises the possibility that some sort of

Fig. 10.7 The passenger pigeon, *Ectopistes migratorius*. After Audubon (1953).

genetic management may also be necessary. For example, a tendency to avoid inbreeding or outcrossing with extremely dissimilar phenotypes could conceivably be sufficiently strong to generate the Allee effect, so genetic principles may be important.

Inbreeding depression

Large outcrossing populations that are suddenly reduced to a few individuals suffer reduced fecundity and viability, or inbreeding depression (Chapter 2). For example, inbreeding depression is virtually ubiquitous in captive populations of mammals (Ralls *et al.* 1988; Fig. 10.8). The average cost in these populations may be as much as a 33 per cent decrease in viability (Ralls & Ballou 1986; Harvey & Read 1988). The costs of inbreeding are also indicated by the large number of natural mechanisms that function to reduce its occurrence (Table 2.5).

The costs of inbreeding will depend first on the past history of the population, or the rate and extent to which deleterious alleles have been exposed to selection. Artificial selection on species used in agriculture over many hundreds of generations has almost certainly had the effect of removing deleterious alleles, though such domesticated populations remain susceptible to at least some inbreeding depression. The second factor influencing risk of exposure of rare alleles will be population size. Clearly a large population size allows greater mate choice, and a lower rate of incest.

Under certain circumstances it may be decided that permanent captivity represents the best hope for maintenance of a species. This would be true if the habitat of the species was totally destroyed, a fate which awaits a growing number of endangered species. Because inbreeding depression is virtually ubiquitous in small captive colonies (Fig. 10.8), it has been suggested that a permanent solution to inbreeding depression should be sought. The solution lies in the careful exposure and elimination of deleterious recessives to produce a homogeneous but stable group of individuals which can interbreed without harm. Speke's gazelle (*Gazella spekei*) is an extremely rare antelope whose

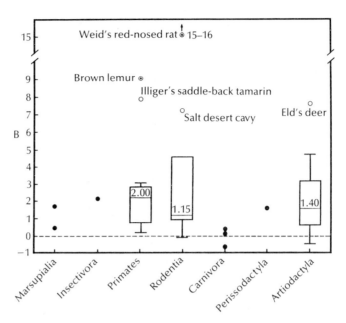

Fig. 10.8 Variation in the cost of inbreeding in 40 captive populations of 38 mammal species. B estimates the number of recessive lethal equivalents per gamete. The average for all mammals is just over two, and virtually all populations show some inbreeding depression. After Ralls *et al.* (1988).

habitat is restricted to a zone of guerilla war between Ethiopia and Somalia. The original habitat suffers from both the effects of warfare and the habitat degradation that has rapidly changed the vegetation of the Sahel region of Africa (Sinclair & Fryxell 1985), so reintroduction is not possible. The entire captive population descends from one male and three females, and the effective population size probably approaches two, or close to the smallest imaginable bottleneck. The population suffered from severe inbreeding depression during its growth to its current size of about 30 animals spread over three zoos. Templeton & Read (1983) have instigated an ambitious breeding design which attempts to combat the inbreeding depression, and current evidence suggests that their efforts are meeting with success. The design involved favouring parents that were already inbred but nonetheless have proven viability. The average genetic contribution of each of the four original founders was also equalised. The progeny of these systematically inbred parents now have greater survivorship than randomly mated progeny (Fig. 10.9).

The problem of outbreeding

The problems associated with extreme outcrossing are only rarely considered by managers. Indeed, in order to minimise inbreeding and to restore alleles lost from subpopulations by drift, individuals from captive populations or their sperm are often transported great distances between isolated subpopulations. The effects of extreme outcrossing

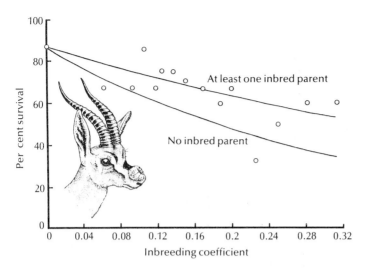

Fig. 10.9 The juvenile 30-day survivorship of Speke's gazelle (*Gazella spekei*) in relation to the intensity of inbreeding. While survival declines with increased inbreeding, offspring from planned matings with at least one inbred parent (○ and upper curve) suffer less than the progeny of parents which were not inbred (lower curve), suggesting that a programme of controlled inbreeding will eventually eliminate the problem of depression. After Ralls & Ballou (1986).

occasionally manifest themselves in the wild. Templeton (1986) reports the extraordinary demise of the Tatra mountain ibex (*Capra ibex ibex*), which was overhunted to extinction in Czechoslovakia but was successfully reintroduced from Austria. In an attempt to increase the vigour of the population, two other subspecies of ibex (*C.i. aegagarus* and *C.i. nubiana*) from the Middle East were introduced. The hybrids were fertile but rutted in the autumn instead of winter like the native ibex, and the offspring were born in the coldest part of the year. Extinction resulted.

Hybridisation also poses a problem where habitat destruction erodes barriers between formerly distinct populations, leading to the genetic subsumption of one form by another, particularly when one is much more numerous than the other. This problem is exaggerated when males of the new species have characteristics favoured by females of the old species (Ryan 1990; Chapter 7).

Nonetheless, it has been argued that extreme outcrossing can be used to capture genes otherwise doomed to disappear. In an ambitious example, the last individual of a morphologically distinct population of seaside sparrow (*Ammospiza maritima*) was hybridised with a neighbouring population in order to capture its genes. An 87.5 per cent copy of the original was claimed (Kale 1987). However, further genetic analysis using comparisons of mitochondrial DNA suggested that the melanism which had led to the assignment of subspecific status to the isolated sparrow population was not indicative of great genetic divergence (Avise 1989). Therefore much of the original effort was wasted on capture of genes promoting melanism, and not on preserving genetically distinct subpopulations. These and other examples emphasise the fundamental role of systematics in the conservation effort.

Loss of alleles by drift

As we have already seen, the random fluctuations in allele frequencies caused by genetic drift will lead to the loss of rare alleles, and can increase homozygosity. When factors which maintain genetic variation are absent (see Chapter 2), the proportion of genetic variance lost per generation is about $1/(2N_e)$ (Lande 1988b). Pursuing arguments developed in Chapter 7, it has been suggested in the context of conservation biology that reduced genetic variance will inhibit the ability of a population to track a changing environment (Pease *et al.* 1989), and increase its susceptibility to catastrophic effects of disease (O'Brien & Evermann 1988).

Not only can an estimate of the population size be of immeasurable value in making a genetic prognosis for the population of concern, but we can also use high levels of homozygosity in present day populations to infer past bottlenecks. An unusually compelling example comes from *Actinonyx jubatus*, the cheetah, which suffers extraordinarily low genetic diversity, even in the typically hypervariable major histocompatibility region (O'Brien *et al.* 1985, 1986). More than half the cheetahs in zoo colonies in the United States were killed by an outbreak of virus which causes feline infectious peritonitis. The same disease is unlikely to kill domestic cats, usually killing only about one per cent of the individuals it infects. O'Brien *et al.* (1985) speculate that the virus acclimated to one cheetah and rapidly spread to other cheetahs with uniform immunological defences, so all succumbed.

Is there a minimum genetically viable population?

Effort has been devoted to defining an effective population size above which drift will be ineffective against the replenishment of genetic variance by new mutations. The true rate of mutational input and its significance remains controversial (Chapter 2; Barton & Turelli 1989), but the limited experimental evidence indicates that the number of new mutations affecting a character (ΔV_M) is in the vicinity of 10^{-3} per character per generation (Lande 1976; Barton & Turelli 1989). Because the rate of loss is balanced by the rate of gain when $\Delta V_M = 1/(2N_e)$, or when $N_e = 500$, Franklin (1980) concluded that 500 was the minimum size adequate to retard the effects of drift. On the basis of empirical observations of the effects of inbreeding he concluded that a N_e of 50 would be necessary to avoid depression, giving rise to the so-called 50/500 rule. These data also indicate that the maintenance of genetic variance requires that the effective population size be an order of magnitude greater than would be required to avoid an effect of inbreeding. However, it is also worth noting that free-living species which have suffered genetic impoverishment such as the cheetah still suffer from inbreeding depression. Juvenile mortality in inbred progeny is about 70 per cent higher than in outcrossed progeny (Hedrick 1987; see Charlesworth *et al.* 1990 for discussion of continued origin of genetic variance in inbred populations).

These results have proved highly influential, and have been explicitly incorporated into a number of management programs (see Lande 1988b). However, there are a number of problems with this enthusiasm. First, we do not know the reliability of the estimates of ΔV_M, which have their basis in an eclectic set of traits such as bristle number in *Drosophila*, which are chosen for their tractability in the laboratory (Barton & Turelli 1989). Second, managers rarely know the effective population size, for reasons discussed in Chapter 2, and even when they attempt to adjust for effects of sex ratio, population fluctuations and variance in reproductive success, they often use formulae appropriate only for populations with discrete generations (Lande & Barrowclough 1987; Koenig 1988). Third, Lande & Barrowclough (1987) show that the results are very different for maintenance of single-locus variation, as ΔV_M may be in the order of 10^{-6}, so that N_e may need to exceed 10^5 (Table 10.2). While the maintenance of some neutral single-locus variation may not be crucial (Simberloff 1988), the effect of single genes of large effect, such as those causing resistance to disease and pesticides, cannot be ignored (Lande 1988b).

Quite apart from the problems caused by the genetic assumptions in the 50/500 rule, there are two dangers inherent in the application of guidelines like these in practical conservation biology. First, there is the simple problem that demography may be far more important than genetics in the majority of cases of crisis management. Lande (1988b) provides two instances where the belief that a population size of 500 is adequate to allow indefinite conservation ignores demographic detail, leading to the implementation of management plans in which extinction is the likely outcome.

Perhaps more pernicious is the tendency to believe that once a

Table 10.2 Effective population size (N_e) for mutation to maintain significant quantities of genetic variation at equilibrium. After Lande & Barrowclough (1987).

Type of variation	Nature of selection	Necessary effective population size	Generations for mutation to replenish variation
Quantitative	Neutral	~500	10^2-10^3
Quantitative	Stabilising, or fluctuating optimum	~500	10^2-10^3
Single-locus	Neutral	10^5-10^6	10^5-10^7
Single-locus	Deleterious (incompletely recessive)	Independent of N_e (always present unless inbred)	~10^2

population has fallen below some critical level, be it defined by genetic or demographic criteria, we might as well give up, and gallop on to the next crisis as it arises. As some of the sternest critics of the 50/500 rule have stressed (e.g. Soulé 1987; Simberloff 1988), there is no such thing as a hopeless case in conservation biology, only expensive cases.

The problem of habitat fragmentation

A common consequence of the degradation of natural environments is habitat fragmentation, increasing the prospect that a species will be scattered between several patches. The effects of such fragmentation are usually portrayed as deleterious, among other reasons because the species–area relation indicates that diversity will be greatest in large areas. It is also common for individual species to require a patch of a certain size in order to persist (Fig. 10.9). This has led to the proposition that a single large reserve will generally be superior to several small reserves with equivalent total area (e.g. Diamond 1975), sometimes called the SLOSS principle. Nonetheless, there are several important reservations about whether a single reserve is really superior. First, there is limited empirical evidence relevant to the issue. Although some studies suggest that archipelagoes contain greater total diversity than an equivalent area of mainland habitat (Soulé & Simberloff 1986), they are not very relevant in helping us understand what would happen if there was no mainland population to act as a source of colonisers for the islands (Simberloff 1988). Second, several small reserves afford greater protection against the catastrophic impact of the introduction of a disease or predator to a particular patch. For example, the tiny islands of Shark Bay in Western Australia harbour several mammals which no longer occur anywhere on the mainland (Strahan 1983), presumably because they have been isolated from the effects of habitat degradation and predation by introduced carnivores.

The contribution of genetics to this debate is even less helpful. We have two aims: to maintain genetic diversity and preserve as many alleles as possible, and to avoid inbreeding depression. The solution to inbreeding depression is simple. The number of individuals which can potentially interbreed should be as large as possible, and hence a single reserve is always best. At first glance a similar result might seem to follow for preservation of alleles. However, a reasonably robust result from population genetic models suggests that fragmented populations are more effective at preserving genetic diversity, provided that occasional migration between the subpopulation occurs naturally or is achieved by direct intervention (Fig. 10.10; Kimura & Crow 1964; Berry 1971; Lacy 1987). This is because the random nature of drift will often preserve a different subset of alleles in each subpopulation, and may allow the fixation of otherwise rare alleles; and because local adaptation will create distinct subpopulations, whose difference might otherwise be eroded by gene flow.

Still further complexity is added when the subpopulations differ in

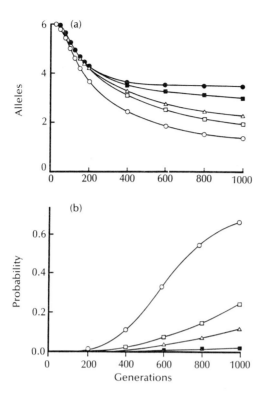

Fig. 10.10 The effects of drift in intact and subdivided populations where the total population contains 250 individuals (○, intact populations; ●, subdivided populations with no migration; □, migration every five generations; △, migration every ten generations; ■, migration every fifty generations). (a) Number of alleles represented. (b) Probabilities of allele fixation. After Boecklen & Bell (1987).

size. Returning to the mating habits of radishes reported in Chapter 7, Ellstrand *et al.* (1989) have recently reported that small isolated populations of radishes are more likely to receive pollen from a distant (>650 m) large population than a nearby (225–400 m) small population. The extent of gene flow was sufficient to play an important role in the evolution of the small populations, and could clearly influence the extent of differentiation between small isolated populations.

Coupled with the Allee effect, balancing these conflicting demands is a major challenge for managers and theoreticians, and may be critical in the case of the management of species whose short- or even long-term survival depends on their maintenance in captivity.

RELEASE OF GENETICALLY ENGINEERED ORGANISMS

There is a tendency among some environmentalists to regard all new technology as a bane invented for the express purpose of destroying nature for the sake of short-term financial gain. Particular umbrage has been directed against the release of new 'organisms' manufactured by recombinant DNA technology, with the obvious analogies between Frankenstein and his creations. The aim of this section is to argue cautiously in favour of recombinant DNA technology, subject to appropriate safety standards being established and observed. There may even be great benefits for the environment, such as the potential use of microorganisms in the destruction of pollutants. However, I also wish to argue that the establishment and monitoring of standards requires a special knowledge of both genetics and ecology, or exactly the type of training I hope is received by evolutionary ecologists.

Stability of recombinant DNA

The most commonly invoked 'evolutionary argument' supporting the inherent safety of newly released engineered organisms is based on the costs of carrying inserted genes. Lenski & Nguyen (1988) call this the Excess Baggage hypothesis, and trace exact parallels in evolutionary arguments advanced by Darwin. Organs are often lost during time in species which have no use for those organs. For example, cave-dwelling fish are often blind and may even lack functioning eyes (Fig. 10.11). Loss of organs in the absence of their use has been attributed to both absence of selection for their function, and active selection against their function. The former effect occurs because deleterious mutations are not eliminated by selection. The latter effect could arise because of the energetic expense of maintenance of function. The costs of maintenance of an organ is likely to be tissue-specific. For example, brain tissue in vertebrates is energetically very expensive, and is reduced in size when not needed (Nottebohm 1981). In *Escherischia coli*, the bacteria most commonly used in recombinant studies, the translational machinery involved in protein synthesis is enormously expensive, and may comprise 40 per cent of the dry weight of the cell (Lewin 1987; Bulmer 1988).

The alternative though less probable outcome is that the temporary reduction in fitness will accelerate adaptive evolution by pushing a population away from one adaptive peak sufficiently far to enable crossing a valley to a new point on the adaptive landscape. There is certainly good evidence that the initial selective disadvantage may be ameliorated by subsequent mutations, and these mutations may occur in completely different parts of the genome (Lenski & Levin 1985; McKenzie, et al. 1982; Clarke & McKenzie 1987). In any case, selection after release will enhance the fitness of the new organism, not reduce it (Tiedje et al. 1989).

The original concern over recombinant organisms centred on the possibility of laboratory escapes of highly virulent pathogenic organisms used in the study of diseases. The response was initiated by geneticists themselves, and has led to an enviable record of industrial safety. Two procedures have proved effective: strict containment adjusted according to the potential risk to human health; and use of physiologically incompetent vectors or hosts for the foreign DNA, rendering survival

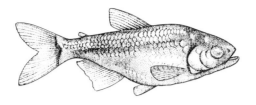

Fig. 10.11 Cave fish like *Anoptichthys jordani* are often blind. After Migdalski & Fichter (1976).

outside the benign conditions of laboratory culture virtually impossible. As Simonsen & Levin (1988) point out, neither safety approach can be extended to the release of genetically engineered organisms. Release and containment are obviously incompatible, and it is also important to note that physiological disablement is likely to prevent the new organism achieving the objective of its release. However, on the positive side, the virulence of new organisms is also totally different.

Rules for introductions

The study of introductions has preoccupied ecologists for many years, and should provide some guidance to the frequency and circumstances in which invasion is likely to be deleterious. Williamson (1988) estimates that introductions are 'successful' for 10 per cent or less of deliberate attempts. Fewer than 10 per cent of these introduced species then pose any problems (become weeds in the broadest sense) so the probability of a new introduction becoming a nuisance may be as little as 0.1 per cent to 1 per cent (see also Simberloff 1981). Even spectacular cases of 'successful' but ultimately catastrophic introductions like the invasion of rabbits in Australia were only successful after several attempts. Genetic engineering has the potential to reduce these dangers, e.g. by enhancing substrate or host specificity. Nonetheless the probability of harm cannot be eliminated (Table 10.3), and the concern of many ecologists centres on the proposed increase in the number of introductions, not on the techniques with which the new organisms are produced.

Pimm (1987) offers three rules for introductions. The impact of introductions are likely to be most severe when:

1 Species are introduced into places when predators are absent, such as islands. The absence of parasites may also facilitate colonisation (Freeland 1983).

2 Species are introduced into relatively simple communities where the removal of a few plant species will cause the collapse of entire food chains. Once again the catastrophic effect of some introductions to islands provide the best example.

3 Highly polyphagous species are introduced.

Table 10.3 The outcomes of introductions of genetically engineered organisms that need to be avoided (Tiedje *et al.* 1989).

---

1   The creation of new pests when an organism normally confined to agricultural land escapes into other habitats as a consequence of its enhanced characteristics
2   Enhancement of the effects of existing pests through hybridisation with related transgenic crop plants
3   Harm to non-target species, either because the new properties affect beneficial species as well as the pests against which they were targeted, or because the new species is an effective predator, competitor, or parasite
4   Adverse effects on ecosystem processes, such as the effects of increased nitrogen assimilation on ecosystem processes
5   Incomplete degradation of a target pollutant. The degradation of trichlorethylene produces the much more toxic vinyl chloride
6   Squandering of natural resources. For example, introduction of bacterial toxins into eukaryotes could accelerate the evolution of resistance to the toxins, rendering the bacteria ineffective as pest control agents

---

Are genetically engineered organisms different?

Beyond this concern over the frequency of introductions lies worry over three aspects of method which distinguish genetically engineered organisms from the products of ordinary artificial selection. First, genomes from completely different lineages can be combined in a single organism (the Fisher/Muller effect carried to extremes). Bacterial genes can be inserted into plant or animals. Second, despite the incorrect impression sometimes held by ecologists that molecular techniques are routine laboratory procedures, there is a possibility that the genetic transfers which prove feasible will also be the genes most likely to transfer between species after the release of the engineered organism. · This might lead, for example, to introduction of a gene causing resistance to insect pests into a weed or potential weed (Table 10.3). Third, the introduction of a completely new gene may have complicated pleiotropic effects that will be overlooked in the excitement of measuring the success and failure of the genetic transfer.

   However, perhaps the overriding concern is the extremely diverse expertise and time scale required to produce the organisms, and then assess the ecological and genetic problems, and the tendency for short-term financial exigencies to override long-term considerations. For example, Tiedje *et al.* (1989) draw attention to the difficulty in assessing whether a microorganism which has completed its function (perhaps degradation of an oil slick) has disappeared from the ecosystem. Indeed, Tiedje *et al.* comment: 'Rather than focus on whether an introduced transgenic organism is likely to disappear completely, emphasis should be placed on whether its population is likely to remain viable or in-crease in size under appropriate environmental conditions.' There is a curious parallel between the problems of conservation biology and concern over the fate of introduced species.

   Tiedje *et al.* go on to list 34 problems that require evaluation on a case-by-case basis in order to judge the safety or harm of a particular introduction, many of which could be applied with considerable benefit to all introductions, regardless of the technology involved. It is a challenge to ecologists to ensure that regulatory authorities demand that this assessment is performed.

GENETIC IMPOVERISHMENT OF AGRICULTURAL SPECIES

To a great extent our agriculture relies on just a few species. Of the 250 000 species of higher plants currently known, just 30 contribute 95 per cent of human nutrition (Sattaur 1989; Fig. 10.12), with a great proportion coming from just four species (wheat, maize, rice and potatoes). Most of the thirty species have been hybridised, inbred and artificially selected for generations. Indeed, one of the most effective ways of bringing a species into cultivation is to select strongly for vegetative reproduction, suppressing sexuality with its potential to produce diverse offspring other than the favoured type (Harlan *et al.* 1973; Harlan 1975; Zohary & Spiegel-Roy 1975; Frankel & Soulé 1981). While a much greater proportion of the world's large herbivorous mammals have been domesticated or exploited for meat, many others have not, despite the availability of several possible candidates (Kemp 1988; Kyle 1990).

We can be proud of our performance in increasing the returns derived from these crops. Recent dramatic improvement in agricultural productivity is in part caused by the success of plant breeders, using artificial selection to enhance output. The generation of new varieties is likely to accelerate as the use of recombinant DNA technology to achieve horizontal transfer of genes between organisms increases.

Although in the past successful cultivation has often initially been associated with an increase in intraspecific diversity as the geographic and climatic range over which the crop is grown expands (Frankel & Soulé 1981), current agricultural practices are tending to work in the opposite direction. The economic success of the agricultural revolution combined with the fundamental economic advantage derived from market monopoly has led to a reduction rather than an increase in the genetic strains used in agriculture. The motivation for governments, farmers and breeders are easy enough to understand. Governments reduce their reliance on exports, farmers boost their cash crops, and breeders achieve market monopoly, particularly when allowed to patent their products, or, in the case of some hybrids, where farmers cannot produce their own seeds. Neither conventional breeding nor the use of DNA technology are cheap or rapid processes, yet the financial rewards from both procedures are potentially great. There have been considerable pressures for the right to patent new varieties in order to preserve the initial investment.

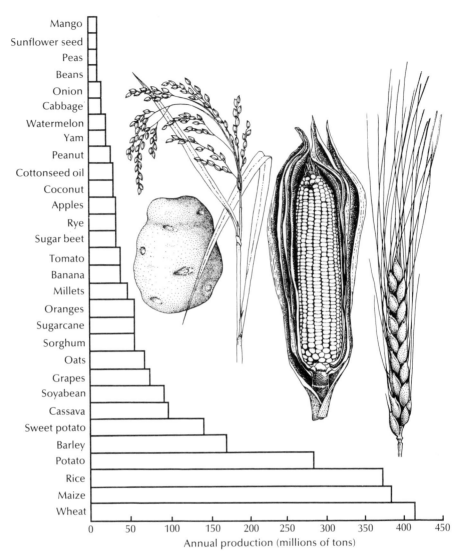

Fig. 10.12 The thirty food crops which contribute 95 per cent of human nutrition. Our diets are dominated by just four species, all of which have been subjected to centuries of hybridisation and inbreeding. The catastrophic impact of potato blight illustrates the consequences of devastation of any one crop by disease. After Sattaur (1989).

Where will the new genes
and agricultural varieties come from?

The strong tendency towards a reduction in species diversity and intraspecific diversity of plants under cultivation leads to three sources of concern, all of which can be directly linked to the theory introduced in the preceding chapters of this text.

1 The prospect of catastrophic failure of production in major food crops increases as the genetic diversity of those crops declines (Chapters 7

and 10). Pathogens and parasites are a special cause for concern (see also Chapter 5), as is the rapidity with which human-induced climate changes are likely to take place. We need only recall the devastating social and economic effects of blight, *Phytophthora infestans* on potatoes, *Solanum tuberosum*, to understand the severity of these problems.

2   Each time a variety of a major food species is taken out of cultivation, or its wild relatives disappear in their original range, a valuable repository of genes which could be used in crop improvement is lost. For example, potato blight apparently evolved in Mexico, where the wild species of *Solanum* are resistant to blight (Frankel & Soulé 1981).

3   In concentrating on only a few species for agriculture, we are substantially underestimating the nutritional warehouse and genetic warehouse provided by the great diversity of life. The vast diversity of plants thus far unstudied by botanists not only provide a source of useful genes (for pathogen resistance, climatic tolerance etc.), but also a pharmacological warehouse which beggars the imagination of the most creative laboratory-based chemist. The reduction in human suffering from malaria caused by the discovery of quinine in the bark of a South American tree (*Cinchona*) is inestimable (Fig. 10.13). The bizarre alkaloids in the rosy periwinkle from Madagascar have greatly increased the chance that human children will survive leukaemia. By any standards, we must assume that these products are scratching the tip of the iceberg (e.g. Fellows 1989). Most of our food and pharmaceutical products come from the tropics, where our ignorance of biodiversity is most pronounced.

Solutions: gene banks and the value of biodiversity

Clearly we should give priority to the preservation of genetic diversity, both among the species on which we currently rely and those on which we might soon rely. Frankel & Soulé (1981) provide an excellent summary of the principles that might be used in maintaining representative collections of seeds or tissue from plants. The increased opportunities for horizontal transfer of genes between species adds even greater strength for their arguments. Obviously the problem with maintaining animals is a greater one, though advances in storage of frozen sperm provide one way of maintaining collections of diversity. The preservation of breeds of livestock which have been virtually discarded from agriculture such as the Gloucester Old Spot pig and the Lakeland Herdwick sheep is obviously costly, but it is increasingly recognised that these oddities have more than just curiosity value, and may contain undetected physiological quirks of enormous value to breeders and genetic engineers (Tudge 1989).

There are many arguments in favour of the preservation of biodiversity (see Wilson 1988 for an excellent review). Unfortunately, the economic pragmatists who dominate our political agenda are prepared

Fig. 10.13 The fruits of *Cinchona*. The bark of this tropical tree gives rise to quinine.

to ignore most of them. The genetic resources of the planet are an economic resource of extreme value. We should lose no opportunity to convince politicians of the importance of preserving those resources.

SUMMARY

As in most pure science, research in evolutionary ecology generates perspectives of great value in applied endeavour. If applied wisely, the base of knowledge should allow us to anticipate the danger or value of our actions, or point out the uncertainties which surround our attempts to manage and manipulate the natural world. Three very new areas of concern illustrate these applications: the design of tactics for the conservation of species in the wild and captivity; the risks of introductions of the products of new technologies like genetic engineering, and the importance of preserving genetic diversity in the species on which we rely for nutrition. It is the responsibility of scientists with relevant knowledge to ensure that these problems are at the forefront of our political agenda.

FURTHER READING

Two texts edited by Soulé (1986, 1987) should be read carefully by all aspiring conservation biologists. The dangers and advantages of recombinant DNA are discussed in a balanced way in a special joint issue of *Trends in Ecology and Evolution* 3, and *Trends in Biotechnology* 6, and in a report commissioned by the Ecological Society of America (Tiedje *et al.* 1989). Frankel & Soulé (1981) is a good account of methods for preserving the genetic diversity of crops.

TOPICS FOR DISCUSSION

[309]
CHAPTER 10
*Applied
evolutionary
ecology*

1 Attempt to classify the avifauna of your state or country using the classification proposed by Rabinowitz *et al.* (1986). Compare your classification with the list of endangered species proposed either by your local ornithological group or government conservation agency. Which type of rarity attracts greatest concern from these groups?

2 The dusky seaside sparrow was deemed worthy of preservation while considered a separate subspecies. That recognition stemmed from its distinctive coloration. Genetic evidence suggests that it is not a separate subspecies (Avise 1989). Debate the criteria which determine whether something is worthy of preservation.

3 Prepare a response to the argument that because extinction is the fate of most organisms, it should not be of overwhelming concern in the management of natural resources.

Chapter 11
The Scope of Evolutionary
Ecology Revisited

Attempting a book of this sort is a salutary task. The rewards arise from being forced to delve into material that would otherwise have escaped my attention, improving the content of my lectures and the state of my mind. The difficulty is that the influx of new ideas and data is causing blurring at the edges of cherished principles that once seemed clearly established, or organisms that once seemed well understood. Despite the enormous progress in evolutionary ecology over the last century, it is no surprise that phrases that recur throughout the text are: 'remain poorly understood', 'await empirical testing' and 'is not clear'. The next generation of evolutionary ecologists are challenged by so many unanswered questions that the field will continue to be a vibrant one. In the following discussion I classify those challenges, and illustrate my classification with some of the problems I have discussed in the text.

The new data sources and tools

Ecology is a science rich in data and method, yet many questions have remained intractable because we lack the methods for probing natural systems at the appropriate level. New technology or methods always offer exciting new directions, and should never be ignored. In almost any area of evolutionary theory, the opportunities offered by the new tools of molecular biology will prove of extraordinary value. Not only will the organisation of the genome be accessible to direct examination, but so will higher order questions such as genealogical organisation of natural populations (Burke 1989). The choice of the cheetah as the vignette for this Chapter represents the fruitful perspectives that have emerged from collaboration between zoo biologists and field ecologists with molecular biologists and virologists (O'Brien *et al.* 1985, 1986; Wildt *et al.* 1987).

The other obvious explosion will come as new species are subjected to scrutiny, when systematists are given the encouragement they rightly deserve, and when the natural history of more species becomes known, rendering them accessible as evolutionary models. *Antechinus stuartii*, the animal to which I have devoted the last ten years of my empirical research, has a population structure more simple than almost any other animal, greatly enhancing its tractability as a study organism (Lee & Cockburn 1985). Although it is the second most common mammal in forests of eastern Australia, nothing was known of its peculiar life history until 25 years ago.

Beyond data to theory

The plethora of data in some areas has clearly outpaced the rate of advancement of theoretical models concerned with explaining that data. The many problems associated with explaining species diversity and range are the best example (Chapter 9).

Beyond theory to experiment

By contrast, in other areas theory is clearly in advance of data and adequate empirical testing. The various hypotheses for the evolution of bizarre male adornment presented in Chapter 7 represent a curious case where the initial interest in the subject was data-driven; an obvious pattern required explanation. However, we are now confronted with a suite of apparently feasible models with no obvious empirical means to distinguish among them. Careful choice of study organism offers the best hope of advancement in such areas.

Assigning frequency

More happily, in certain cases we know that certain evolutionary mechanisms are at least feasible because they have been subjected to empirical and theoretical scrutiny. However, where the feasibility of several important models attempting to explain the same phenomenon has been demonstrated, we need to tackle the question of their relative frequency in nature. Several circumstances plausibly lead to speciation (Chapter 8), or maintain genetic variation in natural populations (Chapter 2), but we know little of their relative importance. Assigning frequencies to their occurrence would be of inestimable value.

Big questions versus little questions

A very great deal of the research in any field is driven by two imperatives: the time it takes to complete a doctoral program, and the duration of funding made available by granting agencies. Brown & Maurer (1989) point out that these influences conspire to prevent attempts at synthesis, at the long-term studies so crucial in ecology (Chapter 6), and in tackling the really big unanswered questions, like why are there so many species (Chapter 9), and can we reduce the diversity of ecological and evolutionary processes into simple classifications (Chapter 5)? We must seek ways to allow researchers to tackle these larger issues, and to encourage the collaboration that ecologists often seem to avoid.

Seeking applications

Last, we must always be prepared to make use of the applications of our knowledge, and beware the temptation to protest ignorance as an excuse for inaction. As Brown & Roughgarden (1989) point out, we may also need to step away from our obvious aesthetic preference for working in beautiful pristine natural ecosystems, and actively consider the evolutionary consequences of human impact in all habitats, before the application of our collective wisdom is no longer possible.

References

Abrahams, M.V. 1986. Patch choice under perceptual constraints: a cause for departures from an ideal free distribution. *Behav. Ecol. Sociobiol.* **19**, 409–15.

Abrams, P.A. 1986a. Adaptive responses of predators to prey and prey to predators: the failure of the arms-race analogy. *Evolution* **40**, 1229–47.

Abrams, P.A. 1986b. Is predator–prey coevolution an arms race? *Trends Ecol. Evol.* **1**, 108–10.

Abrams, P.A. 1987a. On classifying interactions between populations. *Oecologia (Berl).* **73**, 272–81.

Abrams, P.A. 1987b. Alternative models of character displacement and niche shift. 1. Adaptive shifts in resource use when there is competition for nutritionally nonsubstitutable resources. *Evolution* **41**, 651–61.

Abrams, P.A. 1987c. Alternative models of character displacement and niche shift. 2. Displacement when there is competition for a single resource. *Amer. Natur.* **130**, 271–82

Ågren, G. 1984. Incest avoidance and bonding between siblings in gerbils. *Behav. Ecol. Sociobiol.* **14**, 161–9.

Aitkin, M., Anderson, D., Francis, B. & Hinde, J. 1989. *Statistical Modelling in GLIM.* Clarendon Press, Oxford.

Alberch, P. 1980. Ontogenesis and morphological diversification. *Amer. Zool.* **20**, 653–67.

Alberch, P. 1982. Developmental constraints in evolutionary processes. In: J.T. Bonner (ed.) *Evolution and Development*, pp. 313–32. Springer-Verlag, Berlin.

Alberch, P., Gould, S.J., Oster, G.F. & Wake, D.B. 1979. Size and shape in ontogeny and phylogeny. *Paleobiol.* **5**, 296–317.

Albin, R.L. 1988. The pleiotropic gene theory of senescence: supportive evidence from human genetic disease. *Ethol. Sociobiol.* **9**, 371–82.

Alexander, M. 1981. Why microbial predators and parasites do not eliminate their prey and hosts. *Ann. Rev. Microbiol.* **35**, 113–33.

Allee, W.C., Park, O., Emerson, A.E., Park, T. & Scmidt, K.P. 1949. *Principles of Animal Ecology.* W.B. Saunders, Philadelphia.

Allen, J.A. & Anderson, K.P. 1984. Selection by passerine birds is anti-apostatic at high prey density. *Biol. J. Linn. Soc.* **23**, 237–46.

Alvarez, L.W., Alvarez, W., Asaro, F. & Michel, H.V. 1980. Extraterrestrial cause for the Cretaceous–Tertiary mass extinction. *Science* **208**, 1095–108.

Alvarez, W., Kauffman, E.G., Surlyk, F., Alvarez, L.W., Asaro, F. & Michel, H.V. 1984. Impact theory of mass extinctions and the invertebrate fossil record. *Science* **223**, 1135–41.

Anderson, P.R. & Oakeshott, J.G. 1984. Parallel geographical patterns of allozyme variation in two sibling *Drosophila* species. *Nature* **308**, 729–31.

Anderson, R.C. 1972. The ecological relationship of meningeal worm and native cervids in North America *J. Wildl. Dis.* **8**, 304–10.

Anderson, R.M. 1982. Coevolution of hosts and parasites. *Parasitology* **85**, 411–26.

Andersson, M. 1982. Female choice selects for extreme tail length in a widowbird. *Nature* **299**, 818–20.

Andersson, M. 1986. Evolution of condition-dependent sex ornaments and mating preferences: sexual selection based on viability differences. *Evolution* **40**, 804–16.

Andersson, M. 1987. Genetic models of sexual selection: some aims, assumptions and tests. In: J.W. Bradbury & M.B. Andersson (eds.) *Sexual Selection; Testing the Alternatives.* pp. 41–53. John Wiley & Sons, Chichester.

Andrewartha, H.G. & Birch, L.C. 1954. *The Distribution and Abundance of Animals.* University of Chicago Press,

Chicago.

Antonovics, J. 1987. The evolutionary dys-synthesis: which bottles for which wine? *Amer. Natur.* **129**, 321–31.

Antonovics, J. & Ellstrand, N.C. 1984. Experimental studies of the evolutionary significance of sexual reproduction. I. A test of the frequency-dependent selection hypothesis. *Evolution* **38**, 103–15.

Arnold, S.J. 1983. Sexual selection: the interface of theory and empiricism. In: P. Bateson (ed.) *Mate Choice*, pp. 67–107. Cambridge University Press, Cambridge.

Arnold, S.J. & Wade, M.J. 1984a. On the measurement of natural and sexual selection: theory. *Evolution* **38**, 709–19.

Arnold, S.J. & Wade, M.J. 1984b. On the measurement of natural and sexual selection: applications. *Evolution* **38**, 720–34.

Ashmole, N.P. 1963. The regulation of numbers of tropical oceanic birds. *Ibis* **103b**, 458–73.

Atkins, M.D. 1978. *Insects in Perspective*. Macmillan, New York.

Atsatt, P.R. 1981. Lycaenid butterflies and ants: selection for enemy-free space. *Amer. Natur.* **118**, 638–54.

Audubon, J.J. 1953. *The Birds of America*. Macmillan, New York.

Auld, T.D. 1987. Population dynamics of the shrub *Acacia suaveolens* (Sm.) Willd.: survivorship throughout the life cycle, a synthesis. *Aust. J. Ecol.* **12**, 139–51.

Austin, M.P. 1987. Models for the analysis of species' response to environmental gradients. *Vegetatio* **69**, 35–45.

Austin, M.P., Nicholls, A.O. & Margules, C.R. 1990. Measurement of the realized qualitative niche: environmental niches of five *Eucalyptus* species. *Ecol. Monogr.* **60**, 161–77.

Avise, J.C. 1989. A role for molecular genetics in the recognition and conservation of endangered species. *Trends Ecol. Evol.* **4**, 279–81.

Baker, H.G. 1959. Reproductive methods as factors in speciation in flowering plants. Cold Spring Harb. *Symp. quant. Biol.* **24**, 177–91.

Baker, H.G. 1967. Support for Baker's Law—as a rule. *Evolution* **21**, 853–6.

Baker, H.G. 1972. Seed mass in relation to environmental conditions in California. *Ecology* **53**, 997–1010.

Baker, H.G. 1984. Some functions of dioecy in seed plants. *Amer. Natur.* **124**, 149–58.

Bambach, R.K. 1985. Classes and adaptive variety: the ecology of diversification in marine faunas through the Phanerozoic. In: J.W. Valentine (ed.) *Phanerozoic Diversity Patterns: Profiles in Macroevolution*, pp. 191–253. Princeton University Press, Princeton.

Barlow, G.W., Rogers, W. & Fraley, N. 1986. Do Midas cichlids win through prowess or daring? It depends. *Behav. Ecol. Sociobiol.* **19**, 1–8.

Barnes, R.D. 1987. *Invertebrate Zoology*. 5th edn. Saunders College Publishing, Philadelphia.

Barnosky, A.D. 1986. 'Big game' extinction caused by Late Pleistocene climatic change: Irish elk (*Megaloceros giganteus*) in Ireland. *Quat. Res.* **25**, 128–35.

Barrett, S.C.H. 1990. The evolution and adaptive significance of heterostyly. *Trends Ecol. Evol.* **5**, 144–8.

Bartlein, P.J. & Prentice, I.C. 1989. Orbital variations, climate and paleoecology. *Trends Ecol. Evol.* **4**, 195–9.

Barton, N.H. & Charlesworth, B. 1984. Genetic revolutions, founder effects, and speciation. *Ann. Rev. Ecol. Syst.* **15**, 133–64.

Barton, N.H., Jones, J.S. & Mallet, J. 1988. No barriers to speciation. *Nature* **336**, 13–14.

Barton, N.H. & Turelli, M. 1989. Evolutionary quantitative genetics: how little do we know? *Ann. Rev. Genet.* **23**, 337–70.

Bateman, A.J. 1948. Intra-sexual selection in *Drosophila*. *Heredity* **2**, 349–68.

Bateson, P. (ed.) 1983. *Mate Choice*. Cambridge University Press, Cambridge.

Baudinette, R.V. 1989. The biomechanics and energetics of locomotion in Macropodoidea. In: G. Grigg, P. Jarman & I. Hume (eds.) *Kangaroos, Wallabies and Rat-Kangaroos*, pp. 245–53. Surrey Beatty & Son, New South Wales.

Bawa, K. 1980. Evolution of dioecy in flowering plants. *Ann. Rev. Ecol. Syst.*

11, 15–39.

Bawa, K. 1982. Outcrossing and the incidence of dioecy in flowering plants. *Amer. Natur.* **119**, 866–71.

Bazzaz, F.A., Carlson, R.W. & Harper, J.L. 1979. Contribution to the reproductive effort by photosynthesis of flowers and fruits. *Nature* **279**, 554–5.

Beauchamp, G.K., Yamazaki, K. & Boyse, E.A. 1985. The chemosensory recognition of genetic individuality. *Scient. Amer.* **235**(5), 66–72.

Begon, M., Harper, J.L. & Townsend, C.R. 1986. *Ecology: Individuals, Populations and Communities.* Blackwell Scientific Publications, Oxford.

Begon, M. & Mortimer, M. 1986. *Population Ecology: A Unified Study of Plants and Animals.* 2nd edn. Blackwell Scientific Publications, Oxford.

Bell, G. 1982. *The Masterpiece of Nature: the Evolution and Genetics of Sexuality.* Croom Helm, London.

Bell, G. 1984a. Measuring the cost of reproduction. I. The correlation structure of the life table of a planktonic rotifer. *Evolution* **38**, 300–13.

Bell, G. 1984b. Measuring the cost of reproduction. II. The correlation structure of the life tables of five freshwater invertebrates. *Evolution* **38**, 314–26.

Bell, G. 1984c. Evolutionary and non-evolutionary theories of senescence. *Amer. Natur.* **124**, 600–3.

Bell, G. 1987. Two theories of sex and variation. In: S.C. Stearns (ed.) *The Evolution of Sex and its Consequences*, pp. 117–34. Birkhäuser Verlag, Basel.

Bell, G. 1989. A comparative method. *Amer. Natur.* **133**, 553–71.

Bell, G. & Koufopanou, V. 1986. The cost of reproduction. *Oxf. Surv. Evol. Biol.* **3**, 83–131.

Bell, M.A., Baumgartner, J.V. & Olson, E.C. 1985. Patterns of temporal change in single morphological characters of a Miocene stickleback fish. *Paleobiology* **11**, 258–71.

Bender, E.A., Case, T.J. & Gilpin, M.E. 1984. Perturbation experiments in community ecology: theory and practice. *Ecology* **65**, 1–13.

Bender, W., Akam, M., Karch, F., Beachy, P.A., Peifer, M., Spierer, P., Lewis, E.B. & Hogness, D.S. 1983. Molecular genetics of the bithorax complex in *Drosophila melanogaster. Science* **221**, 23–9.

Bengtsson, B.O. 1985. Biased conversion as the primary function of recombination. *Genet. Res.* **47**, 77–80.

Bengtsson, J. 1989. Interspecific competition increases local extinction rate in a metapopulation system. *Nature* **340**, 713–15.

Benton, M.J. 1987. Progress and competition in macroevolution. *Biol. Rev.* **62**, 305–38.

Berenbaum, M. 1983. Coumarins and caterpillars: a case for coevolution. *Evolution* **37**, 163–79.

Berg, R.L. 1960. The ecological significance of correlation pleiades. *Evolution* **14**, 171–80.

Berglund, A. 1986. Sex change by a polychaete: effects of social and reproductive costs. *Ecology* **67**, 837–45.

Bernays, E. & Graham, M. 1988. On the evolution of host specificity in phytophagous arthropods. *Ecology* **69**, 886–92.

Bernstein, H., Hopf, F.A. & Michod, R.E. 1987. The molecular basis for the evolution of sex. *Adv. Genet.* **24**, 323–70.

Bernstein, H., Hopf, F.A. & Michod, R.E. 1988. Is meiotic recombination an adaptation for repairing DNA, producing genetic variation, or both? In: R.E. Michod & B.R. Levin (eds.) *The Evolution of Sex*, pp. 139–60. Sinauer, Sunderland, Massachusetts.

Berry, R.J. 1971. Conservation aspects of the genetical constitution of populations. In: E. Duffey & A.S. Watt (eds.) *The Scientific Management of Animal and Plant Communities for Conservation*, pp. 177–206. Blackwell Scientific Publications, Oxford.

Berven, K.A. 1987. The heritable basis of variation in larval developmental patterns within populations of the wood frog (*Rana sylvatica*). *Evolution* **41**, 1088–97.

Berven, K.A. & Gill, D.E. 1983. Interpreting geographic variation in life-history traits. *Amer. Zool.* **23**, 85–97.

Beverley, S.M. & Wilson, A.C. 1985. Ancient origin for Hawaiian Drosophilinae inferred from protein comparisons. *Proc. Natl. Acad. Sci. USA* **82**, 4753–7.

Bierbaum, T.J., Mueller, L.D. & Ayala,

F.J. 1989. Density-dependent evolution of life-history traits in *Drosophila melanogaster*. *Evolution* **43**, 382–92.

Bierzychudek, P. 1982. The demography of jack-in-the-pulpit, a forest perennial that changes sex. *Ecol. Monogr.* **52**, 335–51.

Bierzychudek, P. 1985. Patterns in plant parthenogenesis. *Experientia* **41**, 1235–45.

Bierzychudek, P. 1987a. Pollinators increase the cost of sex by avoiding female flowers. *Ecology* **68**, 444–7.

Bierzychudek, P. 1987b. Resolving the paradox of sexual reproduction: a review of experimental tests. In: S.C. Stearns (ed.) *The Evolution of Sex and its Consequences*, pp. 163–74. Birkhäuser Verlag, Basel.

Birkhead, T.R. & Hunter, F.M. 1990. Mechanisms of sperm competition. *Trends Ecol. Evol.* **5**, 48–52.

Blackburn, T.M., Harvey, P.H. & Pagel, M.D. 1990. Species number, population density and body size relationships in natural communities. *J. Anim. Ecol.* **59**, 335–45.

Blaustein, A.R., Bekoff, M. & Daniels, T.J. 1987. Kin recognition in vertebrates (excluding primates): empirical evidence. In: D.J. Fletcher & C.D. Michener (eds.) *Kin Recognition in Animals*, pp. 287–331., John Wiley & Sons, Chichester.

Blaustein, A.R., O'Hara, R.K. & Olson, D.H. 1984. Kin preference behaviour is present after metamorphosis in *Rana cascadae* frogs. *Anim. Behav.* **32**, 445–50.

Bodmer, W.F., Trowsdale, J., Young, J. & Bodmer, J. 1986. Gene clusters and the evolution of the major histocompatibility system. *Phil. Trans. R. Soc. Lond.* **B312**, 301–15.

Boecklen, W.J. & Bell, G.W. 1987. Consequences of faunal collapse and genetic drift for the design of nature reserves. In: D.A. Saunders, A.A. Burbidge & A.J.M. Hopkins (eds.) *Nature Conservation: the Role of Natural Remnants*, pp. 141–1. Surrey Beatty & Sons, Sydney.

Bogusz, D., Appleby, C.A., Landsmann, J., Dennis, E.S., Trinick, M.J. & Peacock, W.J. 1988. Functional haemoglobin genes in non-nodulating plants. *Nature* **331**, 178–80.

Bonaventura, J., Bonaventura, C. & Tesh, S. (eds.) 1982. *Physiology and Biology of Horseshoe Crabs: Studies on Normal and Environmentally Stressed Crabs*. Alan R. Liss, New York.

Boorman, E. & Parker, G.A. 1976. Sperm (ejaculate) competition in *Drosophila melanogaster*, and the reproductive value of females to males in relation to female age and mating status. *Ecol. Entomol.* **1**, 145–55.

Borgia, G. 1979. Sexual selection and the evolution of mating systems. In: M.S. & N.A. Blum (eds.) *Sexual Selection and Reproductive Competition in Insects*, pp. 19–80. Academic Press, New York.

Borgia, G. & Collis, K. 1989. Female choice for parasite-free male satin bowerbirds and the evolution of bright male plumage. *Behav. Ecol. Sociobiol.* **25**, 445–54.

Bosbach, K., Hurka, H. & Haase, R. 1982. The soil seed bank of *Capsella bursa-pastoris* (Cruciferae): its influence on population variability. *Flora* **172**, 47–56.

Bowen, G.D. 1980. Mycorrhizal roles in tropical plants and ecosystems. In: P. Mikola (ed.) *Tropical Mycorrhizal Research*, pp. 116–90. Clarendon Press, Oxford.

Bowers, M.A. & Brown, J.H. 1982. Body size and coexistence in desert rodents: chance or community structure. *Ecology* **63**, 391–400.

Boyce, M.S. 1984. Restitution of *r*- and *K*-selection as a model of density-dependent natural selection. *Ann. Rev. Ecol. Syst.* **15**, 427–47.

Boyce, M.S. & Perrins, C.M. 1987. Optimizing great tit clutch size in a fluctuating environment. *Ecology* **68**, 142–57.

Bradbury, J.W. 1977. Lek mating behaviour in the hammer-headed bat. *Z. Tierpsychol.* **45**, 225–55.

Bradbury, J.W. & Andersson, M. (ed.) 1987. *Sexual Selection: Testing the Alternatives*. John Wiley & Sons, Chichester.

Bradbury, J.W., Gibson, R.M. & Tsai, I.M. 1986. Leks and the unanimity of female choice. *Anim. Behav.* **34**, 1694–708.

Bradford, D.F. & Smith, C.C. 1977. Seed predation and seed number in *Scheelea* palm fruits. *Ecology* **58**, 667–73.

Bradshaw, A.D. 1965. Evolutionary significance of phenotypic plasticity in plants. *Adv. Genet.* **13**, 115–55.

Bradshaw, A.D. & McNeilly, T. 1981. *Evolution and Pollution.* Edward Arnold, London.

Bradshaw, W.E. 1986. Pervasive themes in insect life cycle strategies. In: F. Taylor & R. Karban (eds.) *The Evolution of Insect Life Cycles*, pp. 261–75. Springer-Verlag, New York.

Brattsten, L.B., Holyoke, C.W., Leeper, J.R. & Raffa, K.S. 1986. Insecticide resistance: challenge to pest management and basic research. *Science* **231**, 1255–60.

Bremermann, H.J. 1983. Theory of catastrophic diseases of cultivated plants. *J. theor. Biol.* **100**, 255–74.

Bremermann, H.J. 1987. The adaptive significance of sexuality. In: S.C. Stearns (ed.) *The Evolution of Sex and its Consequences*, pp. 135–61. Birkhäuser Verlag, Basel.

Briand, F. & Cohen, J.E. 1987. Environmental correlates of food chain length. *Science* **238**, 956–60.

Briggs, D. & Walters, S.M. 1984. *Plant Variation and Evolution.* 2nd edn. Cambridge University Press, Cambridge.

British Museum of Natural History. 1983. *British Mesozoic Fossils.* British Museum (Natural History), London.

Brockmann, H.J. 1984. The evolution of social behaviour in insects. In: J.R. Krebs & N.B. Davies (eds.) *Behavioural Ecology: An Evolutionary Approach.* 2nd edn., pp. 340–61. Blackwell Scientific Publications, Oxford.

Brooke, M. de L. & Davies, N.B. 1987. Recent changes in host usage by cuckoos *Cuculus canorus* in Britain. *J. Anim. Ecol.* **56**, 873–83.

Brooke, M. de L. & Davies, N.B. 1988. Egg mimicry by cuckoos *Cuculus canorus* in relation to discrimination by hosts. *Nature* **335**, 630–2.

Brouwers, E.M., Clemens, W.A., Spicer, R.A., Ager, T.A., Carter, D. & Sliter, W.V. 1987. Dinosaurs on the North Slope, Alaska: high latitude, latest Cretaceous environments. *Science* **237**, 1608–10.

Brown, A.L. 1987. Positively darwinian molecules? *Nature* **326**, 12–13.

Brown, D. 1988. Components of reproductive success. In: T.H. Clutton-Brock (ed.) *Reproductive Success: Studies on Individual Variation in Contrasting Breeding Systems*, pp. 439–53. University of Chicago Press, Chicago.

Brown, J.H. 1975. Geographical ecology of desert rodents. In: M.L. Cody & J.M. Diamond (eds.) *Ecology and Evolution of Communities*, pp. 315–41. Cambridge University Press, Cambridge.

Brown, J.H. 1981. Two decades of homage to Santa Rosalia: toward a general theory of diversity. *Amer. Zool.* **21**, 877–88.

Brown, J.H. 1984. On the relationship between abundance and distribution of species. *Amer. Natur.* **124**, 255–79.

Brown, J.H. 1987. Variation in desert rodent guilds: patterns, processes and scales. In: J.H.R. Gee & P.S. Giller (eds.) *Organization of Communities: Past and Present*, pp. 185–203. Blackwell Scientific Publications, Oxford.

Brown, J.H. & Gibson, A.C. 1983. *Biogeography.* Mosby, St Louis.

Brown, J.H. & Maurer, B.A. 1986. Body size, ecological dominance and Cope's rule. *Nature* **324**, 248–50.

Brown, J.H. & Maurer, B.A. 1987. Evolution of species assemblages: effects of energetic constraints and species dynamics on the diversification of the North American avifauna. *Amer. Natur.* **130**, 1–17.

Brown, J.H. & Maurer, B.A. 1989. Macroecology: the division of food and space among species on continents. *Science* **243**, 1145–50.

Brown, J.H. & Roughgarden, J. 1989. US ecologists address global change. *Trends Ecol. Evol.* **4**, 255–6.

Brown, J.L. 1987. *Helping and Communal Breeding in Birds: Ecology and Evolution.* Princeton University Press, Princeton.

Brown, J.S. & Venable, D.L. 1986. Evolutionary ecology of seed-bank annuals in temporally varying environments. *Amer. Natur.* **127**, 31–47.

Bryant, E.H., McCommas, S.A. & Combs, L.M. 1986. The effect of an experimental bottleneck on quantitative genetic variation in the housefly. *Genetics* **114**, 1191–211.

Bull, J.J. 1983. *Evolution of Sex Determining Mechanisms.* Benjamin/Cummings, Menlo Park, California.

Bull, J.J. & Harvey, P.H. 1989. A new reason for having sex. *Nature* **339**, 260–1.

Bulmer, M.G. 1988. Evolutionary aspects of protein synthesis. *Oxford Surv. Evol. Biol.* **5**, 1–40.

Bulmer, M.G. 1989. Maintenance of genetic variability by mutation–selection balance: a child's guide through the jungle. *Genome* **31**, 761–7.

Burbidge, A.A., Johnson, K.A., Fuller, P.J. & Southgate, R.I. 1988. Aboriginal knowledge of the mammals of the central deserts of Australia. *Aust. Wildl. Res.* **15**, 9–40.

Burdon, J.J. 1987a. Phenotypic and genetic patterns of resistance to the pathogen *Phaksopora pachyrhizi* in populations of *Glycine clandestina. Oecologia* **73**, 257–67.

Burdon, J.J. 1987b. *Diseases and Plant Population Biology.* Cambridge University Press, Cambridge.

Burdon, J.J. & Jarosz, A.M. 1988. The ecological genetics of plant–pathogen interactions in natural communities. *Phil. Trans. R. Soc. Lond.* **B321**, 349–63.

Burdon, J.J., Jarosz, A.M. & Kirby, G.C. 1989. Pattern and patchiness in plant–pathogen interactions—causes and consequences. *Ann. Rev. Ecol. Syst.* **20**, 119–36.

Burdon, J.J. & Marshall, D.R. 1981. Biological control and the reproductive mode of weeds. *J. Appl. Ecol.* **18**, 649–59.

Bürger, R., Wagner, G.P. & Stettinger, F. 1989. How much heritable variation can be maintained in finite populations by mutation–selection balance. *Evolution* **43**, 1748–66.

Burgman, M.A. 1989. The habitat volumes of scarce and ubiquitous plants: a test of the model of environmental control. *Amer. Natur.* **133**, 228–39.

Burke, T. 1989. DNA fingerprinting and other methods for the study of mating success. *Trends Ecol. Evol.* **4**, 139–44.

Burke, T., Davies, N.B., Bruford, M.W. & Hatchwell, B.J. 1989. Parental care and mating behaviour of polyandrous dunnocks *Prunella modularis* related to paternity by DNA fingerprinting. *Nature* **338**, 249–51.

Burt, A. & Bell, G. 1987a. Mammalian chiasma frequencies: a critical test of two theories of recombination. *Nature* **326**, 803–5.

Burt, A. & Bell, G. 1987b. Red Queen versus Tangled Bank models. Burt and Bell reply. *Nature* **330**, 118.

Bush, G.L. 1966. Taxonomy, cytology and evolution of the genus *Rhagoletis* in North America. *Bull. Harvard Mus. Comp. Zool.* **134**, 431–562.

Bush, G.L. 1974. The mechanism of sympatric host race formation in the true fruit flies. In: M.J.D. White (ed.) *Genetic Mechanisms of Speciation in Insects*, pp. 3–23. Australian and New Zealand Book Company, Sydney.

Bush, G.L. 1975. Modes of animal speciation. *Ann. Rev. Ecol. Syst.* **6**, 339–64.

Buss, L.W. 1983. Evolution, development, and the units of selection. *Proc. Natl. Acad. Sci. USA* **80**, 1387–91.

Buss, L.W. 1987. *The Evolution of Individuality.* Princeton University Press, Princeton.

Buss, L.W. 1988. Diversification and germ-line determination. *Paleobiology* **14**, 313–21.

Butlin, R. 1987. Speciation by reinforcement. *Trends Ecol. Evol.* **2**, 8–13.

Caccone, A. & Powell, J.R. 1989. DNA divergence among hominoids. *Evolution* **43**, 925–42.

Cairns, J., Overbaugh, J. & Miller, S. 1988. The origin of mutants. *Nature* **335**, 142–5

Caisse, M. & Antonovics, J. 1978. Evolution in closely adjacent plant populations. IX. Evolution of reproductive isolation in clinal populations. *Heredity* **40**, 371–84.

Calder, W.A. 1984. *Size, Function and Life History.* Harvard University Press, Cambridge, Massachussets.

Caldwell, M.M., Teramura, A.H. & Tevini, M. 1989. The changing solar ultraviolet climate and the ecological consequences for higher plants. *Trends Ecol. Evol.* **4**, 363–7.

Calow, P. 1979. The cost of reproduction—a physiological approach. *Biol. Rev.* **54**, 23–40.

Caplan, A.L. 1977. Tautology, circularity, and biological theory. *Amer. Natur.* 111, 390–3.

Carson, H.L. 1982. Speciation as a major reorganization of polygenic behaviour. In: C. Borigozzi (ed.) *Mechanisms of Speciation*, pp. 411–33. Alan Liss, New York.

Carson, H.L. 1990. Increased genetic variance after a population bottleneck. *Trends Ecol. Evol.* 5, 228–9.

Carson, H.L. & Templeton, A.R. 1984. Genetic revolutions in relation to speciation phenomena: the founding of new populations. *Ann. Rev. Ecol. Syst.* 15, 97–131.

Caraco, T. & Wolf, L.L. 1975. Ecological determinants of group size of foraging lions. *Amer. Natur.* 109, 343–52.

Caswell, H. 1978. Predator-mediated coexistence: a non-equilibrium model. *Amer. Natur.* 112, 127–54.

Chandler, C.R. & Gromko, M.H. 1989. On the relationship between species concepts and speciation processes. *Syst. Zool.* 38, 116–25.

Chapman, R.F. 1982. Chemoreception: the significance of receptor numbers. *Adv. Insect Physiol.* 16, 247–356.

Charlesworth, B. 1980. *Evolution in Age-Structured Populations*. Cambridge University Press, Cambridge.

Charlesworth, B. 1984a. Evolutionary genetics of life histories. In: B. Shorrocks (ed.) *Evolutionary Ecology*, pp. 117–33. Blackwell Scientific Publications, Oxford.

Charlesworth, B. 1984b. The cost of phenotypic evolution. *Paleobiology* 10, 319–27.

Charlesworth, B. 1987a. The population biology of transposable elements. *Trends Ecol. Evol.* 2, 21–3.

Charlesworth, B. 1987b. The heritability of fitness. In: J.W. Bradbury & M.B. Andersson (eds.) *Sexual Selection: Testing the Alternatives*, pp. 21–40. Wiley, Chichester.

Charlesworth, B. 1988. The evolution of mate choice in a fluctuating environment. *J. theor. Biol.* 130, 191–204.

Charlesworth, B. 1990. Optimization models, quantitative genetics, and mutation. *Evolution* 44, 520–38.

Charlesworth, B., Charlesworth, D. & Morgan, M.T. 1990. Genetic loads and estimates of mutation rates in highly inbred plant populations. *Nature* 347, 380–2.

Charlesworth, B., Lande, R. & Slatkin, M. 1982. A neo-Darwinian commentary on macroevolution. *Evolution* 36, 474–98.

Charlesworth, D. 1985. Distribution of dioecy and self-compatibility in angiosperms. In: P.J. Greenwood, P.H. Harvey & M. Slatkin (eds.) *Evolution: Essays in Honour of John Maynard Smith*, pp. 237–68. Cambridge University Press, Cambridge.

Charlesworth, D. & Charlesworth, B. 1987. Inbreeding depression and its evolutionary consequences. *Ann. Rev. Ecol. Syst.* 18, 237–68.

Charnov, E.L. 1976. Optimal foraging, the marginal value theorem. *Theor. Popul. Biol.* 9, 129–36.

Charnov, E.L. 1979. Natural selection and sex change in a pandalid shrimp: a test of life history theory. *Amer. Natur.* 113, 715–34.

Charnov, E.L. 1982. *The Theory of Sex Allocation*. Princeton University Press, Princeton.

Charnov, E.L. & Krebs, J.R. 1974. On clutch-size and fitness. *Ibis* 116, 217–19.

Charnov, E.L. & Schaffer, W.M. 1973. Life history consequences of natural selection: Cole's result revisited. *Amer. Natur.* 107, 291–303.

Chatterton, B.D.E. & Speyer, S.E. 1989. Larval ecology, life history strategies, and patterns of extinction and survivorship among Ordovician trilobites. *Paleobiology* 15, 118–32.

Cheetham, A.H. 1986. Tempo of evolution in a Neogene bryozoan: rates of morphologic change within and across species boundaries. *Paleobiology* 12, 190–202.

Chesson, P.L. & Case, T.J. 1986. Overview: nonequilibrium community theories: chance, variability, history and coexistence. In: J. Diamond & T. Case (eds.) *Community Ecology*, pp. 229–39. Harper & Row, New York.

Chesson, P.L. & Huntly, N. 1989. Short-term instabilities and long-term community dynamics. *Trends Ecol. Evol.* 4, 293–8.

Cheverud, J.M. 1984. Quantitative genetics and developmental constraints on evolution by selection. *J. Theor. Biol.* **110**, 155–71.

Cheverud, J.M. 1988. A comparison of genetic and phenotypic correlations. *Evolution* **42**, 958–68.

Cheverud, J.M., Dow, M.M. & Leutenegger, W. 1985. The quantitative assessment of phylogenetic constraints in comparative analyses: sexual dimorphism in body weight among primates. *Evolution* **39**, 1335–51.

Clark, C.W. 1987. The lazy, adaptable lions: a Markovian model of group foraging. *Anim. Behav.* **35**, 361–8.

Clark, D.L., Cheng-Yuan, W., Orth, C.J. & Gilmore, J.S. 1986. Conodont survival and low iridium abundances across the Permian–Triassic boundary in south China. *Science* **233**, 984–6.

Clarke, B.C. 1979. The evolution of genetic diversity. *Proc. R. Soc. Lond.* **B205**, 453–74.

Clarke, G.M. & McKenzie, J.A. 1987. Developmental stability of insecticide resistant phenotypes in blowfly; a result of canalizing natural selection. *Nature* **325**, 345–65.

Clarkson, E.N.K. 1979. *Invertebrate Paleontology and Evolution*. George Allen & Unwin, Boston.

Clegg, M.T., Kahler, A.T. & Allard, R.W. 1978. Estimation of life cycle components of selection in an experimental plant population. *Genetics* **89**, 765–92.

Clutton-Brock, T.H. 1983. Selection in relation to sex. In: D.S. Bendall (ed.) *Evolution from Molecules to Men*, pp. 457–81. Cambridge University Press, Cambridge.

Clutton-Brock, T.H. 1984. Reproductive effort and terminal investment in iteroparous animals. *Amer. Natur.* **123**, 212–29.

Clutton-Brock, T.H. (ed.) 1988. *Reproductive Success: Studies on Individual Variation in Contrasting Breeding Systems*. University of Chicago Press, Chicago.

Clutton-Brock, T.H. 1989. Mammalian mating systems. *Proc. R. Soc. Lond.* **B236**, 339–72.

Clutton-Brock, T.H., Albon, S.D. & Guinness, F.E. 1984. Maternal dominance, breeding success and birth sex ratios in red deer. *Nature* **308**, 358–60.

Clutton-Brock, T.H., Albon, S.D. & Guinness, F.E. 1985. Parental investment and sex differences in juvenile mortality in birds and mammals. *Nature* **313**, 131–3.

Clutton-Brock, T.H., Albon, S.D. & Guinness, F.E. 1986. Great expectations: dominance, breeding success and offspring sex ratios in red deer. *Anim. Behav.* **34**, 460–71.

Clutton-Brock, T.H., Guinness, F.E. & Albon, S.D. 1982. *Red Deer – Behavior and Ecology of Two Sexes*. Edinburgh University Press, Edinburgh.

Clutton-Brock, T.H. & Harvey, P.H. 1977. Primate ecology and social organization. *J. Zool.* **183**, 1–33.

Cockburn, A. 1988. *Social Behaviour in Fluctuating Populations*. Croom Helm, London.

Cockburn, A., Lee, A.K. & Martin, R.W. 1983. Macrogeographic variation in litter size in *Antechinus* spp. (Marsupialia: Dasyuridae). *Evolution* **37**, 86–95.

Cockburn, A., Scott, M.P. & Scotts, D.J. 1985. Inbreeding avoidance and male-biased natal dispersal in *Antechinus* spp. (Marsupialia: Dasyuridae). *Anim. Behav.* **33**, 908–15.

Cody, M.L. 1966. A general theory of clutch size. *Evolution* **20**, 174–84.

Cohen, J. 1985. Metamorphosis: introduction, usages and evolution. In: M. Bulls & M. Brownes (eds.) *Metamorphosis*. Clarendon Press, Oxford.

Cohen, J.E., Briand, F. & Newman, C.M. 1990. *Community Food Webs: Data and Theory*. Biomathematics. Springer-Verlag, Berlin.

Cohen, J.E., Newman, C.M. & Briand, F.E. 1985. A stochastic theory of food webs. II. Individual webs. *Proc. R. Soc. Lond.* **B224**, 449–61.

Cole, L.C. 1954. The population consequences of life history phenomena. *Q. Rev. Biol.* **25**, 103–27.

Collett, C. 1988. Recent origin for a thermostable alcohol dehydrogenase allele of *D. melanogaster*. *J. Mol. Evol.* **27**, 142–6.

Colwell, R.K. 1986. Population structure and sexual selection for host fidelity in the speciation of hummingbird flower mites. In: S. Karlin & E. Nevo (eds.) *Evolutionary Processes and Theory*, pp. 475–95. Academic Press,

New York.

Colwell, R.K. & Winkler, D.W. 1984. A null model for null models in biogeography. In: D.R. Strong, D. Simberloff, L.G. Abele & A.B. Thistle (eds.) *Ecological Communities: Conceptual Issues and the Evidence*, pp. 344–59. Princeton University Press, Princeton.

Connell, J.H. 1978. Diversity in tropical rain forests and coral reefs. *Science* 199, 1302–9.

Connell, J.H. 1980. Diversity and the coevolution of competitors, or the ghost of competition past. *Oikos* 35, 131–8.

Connell, J.H. 1983. On the prevalence and relative importance of interspecific competition – evidence from field experiments. *Amer. Natur.* 122, 661–96.

Connor, E.F. 1986. The role of Pleistocene forest refugia in the evolution and biogeography of tropical biotas. *Trends Ecol. Evol.* 1, 165–8.

Conway Morris, S. 1977. A new metazoan from the Cambridge Burgess Shale of British Columbia. *Paleontology* 20, 623–40.

Conway Morris, S. 1979. The Burgess Shale (Middle Cambrian) fauna. *Ann. Rev. Ecol. Syst.* 10, 327–49.

Conway Morris, S. 1985. The middle Cambrian metazoan *Wiwaxia corrugata* (Matthew) from the Burgess Shale and *Ogyopsis* Shale, British Columbia, Canada. *Phil Trans. R. Soc. Lond.* B307, 507–86.

Conway Morris, S. 1989. Burgess Shale faunas and the Cambrian explosion. *Science* 246, 339–46.

Cooke, P.H. & Oakeshott, J.G. 1989. Amino acid polymorphisms for esterase-6 in *Drosophila melanogaster*. *Proc. Natl. Acad. Sci. USA* 86, 1426–30.

Cooke, P.H., Richmond, R.C. & Oakeshott, J.G. 1987. High resolution electrophoretic variation at the esterase-6 locus in a natural population of *Drosophila melanogaster*. *Heredity* 55, 259–64.

Coope, G.R. 1987. The response of Late Quaternary insect communitiies to sudden climatic changes. In: J.H.R. Gee & P.S. Giller (eds.) *Organization of Communities: Past and Present*, pp. 421–38. Blackwell Scientific Publications, Oxford.

Coulson, J.C. & Thomas, C. 1985. Differences in the breeding performance of individual kittiwake gulls, *Rissa tridactyla* (L.). In: R.M. Sibly & R.H. Smith (eds.) *Behavioural Ecology: Ecological Consequences of Adaptive Behaviour*, pp. 489–503. Blackwell Scientific Publications, Oxford.

Courtney, S. 1984. The evolution of batch oviposition by Lepidoptera and other insects. *Amer. Natur.* 123, 276–81.

Courtney, S. 1988. If it's not coevolution, it must be predation? *Ecology* 69, 910–11.

Coyne, J.A. 1990. Endless forms most beautiful. *Nature* 344, 30.

Coyne, J.A. & Orr, H.A. 1989. Patterns of speciation in *Drosophila*. *Evolution* 43, 362–81.

Cox, F.E.G. 1989. Parasites and sexual selection. *Nature* 341, 289.

Crane, P.R. & Lidgard, S. 1989. Angiosperm diversification and paleolatitudinal gradients in Cretaceous floristic diversity. *Science* 246, 675–8.

Crosland, M.W.J. & Crozier, R.H. 1986. *Myrmecia pilosula*, an ant with only one pair of chromosomes. *Science* 231, 1278.

Crow, J.F., Engels, W.R. & Denniston, C. 1990. Phase three of Wright's shifting balance theory. *Evolution* 44, 233–47.

Crowl, T.A. & Covich, A.P. 1990. Predator-induced life-history shifts in a freshwater snail. *Science* 247, 949–51.

Crump, M.L. 1981. Variation in propagule size as a function of environmental uncertainty for tree frogs. *Amer. Natur.* 117, 724–37.

Cruz, Y.P. 1981. A sterile defender morph in a polyembryonic hymenopterous parasite. *Nature* 294, 446–7.

Currie, D.J. & Paquin, V. 1987. Large-scale biogeographical patterns of species richness in trees. *Nature* 329, 326–7.

Damuth, J. 1981. Population density and body size in small mammals. *Nature* 290, 699–700.

Damuth, J. 1987. Interspecific allometry of population density in mammals and other animals: the independence of body mass and population energy use. *Biol. J. Linn. Soc.* 31, 193–246.

Darwin, C. 1859. *On the Origin of Species by means of Natural Selection or the Preservation of Favored Races in the Struggle for Life.* John Murray, London.

Darwin, C. 1862. *On the Various Contrivances by which British and Foreign Orchids are Fertilized by Insects.* John Murray, London.

Darwin, C. 1871. *The Descent of Man and Selection in Relation to Sex.* John Murray, London.

Darwin, C. 1874. *The Descent of Man and Selection in Relation to Sex.* 2nd edn. John Murray, London.

Darwin, C. 1877. *The Different Forms of Flowers on Plants of the Same Species.* John Murray, London.

Davies, N.B. 1983. Polyandry, cloaca-pecking and sperm competition in dunnocks. *Nature* **302**, 334–6.

Davies, N.B. 1985. Cooperation and conflict among dunnocks, *Prunella modularis*, in a variable mating system. *Anim. Behav.* **33**, 628–48.

Davies, N.B. 1986. Reproductive success of dunnocks, *Prunella modularis* in a variable mating system. I. Factors influencing provisioning rate, nestling weight and fledging success. *J. Anim. Ecol.* **55**, 139–54.

Davies, N.B. & Brooke, M de L. 1989. An experimental study of co-evolution between the cuckoo, *Cuculus canorus*, and its hosts. II. Host egg markings, chick discrimination and general discussion. *J. Anim. Ecol.* **58**, 225–36.

Davies, N.B. & Houston, A.I. 1986. Reproductive success of dunnocks, *Prunella modularis* in a variable mating system. II. Conflicts of interest among breeding adults. *J. Anim. Ecol.* **55**, 139–54.

Davies, N.B. & Lundberg, A. 1984. Food distribution and a variable mating system in the dunnock, *Prunella modularis*. *J. Anim. Ecol.* **53**, 895–912.

Dawkins, R. 1976. *The Selfish Gene.* Oxford University Press, Oxford.

Dawkins, R. 1979. Twelve misunderstandings of kin selection. *Z. Tierpsychol.* **51**, 184–200.

Dawkins, R. 1981. *The Extended Phenotype: the Gene as the Unit of Selection.* W.H. Freeman, Oxford.

Dawkins, R. 1986. *The Blind Watchmaker.* Longman Scientific & Technical, Harlow, Essex.

Dayan, T., Simberloff, D., Tchernov, E. & Yom-Tov, Y. 1989. Inter- and intraspecific character displacement in mustelids. *Ecology* **70**, 1526–39.

Dayan, T., Simberloff, D., Tchernov, E. & Yom-Tov, Y. 1990. Feline canines: community-wide character displacement among the small cats of Israel. *Amer. Natur.* **136**, 39–60.

De Salle, R. & Templeton, A.R. 1988. Founder effects and the rate of mitochondrial DNA evolution in Hawaiian *Drosophila*. *Evolution* **42**, 1076–84.

Dempster, J.P. 1983. The natural control of populations of butterflies and moths. *Biol. Rev.* **58**, 461–81.

Diamond, J. 1969. Avifaunal equilibria and species turnover rates on the Channel Islands of California. *Proc. Natl. Acad. Sci. USA* **64**, 57–63.

Diamond, J. 1975. The island dilemma: lessons of modern biogeographic studies for the design of nature reserves. *Biol. Conserv.* **7**, 129–46.

Dingle, H. 1986. The evolution of insect life cycle syndromes. In: F. Taylor & R. Karban (eds.) *The Evolution of Insect Life Cycles*, pp. 187–203. Springer-Verlag, New York.

Dingle, H., Brown, C.K. & Hegmann, J.P. 1977. The nature of genetic variance influencing photoperiodic diapause in a migratory insect, *Oncopeltus fasciatus*. *Amer. Natur.* **111**, 1047–59.

Dirzo, R. & Harper, J.L. 1982a. Experimental studies on slug–plant interactions. IV. The performance of cyanogenic and acyanogenic morphs of *Trifolium repens* in the field. *J. Ecol.* **70**, 119–38.

Dirzo, R. & Harper, J.L. 1982b. Experimental studies on slug–plant interactions. III. Differences in the acceptability of individual plants of *Trifolium repens* to slugs and snails. *J. Ecol.* **70**, 101–17.

Dixon, A.F.G. & Kindlmann, P. 1990. Role of plant abundance in determining the abundance of herbivorous insects. *Oecologia* **83**, 282–3.

Dobson, A.P. & Hudson, P.J. 1986. Parasites, disease and the structure of ecological communities. *Trends Ecol. Evol.* **1**, 11–15.

Dobson, A.P., Jolly, A. & Rubenstein, D.

1989. The Greenhouse effect and biological diversity. *Trends Ecol. Evol.* **4**, 64–8.

Dobzhansky, T. 1937. *Genetics and the Origin of Species.* Columbia University Press, New York.

Dobzhansky, T. 1941. *Genetics and the Origin of Species.* 2nd edn. Columbia University Press, New York.

Dobzhansky, T. 1970. *Genetics of the Evolutionary Process.* Columbia University Press, New York.

Dobzhansky, T. 1973. Nothing in biology makes sense except in the light of evolution. *Amer. Biol. Teacher* **35**, 125–29.

Dobzhansky, T., Ayala, F.J., Stebbins, G.L. & Valentine, J.W. 1977. *Evolution.* W.H. Freeman, San Francisco.

Doolittle, W.F. 1987. The origin and function of intervening sequences in DNA: a review. *Amer. Natur.* **130**, 915–28.

Doolittle, W.F. & Sapienza, C. 1980. Selfish genes, the phenotype paradigm and genome evolution. *Nature* **284**, 601–3.

Dover, G.A. 1982. Molecular drive, a cohesive mode of species evolution. *Nature* **299**, 111–17.

Dover, G.A. 1986a. The spread and success of non-Darwinian novelties. In: S. Karlin & E. Nevo (eds.) *Evolutionary Processes and Theory,* pp. 199–237. Academic Press, London.

Dover, G.A. 1986b. Molecular drive in multigene families: how biological novelties arise, spread and are assimilated. *Trends in Genetics* **2**, 159–65.

Dover, G.A. 1988. Evolving the improbable. *Trends Ecol. Evolution* **3**, 81–4.

Dow, D.D. 1980. Communally breeding Australian birds with an analysis of distributional and altitudinal factors. *Emu* **80**, 121–40.

Downhower, J.F., Blumer, L.S. & Brown, L. 1987. Opportunity for selection: an appropriate measure for evaluating variation in the potential for selection? *Evolution* **41**, 1395–400.

Edmunds, G.F. & Alstad, D.N. 1978. Coevolution in insect herbivores and conifers. *Science* **199**, 941–5.

Edney, E.B. & Gill, R.W. 1968. Evolution of senescence and specific longevity. *Nature* **220**, 281–2.

Egid, A. & Lenington, S. 1985. Responses of male mice to odors of females: effects of T- and H-2-locus genotype. *Behav. Gen.* **15**, 287–95.

Ehlinger, T.J. & Wilson, D.S. 1988. Complex foraging polymorphism in bluegill sunfish. *Proc. Natl. Acad. Sci. USA* **85**, 1878–82.

Ehrlich, P.R. 1984. The structure and dynamics of butterfly populations. *Symp. Roy. ent. Soc. Lond.* **11**, 25–40.

Ehrlich, P.R. & Raven, P.H. 1964. Butterflies and plants: a study in coevolution. *Evolution* **18**, 586–608.

Eichler, W. 1948. Some rules in ectoparasitism. *Ann. Mag. Nat. Hist.* **12**, 588–98.

Eldredge, N. 1985. *Time Frames.* Simon & Schuster, New York.

Eldredge, N. & Gould, S.J. 1972. Punctuated equilibria: an alternative to phyletic gradualism. In: T.J.M. Schopf (ed.) *Models in Palaeobiology,* pp. 82–115. Freeman & Cooper, San Francisco.

Ellstrand, N.C. & Antonovics, J. 1985. Experimental studies of the evolutionary significance of sexual reproduction. II. A test of the density-dependent selection hypothesis. *Evolution* **39**, 657–66.

Ellstrand, N.C., Devlin, B. & Marshall, D.L. 1989. Gene flow by pollen into small populations: data from experimental and natural stands of wild radish. *Proc. Natl. Acad. Sci. USA* **86**, 9044–7.

Ellstrand, N.C. & Marshall, D.L. 1986. Patterns of multiple paternity in populations of *Raphanus sativus. Evolution* **40**, 837–42.

Ellstrand, N.C. & Roose, M.L. 1987. Patterns of genotypic diversity in clonal plant populations. *Amer. J. Bot.* **74**, 123–31.

Endler, J.A. 1977. *Geographic Variation, Speciation, and Clines.* Princeton University Press, Princeton.

Endler, J.A. 1986a. *Natural Selection in the Wild.* Princeton University Press, Princeton.

Endler, J.A. 1986b. The newer synthesis? Some conceptual problems in evolutionary biology. *Oxford Rev. Evol. Biol.* **3**, 224–43.

Endler, J.A. & McLellan, T. 1988. The processes of evolution: towards a newer synthesis. *Ann. Rev. Ecol. Syst.*

19, 395–421.

Engelhard, G., Foster, S.P. & Day, T.H. 1989. Genetic differences in mating success and female choice in seaweed flies (*Coelopa frigida*). *Heredity* **62**, 123–31.

Erwin, D.H. 1989. The end-Permian mass extinction: what really happened and did it really matter. *Trends Ecol. Evol.* **4**, 225–9.

Erwin, D.H., Valentine, J.W. & Sepkoski, J.J. 1987. A comparative study of diversification events: the early Paleozoic versus the Mesozoic. *Evolution* **41**, 1177–86.

Erwin, T.L. 1982. Tropical forests: their richness in Coleoptera and other arthropod species. *Coleopt. Bull.* **36**, 74–5.

Erwin, T.L. 1983. Beetles and other arthropods of the tropical forest canopies at Manaus, Brasil, sampled with insecticidal fogging techniques. In: S.L. Sutton, T.C. Whitmore & A.C. Chadwick (eds.) *Tropical Rain Forests: Ecology and Management*, pp. 59–75. Blackwell Scientific Publications, Oxford.

Erwin, T.L. 1988. The tropical forest canopy: the heart of biotic diversity. In: E.O. Wilson (ed.) *Biodiversity*, pp. 123–9. National Academy Press, Washington, DC.

Eshel, I. & Hamilton, W.D. 1984. Parent–offspring correlation in fitness under fluctuating selection. *Proc. R. Soc. Lond.* **B222**, 1–14.

Etter, R.J. 1988. Asymmetrical developmental plasticity in an intertidal snail. *Evolution* **42**, 322–34.

Ewald, P.W. 1983. Host–parasite relations, vectors, and the evolution of disease severity. *Ann. Rev. Ecol. Syst.* **14**, 465–85.

Ewald, P.W. 1987. Transmission modes and evolution of the parasitism–mutualism controversy. *Ann. NY Acad. Sci.* **503**, 295–306.

Ewald, P.W. 1988. Cultural vectors, virulence, and the emergence of evolutionary epidemiology. *Oxford Surv. Evol. Biol.* **5**, 215–45.

Fagerström, T. 1988. Lotteries in communities of sessile organisms. *Trends Ecol. Evol.* **3**, 303–6.

Falconer, D.S. 1981. *Introduction to Quantitative Genetics*. 2nd edn.

Longman, London.

Feder, J.L., Chilcote, C.A. & Bush, G.L. 1988. Genetic differentiation between sympatric host races of the apple maggot fly. *Nature* **336**, 61–4.

Feder, J.L., Chilcote, C.A. & Bush, G.L. 1990a. The geographic pattern of genetic differentiation between host associated populations of *Rhagoletis pomonella* (Diptera: Tephritidae) in the eastern United States and Canada. *Evolution* **44**, 570–94.

Feder, J.L., Chilcote, C.A. & Bush, G.L. 1990b. Regional, local and microgeographic allele frequency variation between apple and hawthorn populations of *Rhagoletis pomonella* in western Michigan. *Evolution* **44**, 595–608.

Feldman, M.W., Christiansen, F.B. & Brooks, L.D. 1980. Evolution of recombination in a constant environment. *Proc. Natl. Acad. Sci. USA* **77**, 4838–41.

Fellows, L. 1989. Botany breaks into the candy store. *New Scientist* **1679**, 45–8.

Felsenstein, J. 1985. Phylogenies and the comparative method. *Amer. Natur.* **125**, 1–15.

Fenchel, T. & Christiansen, F.B. 1977. Selection and interspecific competition. In: F. Christiansen & T. Fenchel (eds.) *Measuring Selection in Natural Populations*, pp. 477–98. Springer-Verlag, Berlin.

Fenner, F. & Ratcliffe, F.N. 1965. *Myxomatosis*. Cambridge University Press, Cambridge.

Figueroa, F., Günther, E. & Klein, J. 1988. MHC polymorphism predating speciation. *Nature* **335**, 265–7.

Fisher, R.A. 1930. *The Genetical Theory of Selection*. Clarendon Press, Oxford.

Flessa, K.W., Erban, H.K., Hallam, A., Hsü, K.J., Hüsnner, H.M., Jablonski, D., Raup, D.M., Sepkoski, J.J., Soulé, M.E., Sousa, W., Stinnesbeck, W. & Vermeij, G.J. 1986. Causes and consequences of extinction. In: D.M. Raup & D. Jablonski (eds.) *Patterns and Processes in the History of Life*, pp. 235–57. Springer-Verlag, Berlin.

Flessa, K.W. & Jablonski, D. 1985. Declining Phanerozoic background extinction rates: effect of taxonomic structure? *Nature* **313**, 216–18.

Flor, H.H. 1956. The complementary genic systems in flax and rust. *Adv. Genet.* **8**, 29–54.

Ford, H.A., Bell, H., Nias, R. & Noske, R. 1988. The relationship between ecology and the incidence of cooperative breeding in Australian birds. *Behav. Ecol. Sociobiol.* **22**, 239–49.

Forshaw, J.M. & Cooper, W.T. 1977. *The Birds of Paradise and Bowerbirds.* Collins, Sydney.

Fox, L.R. & Morrow, P.A. 1981. Specialization: species property or local phenomenon? *Science* **211**, 887–93.

Frank, S.A. & Slatkin, M. 1990. Evolution in a variable environment. *Amer. Natur.* **136**, 244–60.

Frankel, O.H. & Soulé, M.E. 1981. *Conservation and Evolution.* Cambridge University Press, Cambridge.

Franklin, I.R. 1980. Evolutionary change in small populations. In: M. Soulé & B.A. Wilcox (eds.) *Conservation Biology: an Evolutionary-Ecological Perspective,* pp. 135–50. Sinauer, Sunderland, Massachussets.

Freeland, W.J. 1983. Parasites and the coexistence of animal host species. *Amer. Natur.* **121**, 223–36.

Freeman, D.C., Harper, K.T. & Charnov, E.L. 1980. Sex changes in plants: old and new observations and new hypotheses. *Oecologia (Berl.)* **47**, 222–32.

Fretwell, S.D. 1972. *Populations in a Seasonal Environment.* Princeton University Press, Princeton.

Fretwell, S.D. & Lucas, H.L. 1970. On territorial behaviour and other factors influencing habitat distribution in birds. I. Theoretical development. *Acta Biotheoretica* **19**, 16–36.

Friedman, D.B. & Johnson, T.E. 1988. A mutation in the age-1 gene in *Caenorhabditis elegans* lengthens life and reduces hermaphrodite fertility. *Genetics* **118**, 75–86.

Fürsich, F.T. & Jablonski, D. 1984. Late Triassic naticid drillholes: carnivorous gastropods gain a major advantage but fail to radiate. *Science* **224**, 78–80.

Futuyma, D.J. 1986. *Evolutionary Biology.* 2nd edn. Sinauer, Sunderland, MA.

Futuyma, D.J. 1987. On the role of species in anagenesis. *Amer. Natur.* **130**, 465–73.

Futuyma, D.J. 1988. *Sturm und drang* and the evolutionary synthesis. *Evolution* **42**, 217–26.

Futuyma, D.J. & Moreno, G. 1988. The evolution of ecological specialization. *Ann. Rev. Ecol. Syst.* **19**, 207–33.

Futuyma, D.J. & Phillipi, T.E. 1987. Genetic variation and covariation in response to host plants by *Alsophila pometaria* (Lepidoptera: Geometridae). *Evolution* **41**, 269–79.

Futuyma, D.J. & Slatkin, M. (eds.) 1983. *Coevolution.* Sinauer, Sunderland, Massachussets.

Gadgil, M. & Bossert, W. 1970. Life history consequences of natural selection. *Amer. Natur.* **104**, 1–24.

Gaston, K.J. & Lawton, J.H. 1988a. Patterns in the distribution and abundance of insect populations. *Nature* **331**, 709–12.

Gaston, K.J. & Lawton, J.H. 1988b. Patterns in body size, population dynamics, and regional distribution of bracken herbivores. *Amer. Natur.* **132**, 662–80.

Gaston, K. & Lawton, J.H. 1989. Insect herbivores on bracken do not support the core–satellite hypothesis. *Amer. Natur.* **134**, 761–77.

Gauch, H. & Whittaker, R.H. 1972. Coenocline simulation. *Ecology* **53**, 446–54.

Gaul, H. 1961. Use of induced mutations in seed-propagated species. In: *Mutations and Plant Breeding,* pp. 206–51. National Academy of Sciences USA Publication No. 891, Washington.

Gensler, H.L. & Bernstein, H. 1981. DNA damage as the primary cause of aging. *Q. Rev. Biol.* **56**, 279–303.

Ghiselin, M.T. 1969. The evolution of hermaphroditism among animals. *Q. Rev. Biol.* **44**, 189–208.

Ghiselin, M.T. 1974. *The Economy of Nature and the Evolution of Sex.* University of California Press, California.

Gibbs, H.L. & Grant, P.R. 1987a. Ecological consequences of an exceptionally strong El Niño event on Darwin's finches. *Ecology* **68**, 1735–46.

Gibbs, H.L. & Grant, P.R. 1987b. Oscillating selection on Darwin's finches. *Nature* **327**, 511–13.

Gilbert, L.E. 1980. Food web organization and conservation of neotropical diversity. In: M. Soulé & B.A. Wilcox (eds.) *Conservation Biology: an Evolutionary—Ecological Perspective*, pp. 11–34. Sinauer, Sunderland, Massachussets.

Gill, D.E. 1978. On selection at high population density. *Ecology* **59**, 1289–91.

Gill, D.E. & Halverson, T.G. 1984. Fitness variation among branches within trees. In: B. Shorrocks (ed.) *Evolutionary Ecology*. pp. 105–16. Blackwell Scientific Publications, Oxford.

Gillespie, J.H. 1974. Natural selection for within-generation variance in offspring number. *Genetics* **76**, 601–6.

Gillespie, J.H. 1977. Natural selection for variances in offspring numbers: a new evolutionary principle. *Amer. Natur.* **111**, 1010–14.

Gingerich, P.D. 1983. Rates of evolution: effects of time and temporal scaling. *Science* **222**, 159–61.

Gingerich, P.D. 1985. Species in the fossil record: concepts, trends and transitions. *Paleobiology* **11**, 27–41.

Gish, D.T. 1978. *Evolution? The Fossils Say No.* 3rd edn. Creation-Life Publishers, San Diego, California.

Givnish, T. 1980. Ecological constraints on the evolution of breeding systems in seed plants: dioecy and dispersal in gymnosperms. *Evolution* **34**, 959–72.

Givnish, T. 1982. Outcrossing versus ecological constraints in the evolution of dioecy. *Amer. Natur.* **119**, 849–65.

Godfray, H.C.J. 1987. The evolution of clutch size in invertebrates. *Oxf. Surv. Evol. Biol.* **4**, 117–54.

Goldberg, D.E. & Werner, P.A. 1983. Equivalence of competition in plant communities: a null hypothesis and a field experimental approach. *Amer. J. Bot.* **70**, 1098–104.

Goldsmith, M.R. & Kafatos, F.C. 1984. Developmentally regulated genes in silkmoths. *Ann. Rev. Genet.* **18**, 443–87.

Goodman, D. 1987. The demography of chance extinction. In: M.E. Soulé (ed.) *Viable Populations for Conservation*, pp. 11–34. Cambridge University Press, Cambridge.

Goodnight, C.J. 1987. On the effect of founder events on epistatic genetic variance. *Evolution* **41**, 80–91.

Goodnight, C.J. 1988. Epistasis and the effect of founder events on the additive genetic variance. *Evolution* **42**, 441–54.

Goodwin, B. 1988. Rumbling the replicator, *New Scientist* **1603**, 56–60.

Goodwin, B. 1989. Evolution and the generative order. In: B. Goodwin & P. Saunders (eds.) *Theoretical Biology*, pp. 89–100. Edinburgh University Press, Edinburgh.

Gotelli, N.J. & Simberloff, D. 1987. The distribution and abundance of tallgrass prairie plants: a test of the core—satellite hypothesis. *Amer. Natur.* **130**, 18–35.

Gottlieb, L.D. 1984. Genetics and morphological evolution in plants. *Amer. Natur.* **123**, 681–709.

Gould, S.J. 1975. Allometry in primates with emphasis on scaling and the evolution of the brain. *Contib. Primatol.* **5**, 244–92.

Gould, S.J. 1977. *Ontogeny and Phylogeny.* Belknap Press, Cambridge, Massachusetts.

Gould, S.J. 1980a. *The Panda's Thumb.* Penguin, London.

Gould, S.J. 1980b. Is a new and general theory of evolution emerging? *Paleobiology* **6**, 119–30.

Gould, S.J. 1980c. The promise of paleobiology as a nomothetic, evolutionary discipline. *Paleobiology* **6**, 96–118.

Gould, S.J. 1983. The meaning of punctuated equilibrium and its role in validating a hierarchical approach to evolution. *Scientia* **118**, 135–57.

Gould, S.J. 1985a. The paradox of the first tier: an agenda for paleobiology. *Paleobiology* **11**, 2–12.

Gould, S.J. 1985b. *The Flamingo's Smile.* Penguin, Harmondsworth.

Gould, S.J. 1989a. A developmental constraint in *Cerion*, with comments on the definition and interpretation of constraint in evolution. *Evolution* **43**, 516–39.

Gould, S.J. 1989b. *Wonderful Life: the Burgess Shale and the Nature of Life.* Norton, New York.

Gould, S.J. & Eldredge, N. 1977. Punctuated equilibria: the tempo and mode of evolution reconsidered. *Paleobiology* **3**, 115–51.

Gould, S.J. & Eldredge, N. 1986. Punctuated equilibrium at the third stage. *Syst. Zool.* **35**, 143–8.

Gould, S.J., Gilinsky, N.L. & German, R.Z. 1987. Asymmetry of lineages and the direction of evolutionary time. *Science* **236**, 1437–41.

Gould, S.J. & Lewontin, R.C. 1979. The spandrels of San Marco and the Panglossian paradigm: a critique of the adaptationist programme. *Proc. R. Soc. Lond.* **B205**, 581–98.

Gould, S.J. & Vrba, S.E. 1982. Exaptation — a missing term in the science of form. *Paleobiology* **8**, 4–10.

Gowaty, P.A. & Lennartz, M.R. 1985. Sex ratios of nestling and fledgling red-cockaded woodpeckers (*Picoides borealis*) favor males. *Amer. Natur.* **126**, 347–53.

Grafen, A. 1982. How not to measure inclusive fitness. *Nature* **298**, 425–6.

Grafen, A. 1984. Natural selection, kin selection and group selection. In: J.R. Krebs & N.B. Davies (eds.) *Behavioural Ecology: An Evolutionary Approach*. 2nd edn., pp. 62–84. Blackwell Scientific Publications, Oxford.

Grafen, A. 1987a. The logic of divisively asymmetric contests: respect for ownership and the desperado effect. *Anim. Behav.* **35**, 462–7.

Grafen, A. 1987b. Measuring sexual selection: why bother? In: J.W. Bradbury & M.B. Andersson (eds.) *Sexual Selection: Testing the Alternatives*, pp. 221–33. John Wiley & Sons, Chichester.

Grafen, A. 1988a. On the uses of data on lifetime reproductive success. In: T.H. Clutton-Brock (ed.) *Reproductive Success: Studies on Individual Variation in Contrasting Breeding Systems*, pp. 454–71. University of Chicago Press, Chicago.

Grafen, A. 1988b. A centrosomal theory of the short term evolutionary advantages of sexual reproduction. *J. theor. Biol.* **131**, 163–73.

Grafen, A. 1989. The phylogenetic regression. *Phil. Trans. R. Soc. Lond.* **B326**, 119–57.

Grafen, A. 1990a. Biological signals as handicaps. *J. theor. Biol.* **144**, 517–46.

Grafen, A. 1990b. Sexual selection unhandicapped by the Fisher process. *J. theor. Biol.* **144**, 473–516.

Grant, B.R. 1984. The significance of song variation in a population of Darwin's finches. *Behaviour* **89**, 90–116.

Grant, P.R. 1986. *Ecology and Evolution of Darwin's Finches*. Princeton University Press, Princeton.

Grant, V. 1977. *Organismic Evolution*. Freeman, San Francisco.

Grant, V. 1981. *Plant Speciation*. 2nd edn. Columbia University Press, New York.

Graur, D. & Li, W.-H. 1988. Evolution of protein inhibitors of serine proteases: positive Darwinian selection or compositional effects? *J. Mol. Evol.* **28**, 131–35.

Greene, E. 1989. A diet-induced developmental polymorphism in a caterpillar. *Science* **243**, 643–6.

Greenslade, P.J.M. 1983. Adversity selection and the habitat templet. *Amer. Natur.* **122**, 352–65.

Greenwood, P.H. 1981. *The Haplochromine Fishes of the East African Lakes*. Cornell University Press, New York.

Greenwood, P.J. 1980. Mating systems, philopatry and dispersal in birds and mammals. *Anim. Behav.* **28**, 1140–62.

Greenwood, P.J. & Wheeler, P. 1985. The evolution of sexual size dimorphism in birds and mammals a 'hot blooded' hypothesis. In: P.J. Greenwood, P.H. Harvey & M. Slatkin (eds.) *Evolution: Essays in Honour of John Maynard Smith*, pp. 287–99. Cambridge University Press, Cambridge.

Grime, J.P. 1977. Evidence for the existence of three primary strategies in plants and its relevance to ecological and evolutionary theory. *Amer. Natur.* **111**, 1169–94.

Grime, J.P. 1979. *Plant Strategies and Vegetation Processes*. John Wiley & Sons, Chichester.

Gross, M.R. 1985. Disruptive selection for alternative life histories in salmon. *Nature* **313**, 47–8.

Gross, M.R. & Shine, R. 1981. Parental care and mode of fertilization in ectothermic vertebrates. *Evolution* **35**, 775–93.

Grubb, P.J. 1987. Global trends in species-richness in terrestrial vegetation: a view from the northern hemisphere. In: J.H.R. Gee & P.S. Giller (eds.) *Organization of Communities: Past and Present*, pp. 99–118. Blackwell Scientific

Publications, Oxford.

Guerrant, E.O. 1982. Neotenic evolution of *Delphinium nudicaule* (Ranunculaceae): a hummingbird-pollinated larkspur. *Evolution* **36**, 699–712.

Gullan, P.J. & Cockburn, A. 1986. Sexual dichronism and intersexual phoresy in gall-forming coccoids. *Oecologia (Berl.)* **68**, 632–4.

Gustafsson, L. 1986. Lifetime reproductive success and heritability: empirical support for Fisher's fundamental theorem. *Amer. Natur.* **128**, 761–4.

Gustafsson, L. & Sutherland, W.J. 1988. The cost of reproduction in the collared flycatcher *Ficedula albicollis*. *Nature* **335**, 813–15.

Gwynne, D.T. 1984. Male mating effort, confidence of paternity, and insect sperm competition. In: R.L. Smith (ed.) *Sperm Competition and the Evolution of Animal Mating Systems*, pp. 117–49. Academic Press, New York.

Hafner, M.S. & Nadler, S.A. 1988. Phylogenetic trees support the coevolution of parasites and their hosts. *Nature* **332**, 258–9.

Hairston, N.G. Sr 1990. *Ecological Experiments: Purpose, Design and Execution*. Cambridge University Press, Cambridge.

Hairston, N.G. Sr, Nishikawa, K.C. & Stenhouse, S.L. 1987. The evolution of competing species of terrestrial salamanders: niche partitioning or interference. *Evol. Ecol.* **1**, 247–62.

Hairston, N.G. Jr & De Stasio, B.T. 1988. Rate of evolution slowed by a dormant propagule pool. *Nature* **336**, 239–42.

Hairston, N.G. Sr, Smith, F.E. & Slobodkin, L.B. 1960. Community structure, population control, and competition. *Amer. Natur.* **94**, 421–5.

Haldane, J.B.S. 1949. Disease and evolution. *La Ricerca Scientifica* **19** (Supp), 68–86.

Hallam, A. 1987. End-Cretaceous mass extinction event: argument for terrestrial causation. *Science* **238**, 1237–42.

Hamilton, W.D. 1964. The genetical evolution of social behaviour. *J. theor. Biol.* **7**, 1–52.

Hamilton, W.D. 1967. Extraordinary sex ratios. *Science* **156**, 477–8.

Hamilton, W.D. 1982. Pathogens as causes of genetic diversity in their host organisms. In: R.M. Anderson & R.M. May (eds.) *Population Biology of Infectious Diseases*. pp. 269–96. Springer-Verlag, New York.

Hamilton, W.D., Axelrod, R. & Tanese, R. 1990. Sexual reproduction as an adaptation to resist parasites (a review). *Proc. Natl. Acad. Sci. USA* **87**, 3566–73.

Hamilton, W.D. & Zuk, M. 1982. Heritable true fitness and bright birds: a role for parasites? *Science* **218**, 384–7.

Hammerstein, P. & Riechert, S.E. 1988. Payoffs and strategies in territorial contests: ESS analyses of two ecotypes of the spider *Agelenopsis aperta*. *Evol. Ecol.* **2**, 115–38.

Hansen, T.A. 1980. Influence of larval dispersal and geographic distribution on species longevity in neogastropods. *Paleobiology* **6**, 193–207.

Hanski, I. 1982. Dynamics of regional distribution: the core and satellite species hypothesis. *Oikos* **38**, 210–21.

Harcourt, A.H., Harvey, P.H., Larsen, S.G. & Short, R.V. 1981. Testis weight, body weight and breeding system in primates. *Nature* **293**, 55–7.

Harlan, J.R. 1975. *Crops and Man*. American Society of Agronomy, Madison, Wisconsin.

Harlan, J.R., de Wet, J.M.J. & Price, E.G. 1973. Comparative evolution of cereals. *Evolution* **27**, 311–25.

Harper, J.L. 1977. *Population Biology of Plants*. Academic Press, London.

Harper, J.L., Lovell, P.H. & Moore, K.G. 1970. The shapes and sizes of seeds. *Ann. Rev. Ecol. Syst.* **1**, 327–56.

Hart, J.A. 1988. Rust fungi and host plant coevolution: do primitive hosts harbor primitive parasites? *Cladistics* **4**, 339–66.

Hart, T.B. 1990. Monospecific dominance in tropical rainforests. *Trends Ecol. Evol.* **5**, 6–11.

Hartgerink, A.P. & Bazzaz, F.A. 1984. Seedling-scale environmental heterogeneity influences individual fitness and population structure. *Ecology* **65**, 198–206.

Hartl, D.L. 1981. *A Primer of Population Genetics*. Sinauer, New York.

Hartl, D.L., Medhora, M., Green, L. & Dykhuizen, D.E. 1986. The evolution of DNA sequences in *Escherichia coli*.

Phil. Trans. R. Soc. Lond. **B312**, 191–204.

Harvey, P.H. 1985. Intrademic group selection and the sex ratio. In: R.M. Sibly & R.H. Smith (eds.) *Behavioural Ecology: Ecological Consequences of Adaptive Behaviour*, pp. 59–71. Blackwell Scientific Publications, Oxford.

Harvey, P.H., Birley, N. & Blackstock, T.H. 1975. The effect of experience on the selective behaviour of song thrushes feeding on artificial populations of *Cepaea* (Held.). *Genetica* **45**, 211–16.

Harvey, P.H., Colwell, R.K., Silvertown, J.W. & May, R.M. 1983. Null models in ecology. *Ann. Rev. Ecol. Syst.* **14**, 189–211.

Harvey, P.H., Promislow, D.E.L. & Read, A.F. 1989. Causes and correlates of life history differences among mammals. In: V. Standen & R.A. Foley (eds.) *Comparative Socioecology: The Behavioural Ecology of Humans and Other Mammals*, pp. 305–22. Blackwell Scientific Publications, Oxford.

Harvey, P.H. & Ralls, K. 1985. Homage to the null weasel. In: P.J. Greenwood, P.H. Harvey & M. Slatkin (eds.) *Evolution: Essays in Honour of John Maynard Smith*, pp. 155–71. Cambridge University Press, Cambridge.

Harvey, P.H. & Read, A.F. 1988. When incest is not best. *Nature* **336**, 514–15.

Hebert, P.D.N. 1987. Genotypic characteristics of cyclic parthenogens and their obligately asexual derivatives. In: S.C. Stearns (ed.) *The Evolution of Sex and its Consequences*, pp. 175–95. Birkhäuser Verlag, Basel.

Hedrick, P.W. 1986. Genetic polymorphism in heterogeneous environments: a decade later. *Ann. Rev. Ecol. Syst.* **17**, 535–66.

Hedrick, P.W. 1987. Genetic bottlenecks. *Science* **237**, 963.

Hedrick, P.W. & Thomson, G. 1983. Evidence for balancing selection at HLA. *Genetics* **104**, 449–56.

Hedrick, P.W., Thomson, G. & Klitz, W. 1987. Evolutionary genetics and HLA: another classic example. *Biol. J. Linn. Soc.* **31**, 311–31.

Heinrich, B. 1976. The foraging specializ- ations of individual bumblebees. *Ecol. Monogr.* **46**, 105–28.

Heinrich, B. 1979. 'Majoring' and 'minoring' by foraging bumblebees, *Bombus vagrans*: an experimental analysis. *Ecology* **60**, 245–55.

Heinrich, B. 1984. Learning in invertebrates. In: P.S Marler & H.S. Terrace (eds.) *The Biology of Learning*, pp. 135–47. Springer-Verlag, Berlin.

Heinsohn, R.G., Cockburn, A. & Mulder, R.A. 1990. Avian cooperative breeding: old hypotheses and new directions. *Trends Ecol. Evol.* **5**, 403–7.

Heslop-Harrison, J. 1957. The experimental modification of sex expression in flowering plants. *Biol. Rev.* **32**, 38–90.

Hewitt, G.M. 1988. Hybrid zones— natural laboratories for evolutionary studies. *Trends Ecol. Evol.* **3**, 158–67.

Hildebrand, A.R. & Boynton, W.V. 1990. Proximal Cretaceous–Tertiary boundary impact deposits in the Caribbean. *Science* **248**, 843–7.

Hildrew, A.G. & Townsend, C.R. 1987. Organization in freshwater benthic communities. In: J.H.R. Gee & P.S. Giller (eds.) *Organization of Communities: Past and Present*, pp. 347–72. Blackwell Scientific Publications, Oxford.

Hill, R.E. & Hastie, N.D. 1987. Accelerated evolution in the reactive centre regions of serine protease inhibitors. *Nature* **326**, 96–9.

Hill, W.G. 1982. Predictions of response to artificial selection from new mutations. *Genet. Res. Camb.* **40**, 255–78.

Hillgarth, N. 1990. Pheasant spurs out of fashion. *Nature* **345**, 119–20.

Hitching, F. 1982. *The Neck of the Giraffe, or Where Darwin Went Wrong*. Pan, London.

Ho, M.-W., Saunders, P. & Fox, S. 1986 A new paradigm for evolution. *New Scientist* **1497**, 41–3.

Hoekstra, R.F. 1987. The evolution of sexes. In: S.C. Stearns (ed.) *The Evolution of Sex and its Consequences*, pp. 59–91. Birkhäuser Verlag, Basel.

Hoeprich, P.D. 1977. Host–parasite relationships and the pathogenesis of infectious disease. In: P.D. Hoeprich (ed.) *Infectious Diseases*, pp. 34–45.

Harper & Row, London.

Högstedt, G. 1980. Evolution of clutch size in birds: adaptive variation in relation to territory quality. *Science* **210**, 1148–50.

Högstedt, G. 1981. Should there be a positive or negative correlation between survival of adults in a bird population and their clutch size? *Amer. Natur.* **118**, 568–71.

Hoffman, A. & Kitchell, J.A. 1984. Evolution in a pelagic planktic system: a paleobiologic test of models of multispecies evolution. *Paleobiology* **10**, 9–33.

Horn, H.S., Bonner, J.T., Dohle, W., Katz, M.J., Koehl, M.A.R., Meinhardt, H., Ragg, R.A., Reif, W-E., Stearns, S.C. & Strathmann, R.R. 1982. Adaptive aspects of development: group report. In: J.T. Bonner (ed.) *Evolution and Development*, pp. 215–35. Springer-Verlag, Berlin.

Houde, A.E. & Endler, J.A. 1990. Correlated evolution of female mating preferences and male color patterns in the guppy *Poecilia reticulata. Science* **248**, 1405–8.

Houle, D. 1989. The maintenance of polygenic variation in finite populations. *Evolution* **43**, 1767–80.

Houston, A., Clark, C., McNamara, J. & Mangel, M. 1988. Dynamic models in behavioural ecology and evolutionary ecology. *Nature* **332**, 29–34.

Houston, A.I. & McNamara, J.M. 1988. Fighting for food: a dynamic version of the Hawk–Dove game. *Evol. Ecol.* **2**, 51–64.

Howard, R.D. 1983. Sexual selection and variation in reproductive success in a long-lived organism. *Amer. Natur.* **122**, 301–25.

Howard, R.D. 1988. Reproductive success in two species of anurans. In: T.H. Clutton-Brock (ed.) *Reproductive Success: Studies on Individual Variation in Contrasting Breeding Systems*, pp. 99–113. University of Chicago Press, Chicago.

Howard, R.D. & Minchella, D.J. 1990. Parasitism and male competition. *Oikos* **58**, 120–2.

Hoyle, F. & Wickramasinghe, N.C. 1981. *Evolution from Space*. J.M. Dent, London.

Hubbell, S.P. & Foster, R.B. 1986. Biology, chance and history and the structure of tropical rainforest tree communities. In: J. Diamond & T. Case (eds.) *Community Ecology*, pp. 314–30. Harper & Row, New York.

Hudson, R.R., Kreitman, M. & Aguadé, M. 1987. A test of neutral molecular evolution based on nucleotide data. *Genetics* **116**, 153–9.

Hughes, A.L. & Nei, M. 1988. Pattern of nucleotide substitution at major histocompatibility complex class I loci reveals overdominant selection. *Nature* **335**, 167–70.

Hughes, A.L. & Nei, M. 1989. Nucleotide substitution at major histocompatibility complex class II loci: evidence for overdominant selection. *Proc. Natl. Acad. Sci. USA* **86**, 958–62.

Hughes, T.P. 1990. Recruitment limitation, mortality, and population regulation in open systems. *Ecology* **71**, 12–20.

Hurka, H. & Haase, R. 1982. The soil seed bank of *Capsella bursa-pastoris* (Cruciferae): dispersal mechanism and the soil seed bank. *Flora* **172**, 35–46.

Hurlbert, S. 1984. Pseudoreplication and the design of ecological field experiments. *Ecol. Monogra.* **54**, 187–211.

Huston, M. 1979. A general hypothesis of species diversity. *Amer. Natur.* **113**, 81–101.

Hut, P., Alvarez, W., Elder, W.P., Hansen, T., Kauffman, E.G., Keller, G., Shoemaker, E.M. & Weissman, P.R. 1987. Comet showers as a cause of mass extinctions. *Nature* **329**, 118–26.

Hutchings, M.J. & Slade, A.J. 1988. Morphological plasticity, foraging and integration in clonal perennial herbs. In: A.J. Davy, M.J. Hutchings & A.R. Watkinson (eds.) *Plant Population Ecology*, pp. 83–109. Blackwell Scientific Publications, Oxford.

Hutchinson, G.E. 1959. Homage to Santa Rosalia, or why are there so many kinds of animals? *Amer. Natur.* **93**, 145–59.

Hutchinson, G.E. 1978. *An Introduction to Population Ecology*. Yale University Press, New Haven.

Huxley, J.S. 1942. *Evolution: The Modern Synthesis*. Harper, New York.

Istock, C.A. 1967. The evolution of

complex life cycles: an ecological perspective. *Evolution* **21**, 592–605.

Istock, C.A. 1983. The extent and consequence of heritable variation for fitness characters. In: C.R. King & P.S. Dawson (eds.) *Population Biology: Retrospects and Prospects*. Columbia University Press, New York.

Istock, C.A. 1984. Boundaries to life history variation and evolution. In: P.W. Price, C.N. Slobodchikoff & W.S. Gaud (eds.) *A New Ecology: Novel Approaches to Interactive Systems*, pp. 143–68. Wiley, New York.

Jablonski, D. 1986a. Evolutionary consequences of mass extinctions. In: D.M. Raup & D. Jablonski (eds.) *Patterns and Processes in the History of Life*, pp. 313–29. Springer-Verlag, Berlin.

Jablonski, D. 1986b. Background and mass extinctions: the alternation of macroevolutionary regimes. *Science* **231**, 129–33.

Jablonski, D. 1987. Heritability at the species level: analysis of geographic races of Cretaceous mollusks. *Science* **238**, 360–3.

Jablonski, D. & Lutz, R.A. 1983. Larval ecology of marine benthic invertebrates: paleobiological implications. *Biol. Rev.* **58**, 21–89.

Jackson, J.B.C. & Cheetham, A.H. 1990. Evolutionary significance of morphospecies: a test with cheilostome Bryozoa. *Science* **248**, 579–83.

Jaenike, J. 1978. An hypothesis to account for the maintenance of sex in populations. *Evol. Theor.* **3**, 191–4.

Jaenike, J. 1985. Parasite pressure and the evolution of amanitin tolerance in *Drosophila*. *Evolution* **39**, 1295–301.

Jain, S.K. 1978. Inheritance of phenotypic plasticity in soft chess, *Bromus mollis* L. (Gramineae). *Experientia* **34**, 835–6.

Janssen, G.M., de Jong, G., Joosse, E.N.G. & Scharloo, W. 1988. A negative maternal effect in springtails. *Evolution* **42**, 828–34.

Janzen, D.H. 1976. Why bamboo wait so long to flower. *Ann. Rev. Ecol. Syst.* **7**, 347–91.

Janzen, D.H. 1977. Why fruits rot, seeds mold, and meat spoils. *Amer. Natur.* **111**, 691–713.

Janzen, D.H. 1979. New horizons in the biology of plant defenses. In: G.A. Rosenthal & D.H. Janzen (eds.) *Herbivores. Their Interaction with Secondary Plant Metabolites*, pp. 331–50. Academic Press, New York.

Janzen, D.H. 1980. When is it coevolution? *Evolution* **34**, 611–12.

Janzen, D.H. 1985. On ecological fitting. *Oikos* **45**, 308–10.

Janzen, D.H. 1987. Insect diversity of a Costa Rican dry forest: why keep it, and how? *Biol. J. Linn. Soc.* **30**, 343–56.

Jeffreys, A.J., Royle, N.J., Wilson, V. & Wong, Z. 1988. Spontaneous mutation rates to new length alleles at tandem-repetitive hypervariable loci in human DNA. *Nature* **332**, 278–80.

Jeffries, M.J. & Lawton, J.H. 1984. Enemy free space and the structure of ecological communities. *Biol. J. Linn. Soc.* **23**, 269–86.

Jermy, T. 1987. The role of experience in the host selection of phytophagous insects. In: R.F. Chapman, E.A. Bernays & J.G. Stoffolano (eds.) *Perspectives in Chemoreception and Behavior*, pp. 143–57. Springer-Verlag, New York.

Jermy, T. 1988. Can predation lead to narrow food specialization in phytophagous insects? *Ecology* **69**, 902–4.

Johanssen, W. 1911. The genotype concept of heredity. *Amer. Natur.* **45**, 129–59.

Johnson, C.M. & Mulcahy, D.L. 1978. Male gametophyte in maize. II. Pollen vigor in inbred plants. *Theor. Appl. Genet.* **51**, 211–15.

Johnson, C.N. 1986. Philopatry, reproductive success of females, and maternal investment in the red-necked wallaby. *Behav. Ecol. Sociobiol.* **19**, 143–50

Johnson, C.N. 1988. Dispersal and the sex ratio at birth in primates. *Nature* **332**, 726–8.

Johnson, K.R., Nichols, D.J., Attrep, M. & Orth, C.J. 1989. High-resolution leaf-fossil record spanning the Cretaceous/Tertiary boundary. *Nature* **340**, 708–11.

Johnson, T.E. 1987. Aging can be genetically dissected into component processes using long-lived lines of *Caenorhabditis elegans*. *Proc. Natl. Acad. Sci. USA* **84**, 3777–81.

Jones, C.G. & Firn, R.D. 1978. The role of phytoecdysteroids in bracken fern, *Pteridium aquilinum* (L.) as a defense against phytophagous insect attack. *J. Chem. Ecol.* **4**, 117–38.

Jones, J.S., Leith, B.H. & Rawlings, P. 1977. Polymorphism in *Cepaea*: a problem with too many solutions? *Ann. Rev. Ecol. Syst.* **8**, 109–43.

Kale, H.W. 1987. The dusky seaside sparrow: have we learned anything? *Florida Natur.* **60**(3), 2–3.

Kaneshiro, K.Y. 1976. Ethological isolation and phylogeny in the *plantibia* subgroup of Hawaiian *Drosophila*. *Evolution* **30**, 740–5.

Kaneshiro, K.Y. 1980. Sexual isolation, speciation and the direction of evolution. *Evolution* **34**, 437–44.

Kaneshiro, K.Y. & Boake, C.R.B. 1987. Sexual selection and speciation: issues raised by Hawaiian *Drosophila*. *Trends Ecol. Evol.* **2**, 207–12.

Kaplan, R.H. & Cooper, W.S. 1984. The evolution of developmental plasticity in reproductive characteristics: an application of the 'adaptive coin-flipping' principle. *Amer. Natur.* **123**, 393–410.

Karban, R. 1989. Fine-scale adaptation of herbivorous thrips to individual host plants. *Nature* **340**, 60–1.

Karron, J.D. & Marshall, D.L. 1990. Fitness consequences of multiple paternity in wild radish, *Raphanus sativus*. *Evolution* **44**, 260–8.

Kasuya, T. & Marsh, H. 1984. Life history and reproductive biology of the short-finned pilot whale, *Globicephala macrorhynchus*, off the Pacific coast of Japan. *Rep. Int. Whal. Comm. Special* **6**, 259–310.

Keith, T.P. 1983. Frequency distribution of esterase-5 alleles in two populations of *Drosophila pseudoobscura*. *Genetics* **105**, 135–55.

Kelley, S.E. 1989a. Experimental studies of the evolutionary significance of sexual reproduction. V. A field test of the sib-competition lottery hypothesis. *Evolution* **43**, 1054–65.

Kelley, S.E. 1989b. Experimental studies of the evolutionary significance of sexual reproduction. VI. A greenhouse test of the sib-competition hypotheses. *Evolution* **43**, 1066–74.

Kelley, S.E., Antonovics, J. & Schmitt, J.

1988. A test of the short-term advantages of sexual reproduction. *Nature* **331**, 714–17.

Kemp, E. 1988. Fighting for the forest ox. *New Scientist* **1619**, 51–3.

Kenagy, G.J., Masman, D., Sharbaugh, S.M. & Nagy, K.A. 1990. Energy expenditure during lactation in relation to litter size in free-living golden-mantled ground squirrels. *J. Anim. Ecol.* **59**, 73–88.

Kennedy, C.E.J., Endler, J.A., Poynton, S.L. & McMinn, H. 1987. Parasite load predicts mate choice in guppies. *Behav. Ecol. Sociobiol.* **21**, 291–5.

Kennedy, C.E.J. & Southwood, T.R.E. 1984. The number of species of insects associated with British trees: a re-analysis. *J. Anim. Ecol.* **53**, 455–78.

Kent, E.B. 1981. Life history responses to resource variation in a sessile predator, the ciliate protozoan *Tokophrya lemnarum* Stein. *Ecology* **62**, 296–302.

Kent, G.M., Harding, A.J. & Orcutt, J.A. 1990. Evidence for a small magma chamber beneath the East Pacific Rise at 9°30'N. *Nature* **344**, 650–3.

Kerr, J.B. & Hedger, M.P. 1983. Spontaneous spermatogenic failure in the marsupial mouse *Antechinus stuartii* Macleay (Dasyuridae: Marsupialia). *Aust. J. Zool.* **31**, 445–66.

Kershaw, A.P. 1986. Climatic change and aboriginal burning in north-east Australia during the last two glacial/interglacial cycles. *Nature* **322**, 47–9.

Kettlewell, H.B.D. 1973. *The Evolution of Melanism*. Clarendon Press, Oxford.

Kidwell, M.G. 1983. Evolution of hybrid dysgenesis determinants in *Drosophila melanogaster*. *Proc. Natl. Acad. Sci. USA.* **80**, 1655–9.

Kiester, A.R., Lande, R. & Schemske, D.W. 1984. Models of coevolution and speciation in plants and their pollinators. *Amer. Natur.* **124**, 220–43.

Kimura, M. 1983. *The Neutral Theory of Molecular Evolution*. Cambridge University Press, Cambridge.

Kimura, M. & Crow, J.F. 1964. The number of alleles that can be maintained in a finite population. *Genetics* **44**, 725–38.

Kirkpatrick, M. 1986. The handicap mechanism of sexual selection does not work. *Amer. Natur.* **127**, 222–40.

Kirkpatrick, M. 1987. Sexual selection

by female choice in polygynous mammals. *Ann. Rev. Ecol. Syst.* **18**, 43–70.

Kirkpatrick, M. & Jenkins, C.D. 1989. Genetic segregation and the maintenance of sexual reproduction. *Nature* **339**, 300–1.

Kirkpatrick, M., Price, T. & Arnold, S.J. 1990. The Darwin–Fisher theory of sexual selection in monogamous birds. *Evolution* **44**, 180–93.

Kitchell, J.A. & MacLeod, N. 1988. Macroevolutionary interpretations of symmetry and synchroneity in the fossil record. *Science* **240**, 1190–3.

Kitcher, P. 1982. *Abusing Science: the Case Against Creationism.* MIT Press, Cambridge, Massachusetts.

Kjellberg, F., Gouyon, P.-H. & Ibrahim, M., Raymond, M. & Valdeyron, G. 1987. The stability of the symbiosis between dioecious figs and their pollinators: a study of *Ficus carica* L. and *Blastophaga psenes* L. *Evolution* **41**, 693–704.

Klekowski, E.J. 1988. Progressive cross- and self-sterility associated with aging in fern clones and perhaps other plants. *Heredity* **61**, 247–53.

Klekowski, E.J. & Godfrey, P.J. 1989. Ageing and mutation in plants. *Nature* **340**, 389–91.

Knoll, A.H. 1984. Patterns of extinction in the fossil record of vascular plants. In: M.H. Nitecki (ed.) *Extinctions*, pp. 21–68. University of Chicago Press, Chicago.

Knolle, H. 1989. Host density and the evolution of parasite virulence. *J. theor. Biol.* **136**, 199–207.

Koch, P.L. 1986. Clinal geographic variation in mammals: implications for the study of chronoclines. *Paleobiology* **12**, 269–81.

Kodric-Brown, A.K. & Brown, J.H. 1984. Truth in advertising: the kind of traits favoured by sexual selection. *Amer. Natur.* **124**, 309–23.

Koehn, R.K. & Eanes, W.F. 1978. Molecular structure and protein variation within and among populations. *Evol. Biol.* **11**, 39–100.

Koehn, R.K. & Hilbish, T.J. 1987. The adaptive importance of genetic variation. *Amer. Scient.* **75**, 134–41.

Koenig, W.D. 1988. On determination of viable population size in birds and mammals. *Wildl. Soc. Bull.* **16**, 230–4.

Kolata, G. 1984. Steroid systems found in yeast. *Science* **225**, 913–14.

Kondrashov, A.S. 1988. Deleterious mutations and the evolution of sexual reproduction. *Nature* **336**, 435–40.

Koptur, S. 1979. Facultative mutualism between weedy vetches bearing extrafloral nectaries and weedy ants in California. *Amer. J. Bot.* **66**, 1016–20.

Kosuda, K. 1985. The aging effect on male mating activity in *Drosophila melanogaster*. *Behav. Genet.* **15**, 297–303.

Krebs, C.J. 1989. *Ecological Methodology*. Harper & Row, New York.

Krebs, J.R. 1982. Territorial defense in the great tit (*Parus major*): do residents always win? *Behav. Ecol. Socbiol.* **11**, 185–94.

Krebs, J.R. & Davies, N.B. 1987. *An Introduction to Behavioural Ecology.* 2nd edn. Blackwell Scientific Publications, Oxford.

Kreitman, M. 1983. Nucleotide polymorphism at the alcohol dehydrogenase locus of *D. melanogaster*. *Nature* **304**, 412–17.

Krieber, M. & Rose, M.R. 1986. Molecular aspects of the species barrier. *Ann. Rev. Ecol. Syst.* **17**, 465–85.

Kyle, R. 1990. An antelope for all seasonings. *New Scientist* **1711**, 54–7.

Lack, D. 1947a. The significance of clutch-size. *Ibis* **89**, 302–52.

Lack, D. 1947b. *Darwin's Finches.* Cambridge University Press, Cambridge.

Lack, D. 1948. The significance of clutch size. *Ibis* **90**, 25–45.

Lack, D. 1954. *The Natural Regulation of Animal Numbers.* Oxford University Press, Oxford.

Lack, D. & Moreau, R.E. 1965. Clutch-size in tropical passerine birds of forest and savanna. *L'Oiseau* **35**, 76–89.

Lacy, R.C. 1978. Dynamics of t-alleles in *Mus musculus* populations: review and speculation. *Biologist* **60**, 41–67.

Lacy, R.C. 1987. Loss of genetic diversity from managed populations: interacting effects of drift, mutation, immigration, selection, and population subdivision. *Conserv. Biol.* **1**, 143–58.

Lanciani, C.A. 1987. Teaching quantitative concepts of population ecology in general biology courses. *Bull. Ecol. Soc. Amer.* **68**(4), 492–5.

Lande, R. 1976. The maintenance of genetic variability by mutation in a polygenic character with linked loci. *Genet. Res. Camb.* **26**, 221–35.

Lande, R. 1979. Quantitative genetic analysis of multivariate evolution, applied to brain: body size allometry. *Evolution* **33**, 402–16.

Lande, R. 1980. Genetic variation and phenotypic evolution during allopatric speciation. *Amer. Natur.* **116**, 463–79.

Lande, R. 1981. Models of speciation by sexual selection on polygenic traits. *Proc. Natl. Acad. Sci. USA* **78**, 3721–5.

Lande, R. 1982a. A quantitative genetic theory of life history evolution. *Ecology* **63**, 607–15.

Lande, R. 1982b. Rapid origin of sexual isolation and character divergence in a cline. *Evolution* **36**, 213–23.

Lande, R. 1987. Genetic correlations between the sexes in the evolution of sexual dimorphism and mating preferences. In: J.W. Bradbury & M.B. Andersson (eds.) *Sexual Selection; Testing the Alternatives*, pp. 83–94, John Wiley & Sons, Chichester.

Lande, R. 1988a. Quantitative genetics and evolutionary theory. In: B.S. Weir, E.J. Eisen, M. Goodman & G. Namkoong (eds.) *Proceedings of the Second International Symposium on Quantitative Genetics*, pp. 71–84. Sinauer, Sunderland, Massachussets.

Lande, R. 1988b. Genetics and demography in biological conservation. *Science* **241**, 1455–60.

Lande, R. & Arnold, S.J. 1983. The measurement of selection on correlated characters. *Evolution* **37**, 1210–26.

Lande, R. & Barrowclough, G.F. 1987. Effective population size, genetic variation, and their use in population management. In: M.E. Soulé (ed.) *Viable Populations for Conservation*, pp. 87–123. Cambridge University Press, Cambridge.

Larson, A., Prager, E.M. & Wilson, A.C. 1984. Chromosomal evolution, speciation and morphological change in vertebrates: the role of social behaviour. *Chromosomes Today* **8**, 215–28.

Law, R. 1979a. Ecological determinants in the evolution of life histories. In: R.M. Anderson, B.D. Turner & L.R. Taylor (eds.) *Population Dynamics*, pp. 81–103. Blackwell Scientific Publications, Oxford.

Law, R. 1979b. The cost of reproduction in annual meadow grass. *Amer. Natur.* **113**, 3–16.

Law, R. & Lewis, D.H. 1983. Biotic environments and the maintenance of sex — some evidence from mutualistic symbioses. *Biol. J. Linn. Soc.* **20**, 249–76.

Lawlor, D.A., Ward, F.E., Ennis, P.D., Jackson, A.P. & Parham, P. 1988. HLA-A and B polymorphisms predate the divergence of humans and chimpanzees. *Nature* **335**, 268–71.

Lawlor, L.R. 1976. Molting, growth and reproductive strategies in the terrestrial isopod, *Armadillidium vulgare*. *Ecology* **57**, 1179–94.

Lawton, J.H. 1988. More time means more variation. *Nature* **334**, 563.

Lawton, J.H. 1989. What is the relationship between population density and body size in animals? *Oikos* **55**, 429–34.

Lawton, J.H. & Brown, K.C. 1986. The population and community ecology of invading insects. *Phil. Trans. R. Soc. Lond.* **B314**, 607–17.

Lawton, J.H. & Hassell, M.P. 1981. Asymmetrical competition in insects. *Nature* **289**, 793–5.

Lawton, J.H. & Strong, D.R. 1981. Community patterns and competition in folivorous insects. *Amer. Natur.* **118**, 317–38.

Lee, A.K. & Cockburn, A. 1985. *Evolutionary Ecology of Marsupials.* Cambridge University Press, Cambridge.

Lendrem, D. 1986. *Modelling in Behavioural Ecology: An Introductory Text.* Croom Helm, Beckenham, Kent.

Lenski, R.E. 1989. Are some mutations directed? *Trends Ecol. Evol.* **4**, 148–50.

Lenski, R.E. & Levin, B.R. 1985. Constraints on the coevolution of bacteria and virulent phage: a model, some experiments and predictions for natural communities. *Amer. Natur.* **125**, 585–602.

Lenski, R.E. & Nguyen, T.T. 1988. Stability of recombinant DNA and its effects on fitness. *Trends Ecol. Evol.*

3, and *Trends Biotechnol.* **6**, S18–20.

Leutnegger, W. 1979. Evolution of litter size in primates. *Amer. Natur.* **114**, 525–31.

Leverich, W.J. & Levin, D.A. 1979. Age-specific survivorship and reproduction in *Phlox drummondii. Amer. Natur.* **113**, 881–903.

Levin, D.A. 1975. Pest pressure and recombination systems in plants. *Amer. Natur.* **109**, 437–51.

Levin, D.A. 1990. The seed bank as a source of genetic novelty in plants. *Amer. Natur.* **135**, 563–72.

Levin, S.A. 1988. Safety standards for the environmental release of genetically engineered organisms. *Trends Ecol. Evol.* **3**, and *Trends Biotechnol.* **6**, S47–9.

Levins, R. 1968. *Evolution in Changing Environments.* Princeton University Press, Princeton.

Levinton, J.S. 1988. *Genetics, Paleontology and Macroevolution.* Cambridge University Press, Cambridge.

Levinton, J.S., Bandel, K., Charlesworth, B., Muller, G., Nagl, W., Runnegar, B., Selander, R.K., Stearns, S.C., Turner, J.R.G., Urbanek, A.J. & Valentine, J.W. 1986. Organismic evolution: the interaction of microevolutionary and macroevolutionary processes. In: D.M. Raup & D. Jablonski (eds.) *Patterns and Processes in the History of Life,* pp. 167–82. Springer-Verlag, Berlin.

Lewin, B. 1987. *Genes III.* Wiley, New York.

Lewis, A.C. 1986. Memory constraints and flower choice in *Pieris rapae. Science* **232**, 863–5.

Lewis, W.M. 1987. The cost of sex. In: S.C. Stearns (ed.) *The Evolution of Sex and its Consequences,* pp. 33–57. Birkhäuser Verlag, Basel.

Lewontin, R.C. 1965a. Selection for colonizing ability. In: H.G. Baker & G.L. Stebbins (eds.) *The Genetics of Colonizing Species,* pp. 77–94. Academic Press, New York.

Lewontin, R.C. 1965b. Comment. In: H.G. Baker & G.L. Stebbins (eds.) *The Genetics of Colonizing Species,* pp. 481–4. Academic Press, New York.

Lewontin, R.C. 1970. The units of selection. *Ann. Rev. Ecol. Syst.* **1**, 1–18.

Lewontin, R.C. 1974. *The Genetic Basis of Evolutionary Change.* Columbia University Press, New York.

Lewontin, R.C. 1983. The organism as the subject and object of evolution. *Scientia* **118**, 65–82.

Lewontin, R.C. 1985a. Population genetics. *Ann. Rev. Genet.* **19**, 81–102.

Lewontin, R.C. 1985b. Population genetics. In: P.J. Greenwood, P.H. Harvey & M. Slatkin (eds.) *Evolution: Essays in Honour of John Maynard Smith,* pp. 3–18. Cambridge University Press, Cambridge.

Lewontin, R.C. 1987. The shape of optimality. In: J. Dupré (ed.) *The Latest on the Best,* pp. 151–9. MIT Press, Cambridge, Massachussets.

Lill, A.R. 1974. Sexual behaviour of the lek-forming white-bearded manakin (*Manacus manacus trinitatis*). *Z. Tierpsychol.* **36**, 1–36.

Lindén, M. & Møller, A.P. 1989. Cost of reproduction and covariation of life history traits in birds. *Trends Ecol. Evol.* **4**, 367–71.

Lindstedt, S.L. 1985. Birds. In: F.A. Lints (ed.) *Non-Mammalian Models for Research in Aging,* pp. 1–21. Karger, Basel.

Liston, A., Rieseberg, L.H. & Elias, T.S. 1990. Functional androdioecy in the flowering plant *Datisca glomerata. Nature* **343**, 641–2.

Littlejohn, M.J & Watson, H.F. 1985. Hybrid zones and homogamy in Australian frogs. *Ann. Rev. Ecol. Syst.* **16**, 85–112.

Lively, C.M. 1986. Canalization versus developmental conversion in a spatially variable environment. *Amer. Natur.* **128**, 561–72.

Lively, C.M. 1987. Evidence from a New Zealand snail for the maintenance of sex by parasitism. *Nature* **328**, 519–21.

Lively, C.M. 1989. Adaptation by a parasitic trematode to local populations of its snail host. *Evolution* **43**, 1663–71.

Lloyd, D.G. 1984a. Variation strategies of plants in heterogeneous environments. *Biol. J. Linn. Soc.* **21**, 357–85.

Lloyd, D.G. 1984b. Gender allocations in outcrossing cosexual plants. In: R. Dirzo & J. Sarukhan (eds.) *Perspectives on Plant Population Ecology,* pp. 277–300. Sinauer, Sunderland, Massachusetts.

Lloyd, D.G. 1988. Benefits and costs of

biparental and uniparental reproduction in plants. In: R.E, Michod & B.R. Levin (eds.) *The Evolution of Sex,* pp. 233–52. Sinauer, Sunderland, Massachusetts.

Lovett Doust, J. & Lovett Doust, L. 1988. Modules of production and reproduction in a dioecious clonal shrub, *Rhus typhina. Ecology* **69**, 741–50.

Lloyd, D.G. & Bawa, K.S. 1984. Modification of the gender of seed plants in varying conditions. *Evol. Biol.* **17**, 255–338.

Ludwig, D. & Rowe, L. 1990. Life-history strategies for energy gain and predator avoidance under time constraints. *Amer. Natur.* **135**, 686–707.

Lupton, F.G.H. 1977. The plant breeders' contribution to the origin and solution of pest and disease problems. In: J.M. Cherrett & G.R. Sagar (eds.) *Origins of Pest, Parasite, Disease and Weed Problems,* pp. 71–81. Blackwell Scientific Publications, Oxford.

Lynch, M. 1985. Spontaneous mutations for life-history characters in an obligate parthenogen. *Evolution* **39**, 804–18.

Lynch, M. 1988. The rate of polygenic mutation. *Genet. Res. Camb.* **51**, 137–48.

MacArthur, R.H. 1955. Fluctuations of animal populations and a measure of community stability. *Ecology* **36**, 533–6.

MacArthur, R.H. & Levins, R. 1964. Competition, habitat selection, and character displacement in a patchy environment. *Proc. Natl. Acad. Sci. USA* **51**, 1207–10.

MacArthur, R.H. & Levins, R. 1967. The limiting similarity, convergence, and divergence of coexisting species. *Amer. Natur.* **101**, 377–85.

MacArthur, R.H. & Wilson, E.O. 1967. *The Equilibrium Theory of Island Biogeography.* Princeton University Press, Princeton.

Mace, G.M., Harvey, P.H. & Clutton-Brock, T.H. 1981. Brain size and ecology in small mammals. *J. Zool. Lond.* **193**, 333–54.

MacFadden, B.J. & Hulbert, R.C. 1988. Explosive speciation at the base of the adaptive radiation of Miocene grazing horses. *Nature* **336**, 466–8.

Mackay, T.F.C. 1986. Transposable element-induced fitness mutations in *Drosophila melanogaster. Genet. Res.* **48**, 77–87.

Mackay, T.F.C. 1987. Transposable element-induced polygenic mutation in *Drosophila melanogaster. Genet. Res.* **49**, 225–33.

Majerus, M.E.N. 1986. The genetics and evolution of female choice. *Trends Ecol. Evol.* **1**, 1–7.

Mangel, M. & Clark, C.W. 1988. *Dynamic Modeling in Behavioral Ecology.* Princeton University Press, Princeton.

Manly, B.F.J. 1985. *The Statistics of Natural Selection on Animal Populations.* Chapman & Hall, London.

Margulis, L. 1981. *Symbiosis in Cell Evolution: Life and its Environment on the Early Earth.* Freeman, San Francisco.

Markgraf, V. 1985. Late Pleistocene faunal extinctions in southern Patagonia. *Science* **228**, 1110–12.

Marsh, H. & Kaysuya, T. 1984. Changes in the ovaries of the short-finned pilot whale, *Globicephala macrorhyncus,* with age and reproductive activity. *Rep. Int. Whal. Comm. Special* **6**, 311–35.

Marshall, D.L. 1988. Postpollination effects on seed paternity: mechanisms in addition to microgametophyte competition operate in wild radish. *Evolution* **42**, 1256–66.

Marshall, D.L. & Ellstrand, N.C. 1986. Sexual selection in *Raphanus sativus*: experimental data on non-random fertilization, maternal choice, and consequences of multiple paternity. *Amer. Natur.* **127**, 446–61.

Marshall, D.L. & Ellstrand, N.C. 1988. Effective mate choice in wild radish: evidence for selective seed abortion and its mechanism. *Amer. Natur.* **131**, 739–56.

Marshall, D.L. & Ellstrand, N.C. 1989. Regulation of mate number in fruits of wild radish. *Amer. Natur.* **133**, 751–65.

Martin, A.P. & Simon, C. 1988. Anomalous distribution of nuclear and mitochondrial DNA markers in periodical cicadas. *Nature* **336**, 237–9.

Martin, M.J., Pérez-Tomé, J.M. & Toro, M.A. 1988. Competition and genotypic variability in *Drosophila melanogaster. Heredity* **60**, 119–23.

Martin, P.S. & Klein, R.G. (eds.) 1984. *Quartenary Extinctions. A Prehistoric Revolution.* University of Arizona Press, Tucson.

Martin, R.D. 1981. Relative brain size and basic metabolic rate in terrestrial vertebrates. *Nature* **293**, 57–60.

May, R.M. 1978. The dynamics and diversity of insect faunas. In: L.A. Mound & N. Waloff (eds.) *Diversity of Insect Faunas.* Symposia of the Royal Entomological Society of London No. 9, pp. 188–204. Blackwell Scientific Publications, London.

May, R.M. 1986. How many species are there? *Nature* **324**, 514–15.

May, R.M. 1988. How many species are there on earth? *Science* **241**, 1441–9.

May, R.M. & Anderson, R.M. 1983. Epidemiology and genetics in the co-evolution of parasites and hosts. *Proc. R. Soc. Lond.* **B219**, 2381–413.

Maynard Smith, J. 1978. *The Evolution of Sex.* Cambridge University Press, Cambridge.

Maynard Smith, J. 1980. A new theory of sexual investment. *Behav. Ecol. Sociobiol.* **7**, 247–51.

Maynard Smith, J. 1982. *Evolution and the Theory of Games.* Cambridge University Press, Cambridge.

Maynard Smith, J. 1984. The ecology of sex. In: J.R. Krebs & N.B. Davies (eds.) *Behavioural Ecology: An Evolutionary Approach.* 2nd edn. pp. 201–21. Blackwell Scientific Publications, Oxford.

Maynard Smith, J. 1986. Contemplating life without sex. *Nature* **324**, 300–1.

Maynard Smith, J. 1987a. Darwinism stays unpunctured. *Nature* **330**, 516.

Maynard Smith, J. 1987b. When learning guides evolution. *Nature* **329**, 761–2.

Maynard Smith, J. 1988. The evolution of recombination. In: R.E, Michod & B.R. Levin (eds.) *The Evolution of Sex*, pp. 106–25. Sinauer, Sunderland, Massachusetts.

Maynard Smith, J. 1989. *Evolutionary Genetics.* Oxford University Press, Oxford.

Maynard Smith, J., Burian, R., Kauffman, S., Alberch, P., Campbell, J., Goodwin, B., Lande, R., Raup, D. & Wolpert, L. 1985. Developmental constraints and evolution. *Quart. Rev. Biol.* **60**, 265–87.

Maynard Smith, J. & Riechert, S.E. 1984. A conflicting tendency model of spider agonistic behaviour: hybrid-pure population line comparisons. *Anim. Behav.* **32**, 564–78.

Mayr, E. 1942. *Systematics and the Origin of Species.* Columbia University Press, New York.

Mayr, E. 1954. Change of genetic environment and evolution. In: J. Huxley, A.C. Hardy & E.B. Ford (eds.) *Evolution as a Process*, pp. 157–80. George Allen & Unwin, London.

Mayr, E. 1963. *Animal Species and Evolution.* Harvard University Press, Cambridge, Massachusetts.

Mayr, E. 1970. *Populations, Speciation and Evolution.* Harvard University Press, Cambridge, Massachussets.

Mayr, E. 1982a. *The Growth of Biological Thought.* Harvard University Press, Cambridge, Massachussets.

Mayr, E. 1982b. Processes of speciation in animals. In: C. Barigozzi (ed.) *Mechanisms of Speciation*, pp. 1–19. Liss, New York.

Mazer, S.J. 1989. Ecological, taxonomic, and life history correlates of seed mass among Indiana dune angiosperms. *Ecol. Monogr* **59**, 153–75.

Mazzini, M. 1976. Giant spermatozoa in *Divales bipustulatus* (Cleridae). *Inst. J. Inst. Morph. Embryol.* **5**, 107–16.

McArdle, B.H., Gaston, K.J. & Lawton, J.H. 1990. Variation in the size of animal populations: patterns, problems and artefacts. *J. Anim. Ecol.* **59**, 439–54.

McCall, C.T., Mitchell-Olds, T. & Waller, D. 1989. Fitness consequences of outcrossing in *Impatiens capensis* tests of the frequency-dependent and sib-competition models. *Evolution* **43**, 1075–84.

McCauley, D.E. & Wade, M.J. 1980. Group selection: the genetic and demographic basis for the phenotypic differentiation of small populations of *Tribolium castaneum. Evolution* **34**, 813–21.

McClintock, B. 1956. Controlling elements and the gene. *Cold Spring Harbor Symp. Quant. Bio.* **21**, 197–216.

McConnell, T.J., Talbot, W.S., McIndoe, R.A. & Wakeland, E.K. 1988. The origin of MHC class II gene polymorphism within the genus *Mus*. *Nature* **332**, 651–4.

McCullagh, P. & Nelder, J.A. 1983. *Generalized Linear Models*. Chapman & Hall, London.

McKenzie, J.A., Whitten, M.J. & Adena, M.A. 1982. The effect of genetic background on the fitness of diazinon resistance genotypes of the Australian sheep blowfly, *Lucilia cuprina*. *Heredity* **49**, 1–9.

McMenamin, M.A.S. 1988. Paleoecological feedback and the Vendian–Cambrian transition. *Trends Ecol. Evol.* **3**, 205–8.

McNamara, K.J. 1986. The role of heterochrony in the evolution of Cambrian trilobites. *Biol. Rev.* **61**, 121–56.

McNamara, K.J. 1988. Patterns of heterochrony in the fossil record. *Trends Ecol. Evol.* **3**, 176–80.

McPheron, B.A., Smith, D.C. & Berlocher, S.H. 1988. Genetic differences between host races of *Rhagoletis pomonella*. *Nature* **336**, 64–6.

Medawar, P.B. 1952. *An Unsolved Problem in Biology*. H.K. Lewis, London.

Menge, B.A., Lubchenco, J., Gaines, S.D. & Ashkenas, L.R. 1986. A test of the Menge–Sutherland model of community organisation in a tropical rocky intertidal food web. *Oecologia (Berl.)* **71**, 75–89.

Menge, B.A. & Sutherland, J.P. 1976. Species diversity gradients: synthesis of the roles of predation, competition and temporal heterogeneity. *Amer. Natur.* **110**, 351–69.

Menge, B.A. & Sutherland, J.P. 1987. Community regulation: variation in disturbance, competition, and predation in relation to environmental stress and recruitment. *Amer. Natur.* **130**, 730–57.

Meeuse, B. & Morris, S. 1984. *The Sex Life of Flowers*. Faber & Faber, London.

Meyer, A., Kocher, T.D., Basasibwaki, P. & Wilson, A.C. 1990. Monophyletic origin of Lake Victoria cichlid fishes suggested by mitochondrial DNA sequences. *Nature* **347**, 550–3.

Michaels, H.J. & Bazzaz, F.A. 1986. Resource allocation and demography of sexual and apomictic *Antennaria parlinii*. *Ecology* **67**, 27–36.

Michaels, H.J., Benner, B., Hartgerink,

A.P., Lee, T.D., Rice, S., Willson, M.F. & Bertin, R.I. 1988. Seed size variations, magnitude, distribution, and ecological correlates. *Evol. Ecol.* **2**, 157–66.

Michod, R.E. & Hasson, O. 1990. On the evolution of reliable indicators of fitness. *Amer. Natur.* **135**, 788–808.

Michod, R.E. & Levin, B.R. (eds.) 1988. *The Evolution of Sex*. Sinauer, Sunderland, Massachusetts.

Migdalski, E.C. & Fichter, G.S. 1976. *The Fresh and Salt Water Fishes of the World*. Alfred A. Knopf, New York.

Milinski, M. 1979. An evolutionarily stable feeding strategy in sticklebacks. *Z. Tierpsychol.* **51**, 36–40.

Milinski, M. & Bakker, T.C.M. 1990. Female sticklebacks use male coloration in mate choice and hence avoid parasitized males. *Nature* **344**, 330–2.

Miller J.S. 1987. Host-plant relationships in the Papilionidae (Lepidoptera): parallel cladogenesis or colonization? *Cladistics* **3**, 105–20.

Mitchell-Olds, T. 1986. Quantitative genetics of survival and growth in *Impatiens capensis*. *Evolution* **40**, 107–16.

Mitchell-Olds, T. & Rutledge, J.J. 1986. Quantitative genetics in natural populations: a review of the theory. *Amer. Natur.* **127**, 379–402.

Mitchell-Olds, T. & Shaw, R.G. 1987. Regression analysis of natural selection: Statistical inference and biological interpretation. *Evolution* **41**, 1149–61.

Mitter, C. & Brooks, D.R. 1983. Phylogenetic aspects of coevolution. In: D.J. Futuyma & M. Slatkin (eds.) *Coevolution*, pp. 65–98. Sinauer, Sunderland, Massachusetts.

Møller, A.P. 1990. Effects of a haematophagous mite on the barn swallow (*Hirundo rustica*): a test of the Hamilton and Zuk hypothesis. *Evolution* **44**, 771–84.

Moran, N.A. 1988. The evolution of host-plant alternation in aphids: evidence for specialization as a dead end. *Amer. Natur.* **132**, 681–706.

Moran, N.A. 1989. A 48-million-year-old aphid–host plant association and complex life cycle: biogeographic evidence. *Science* **245**, 173–5.

Moran, N.A. 1990. Aphid life cycles: two

evolutionary steps. *Amer. Natur.* **136**, 135–8.

Moran, N.A. & Whitham, T.G. 1988. Evolutionary reduction of complex life cycles: loss of host alternation in *Pemphigus* (Homoptera: Aphididae). *Evolution* **42**, 717–28.

Moran, V.C. 1980. Interactions between phtyophagous insects and their *Opuntia* hosts. *Ecol. Entomol.* **5**, 153–64.

Morse, D.R., Stork, N.E. & Lawton, J.H. 1988. Species number, species abundance and body length relationships of arboreal beetles in Bornean lowland rain forest trees. *Ecol. Entomol.* **13**, 25–37.

Morton, S.R. 1990. The impact of European settlement on the vertebrate animals of arid Australia: a conceptual model. *Proc. Ecol. Soc. Aust.* **16**, 201–13.

Mousseau, T.A. & Roff, D.A. 1987. Natural selection and the heritability of fitness components. *Heredity* **59**, 181–97.

Mueller, L.D. 1987. Evolution of accelerated senescence in laboratory populations of *Drosophila. Proc. Natl. Acad. Sci. USA* **84**, 1974–7.

Mukai, T. 1985. Experimental verification of the neutral theory. In: T. Ohta & K.-I. Aoki (eds.), *Population Genetics and Molecular Evolution*, pp. 125–45. Springer-Verlag, Berlin.

Mukai, T. & Nagano, S. 1983. The genetic structure of natural populations of *Drosophila melanogaster*. XVI. Excess additive genetic variance of viability. *Genetics* **105**, 115–34.

Mukai, T., Schaffer, H.E., Watanabe, T.K. & Crow, J.F. 1974. The genetic variance for viability and its components in a population of *Drosophila melanogaster. Genetics* **72**, 763–9.

Mulcahy, D.L. 1974. Adaptive significance of gametic competition. In: H.F. Linskens (ed.) *Fertilization in Higher Plants*, pp. 27–30. North-Holland, Amsterdam.

Muller, H.J. 1964. The relation of recombination to mutational advance. *Mutat. Res.* **1**, 2–9.

Nadeau, J.H., Britton-Davidian, J., Bonhomme, F. & Thaler, L. 1988. H-2 polymorphisms are more uniformly distributed than allozyme polymorphisms in natural populations of house mice. *Genetics* **118**, 131–40.

Nei, M. 1987. *Molecular Evolutionary Genetics.* Columbia University Press, New York.

Nei, M., Maruyama, T. & Chakraborty, R. 1975. The bottleneck effect and genetic variability in populations. *Evolution* **29**, 1–10.

Newell, N.D. 1952. Periodicity in invertebrate evolution. *J. Paleontol.* **26**, 371–85.

Newman, C.M., Cohen, J.E. & Kipnis, C. 1985. Neo-Darwinian evolution implies punctuated equilibria. *Nature* **315**, 400–1.

Newton, I. (ed.) 1989. *Lifetime Reproductive Success in Birds.* Academic Press, London.

Nicholas, W.L. 1983. *The Biology of Free-Living Nematodes.* 2nd edn. Oxford University Press, Oxford.

Niklas, K.J. 1986. Large scale changes in animal and plant terrestrial communities. In: D.M. Raup & D. Jablonski (eds.) *Patterns and Processes in the History of Life*, pp. 383–405. Springer-Verlag, Berlin.

Niklas, K.J., Tiffney, B.H. & Knoll, A.H. 1983. Patterns in vascular plant diversification. *Nature* **303**, 614–16.

Nilsson, L.A. 1988. The evolution of flowers with deep corolla tubes. *Nature* **334**, 147–9.

Nisbet, I.C.T. 1973. Courtship-feeding, egg-size and breeding success in common terns *Nature* **241**, 141–2.

de Nooij, M.P. & van Damme, J.M.M. 1988. Variation in pathogenicity among and within populations of the fungus *Phomopsis subordinaria* infecting *Plantago lanceolata. Evolution* **42**, 1166–71.

Nottebohm, F. 1981. A brain for all seasons: cyclical anatomical changes in song control nuclei of the canary brain. *Science* **214**, 1368–70.

Nur, N. 1988. The consequences of brood size for breeding blue tits. III. Measuring the cost of reproduction: survival, future reproduction, and differential survival. *Evolution* **42**, 351–62.

Nur, U., Werren, J.H., Eickbush, D.G., Burke, W.D. & Eickbush, T.H. 1988. A 'selfish' B chromosome that enhances its transmission by eliminating the paternal genome. *Science* **240**,

512–14.

Oakeshott, J.G., Gibson, J.G., Anderson, P.R., Knibb, W.R., Anderson, D.G. & Chambers, G.K. 1982. Alcohol dehydrogenase and glycerol-3-phosphate dehydrogenase clines in *Drosophila melanogaster* on different continents. *Evolution* **36**, 86–96.

O'Brien, S.J. & Evermann, J.F. 1988. Interactive influence of infectious disease and genetic diversity in natural populations. *Trends Ecol. Evol.* **3**, 254–9.

O'Brien, S.J., Wildt, D.E. & Bush, M. 1986. The cheetah in genetic peril. *Sci. Amer.* **254**(5), 68–76.

O'Brien, S.J., Roelke, M.E., Marker, L., Newman, A., Winkler, C.A., Meltzer, D., Colly, L., Evermann, J.F., Bush, M. & Wildt, D.E. 1985. Genetic basis for species vulnerability in the cheetah. *Science* **227**, 1428–34.

Officer, C.B., Hallam, A., Drake, C.L. & Devine, J.D. 1987. Late Cretaceous and paroxysmal Cretaceous/Tertiary extinctions. *Nature* **326**, 143–9.

Oka, H.I. & Chang, W-T. 1964. Evolution of responses to growing conditions in wild and cultivated rice forms. *Bot. Bull. Acad. Sci.* **5**, 120–38.

Olson, R.R. & Olson, M.H. 1989. Food limitation of planktotrophic marine invertebrate larvae: does it control recruitment success. *Ann. Rev. Ecol. Syst.* **20**, 225–47.

Orgel, L.E. & Crick, F.H.C. 1980. Selfish DNA: the ultimate parasite. *Nature* **284**, 604–7.

Oring, L.W. 1982. Avian mating systems. *Avian Biology,* **6**, 1–92.

Orzack, S.H. & Tuljapurkar, S. 1989. Population dynamics in variable environments. VII. The demography and evolution of iteroparity. *Amer. Natur.* **133**, 901–23.

Otte, D. & Endler, J.A. (eds.) 1989. *Speciation and its Consequences.* Sinauer, Sunderland, Massachussets.

Pacala, S.W. 1988. Competitive equivalence: the coevolutionary consequences of sedentary habit. *Amer. Natur.* **132**, 576–93.

Packer, C. 1986. The ecology of sociality in felids. In: D.I. Rubenstein & R.W. Wrangham (eds.) *Ecological Aspects of Social Evolution,* pp. 429–51. Princeton University Press, Princeton.

Packer, C., Herbst, L., Pusey, A.E., Bygott, J.D., Cairns, S.J., Hanby, J.P. & Borgerhoff-Mulder, M. 1988. Reproductive success of lions. In: T.H. Clutton-Brock (ed.) *Reproductive Success,* pp. 363–83. University of Chicago Press, Chicago.

Packer, C. & Ruttan, L. 1988. The evolution of cooperative hunting. *Amer. Natur.* **132**, 159–98.

Packer, C., Scheel, D. & Pusey, A.E. 1990. Why lions form groups: food is not enough. *Amer. Natur.* **136**, 1–19.

Pagel, M.D. & Harvey, P.H. 1988a. Recent developments in the analysis of comparative data. *Q. Rev. Biol.* **63**, 413–40.

Pagel, M.D. & Harvey, P.H. 1988b. The taxon-level problem in the evolution of mammalian brain size: facts and artifacts. *Amer. Natur.* **132**, 344–59.

Pagel, M.D. & Harvey, P.H. 1989a. Comparative methods of examining adaptation depend on evolutionary models. *Folia Primatol.* **53**, 203–20.

Pagel, M.D. & Harvey, P.H. 1989b. Taxonomic differences in the scaling of brain on body size among mammals. *Science* **244**, 1589–93.

Pagel, M.D., May, R.M. & Collie, A. 1991. Ecological aspects of the geographic distribution and diversity of mammal species. *Amer. Natur.* in press. **137**, 791–815.

Paine, R.T. 1966. Food web complexity and species diversity. *Amer. Natur.* **100**, 65–74.

Paley, W. 1828. *Natural Theology.* 2nd edn. J. Vincent, London.

Palmer, A.R. 1985. Quantum changes in gastropod shell morphology need not reflect speciation. *Evolution* **39**, 699–705.

Palmer, E. & Pitman, N. 1972. *Trees of Southern Africa.* A.A. Balkema, Capetown.

Paris, O.H. & Pitelka, F.A. 1962. Population characteristics of the terrestrial isopod *Armadillidium vulgare* in California grassland. *Ecology* **43**, 229–48.

Parker, G.A. 1978. Evolution of competitive mate searching. *Ann. Rev. Entomol.* **23**, 173–96.

Parker, M.A. 1988. Genetic uniformity and disease resistance in a clonal plant. *Amer. Natur.* **132**, 538–49.

Partridge, L. 1980. Mate choice increases a component of offspring fitness in fruit flies. *Nature* **283**, 290–1.

Partridge, L. 1983. Non-random mating and offspring fitness. In: P. Bateson (ed.) *Mate Choice*, pp. 227–55. Cambridge University Press, Cambridge.

Partridge, L. 1989. An experimentalist's approach to the role of costs of reproduction in the evolution of life-histories. In: P.J. Grubb & J.B. Whittaker (eds.) *Toward a More Exact Ecology*, pp. 231–46. Blackwell Scientific Publications, Oxford.

Partridge, L. & Andrews, R. 1985. The effect of reproductive activity on the longevity of male *Drosophila melanogaster* is not caused by an acceleration of ageing. *J. Insect Physiol.* **31**, 393–5.

Partridge, L., Green, A. & Fowler, K. 1987. Effects of egg-production and of exposure to males on the survival and egg-production rates of female *Drosophila melanogaster*. *J. Insect Physiol.* **33**, 745–9.

Partridge, L. & Farquhar, M. 1981. Sexual activity reduces lifespan of male fruitflies. *Nature* **294**, 580–2.

Partridge, L. & Harvey, P.H. 1988. The ecological context of life history evolution. *Science* **241**, 1449–55.

Patterson, C. & Smith, A.B. 1987. Is the periodicity of extinctions a taxonomic artefact? *Nature* **330**, 248–51.

Paterson, H.E.H. 1978. More evidence against speciation by reinforcement. *South Afr. J. Sci.* **74**, 369–71.

Paterson, H.E.H. 1982. Perspective on speciation by reinforcement. *South Afr. J. Sci.* **78**, 53–7.

Patton, J.L., Berlin, B. & Berlin, E.A. 1982. Aboriginal perspectives of a mammal community in Amazonian Perú: knowledge and utilization patterns among the Aguaruna Jívaro. In: M.A. Mares & H.H. Genoways (eds.) *Mammalian Biology in South America*. Pymatuning Symposia in Ecology. Vol. 6, pp. 111–28. Pymatuning Laboratory of Ecology, Linesville, Pennsylvania.

Pease, C.M., Lande, R. & Bull, J.J. 1989. A model of population growth, dispersal, and evolution in a changing environment. *Ecology* **70**, 1657–64.

Pemberton, J.M., Albon, S.D., Guinness, F.E., Clutton-Brock, T.H. & Berry, R.J. 1988. Genetic variation and juvenile survival in red deer. *Evolution* **42**, 921–34.

Peters, R.H. 1976. Tautology in evolution and ecology. *Amer. Natur.* **110**, 1–12.

Peters, R.H. 1983. *The Ecological Implications of Body Size*. Cambridge University Press, Cambridge.

Petersen, J.J., Chapman, H.C. & Woodward, D.B. 1968. The bionomics of a mermithid nematode of larval mosquitoes in southwestern Louisiana. *Mosquito News* **28**, 346–52.

Petraitis, P.S., Latham, R.E. & Niesenbaum, R.A. 1989. The maintenance of species diversity by disturbance. *Q. Rev. Biol.* **64**, 393–418.

Pettifor, R.A., Perrins, C.M. & McCleery, R.H. 1988. Individual optimization of clutch size in great tits. *Nature* **336**, 160–2.

Phillipi, T. & Seger, J. 1989. Hedging one's evolutionary bets, revisited. *Trends Ecol. Evol.* **4**, 41–4.

Phillips, P.C. & Arnold, S.J. 1989. Visualizing mutlivariate selection. *Evolution* **43**, 1209–22.

Pianka E.R. 1970. On *r*- and *K*-selection. *Amer. Natur.* **104**, 592–7.

Pierce, N.E. 1985. Lycaenid butterflies and ants: selection for nitrogen-fixing and other protein-rich food plants. *Amer. Natur.* **125**, 888–95.

Pierce, N.E. & Elgar, M.A. 1985. The influence of ants on host plant selection by *Jalmenus evagoras*, a myrmecophilous lycaenid butterfly. *Behav. Ecol. Sociobiol.* **16**, 209–22.

Pierce, N.E. & Mead, P.S. 1981. Parasitoids as selective agents in the symbiosis between lycaenid butterfly caterpillars and ants. *Science* **211**, 1185–7.

Pierson, E.D., Sarich, V.M., Lowenstein, J.M., Daniel, M.J. & Rainey, W.E. 1986. A molecular link between the bats of New Zealand and South America. *Nature* **323**, 60–3.

Pimm. S.L. 1987. Determining the effects of introduced species. *Trends Ecol. Evol.* **2**, 106–8.

Pimm, S.L. & Gittleman, J.L. 1990. Carnivores and ecologists on the road to Damascus. *Trends Ecol. Evol.* **5**, 70–3.

Pimm, S.L., Jones, H.L. & Diamond, J.

1988. On the risk of extinction. *Amer. Natur.* **132**, 757–85.

Pimm, S.L. & Redfearn, A. 1988. The variability of population densities. *Nature* **334**, 613–14.

Policansky, D. 1981. Sex choice and the size advantage model in jack-in-the-pulpit (*Arisaema triphyllum*). *Proc. Natl. Acad. Sci. USA* **78**, 1306–8.

Pomiankowski, A. 1987a. The costs of choice in sexual selection. *J. theor. Biol.* **128**, 195–218.

Pomiankowski, A. 1987b. Sexual selection: the handicap principle does work—sometimes. *Proc. R. Soc. Lond.* **B231**, 123–45.

Popper, K. 1974. Darwinism as a metaphysical research program. In: P.A. Schilpp (ed.) *The Philosophy of Karl Popper*. Open Court, La Salle.

Potts, W.K. & Wakeland, E.K. 1990. Evolution of diversity at the major histocompatibility complex. *Trends Ecol. Evol.* **5**, 181–7.

Price, P.W., Westoby, M. & Rice, B. 1988. Parasite-mediated competition: some predictions and tests. *Amer. Natur.* **131**, 544–55.

Price, P.W., Westoby, M., Rice, B., Atsatt, P.R., Fritz, R.S., Thompson, J.N. & Mobley, K. 1986. Parasite mediation in ecological interactions. *Ann. Rev. Ecol. Syst.* **17**, 487–505.

Primack, R.B. & Antonovics, J. 1981. Experimental ecological genetics in *Plantago*. V. Components of seed yield in the ribwort plantain, *Plantago lanceolata* L. *Evolution* **35**, 1069–79.

Prokopy, R.J., Averill, A.L., Cooley, S.S. & Roitberg, B.D. 1982. Associative learning in egg-laying site selection by apple maggot flies. *Science* **218**, 76–77.

Prokopy, R.J., Diehl, S.R. & Cooley, S.S. 1988. Behavioral evidence for host races in *Rhagoletis pomonella Oecologia* (Berl.) **76**, 138–47.

Pryor, L.D. 1976. *Biology of Eucalypts*. Studies in Biology no. 61. Edward Arnold, London.

Qin, T.K. & Gullan, P.J. 1989. *Cryptostigma* Ferris—a coccoid genus with a strikingly disjunct distribution. *Syst. Entomol.* **14**, 221–32.

Queller, D.C. 1987. Sexual selection in flowering plants. In: J.W. Bradbury & M.B. Andersson (eds.) *Sexual Selection; Testing the Alternatives*, pp. 165–79. Wiley & Sons, Chichester.

Quinn, J.F. 1987. On the statistical detection of cycles in extinctions in the fossil record. *Paleobiology* **13**, 465–78.

Quinn, J.F. & Signor, P.W. 1989. Death stars, ecology, and mass extinctions. *Ecology* **70**, 824–34.

Rabinowitz, R. 1981. Seven forms of rarity. In: H. Synge (ed.) *The Biological Aspects of Rare Plant Conservation*, pp. 205–17. Wiley & Sons, Chichester.

Rabinowitz, D., Cairns, S. & Dillon, T. 1986. Seven forms of rarity and their frequency in the flora of the British Isles. In: M. Soulé (ed.) *Conservation Biology: the Science of Scarcity and Diversity*, pp. 182–204. Sinauer, Sunderland, Massachussets.

Ralls, K. & Ballou, J. 1986. Captive breeding programs for populations with a small number of founders. *Trends Ecol. Evol.* **1**, 19–22.

Ralls, K., Ballou, J.D. & Templeton, A. 1988. Estimates of lethal equivalents and the cost of inbreeding in mammals. *Conserv. Biol.* **2**, 185–93.

Ralls, K. & Harvey, P.H. 1985. Geographic variation in size and sexual dimorphism of North American weasels. *Biol. J. Linn. Soc.* **25**, 119–67.

Rampino, M.R. & Volk, T. 1988. Mass extinctions, atmospheric sulphur and climatic warming at the K/T boundary. *Nature* **332**, 63–5.

Ratcliffe, D. 1979. The end of the large blue butterfly. *New Scientist* **8**, 457–8.

Raup, D.M. 1978. Cohort analysis of generic survivorship. *Paleobiology* **4**, 1–15.

Raup, D.M. 1986. *The Nemesis Affair*. Norton, New York.

Raup, D.M. 1987. Major features of the fossil record and their implications for evolutionary rate studies. In: K.S.W. Campbell & M.F. Day (eds.) *Rates of Evolution*, pp. 1–14. George Allen & Unwin, London.

Raup, D.M. & Boyajian, G.E. 1988. Patterns of generic extinction in the fossil record. *Paleobiology* **14**, 109–25.

Raup, D.M. & Jablonski, D. (eds.) 1986. *Patterns and Processes in the History of Life*. Springer-Verlag, Berlin.

Raup, D.M. & Sepkoski, J.J. 1984. Periodicity of extinctions in the geological past. *Proc. Natl. Acad. Sci. USA* **81**, 801–5.

Raup, D.M. & Sepkoski, J.J. 1986. Periodic extinction of families and genera. *Science* **231**, 833–6.

Rausher, M.D. 1978. Search image for leaf shape in a butterfly. *Science* **200**, 1071–3.

Rausher, M.D. 1988. Is coevolution dead? *Ecology* **69**, 898–901.

Read, A.F. 1987. Comparative evidence supports the Hamilton and Zuk hypothesis on parasites and sexual selection. *Nature* **328**, 68–70.

Read, A.F. 1988. Sexual selection and the role of parasites. *Trends Ecol. Evol.* **3**, 97–102.

Read, A.F. & Harvey, P.H. 1989a. Life history differences among the eutherian radiations. *J. Zool. Lond.* **219**, 329–53.

Read, A.F. & Harvey, P.H. 1989b. Reassessment of the comparative evidence for Hamilton and Zuk's theory on the evolution of secondary sexual characters. *Nature* **339**, 618–20.

Recher, H.F. 1969. Bird species diversity and habitat diversity in Australia and North America. *Amer. Natur.* **103**, 75–80.

Redfearn, A. & Pimm, S.L. 1988. Population variability and polyphagy in herbivorous insect communities. *Ecol. Monogr.* **58**, 39–55.

Reekie, E.G. & Bazzaz, F.A. 1987a. Reproductive effort in plants. 1. Carbon allocation to reproduction. *Amer. Natur.* **129**, 876–96.

Reekie, E.G. & Bazzaz, F.A. 1987b. Reproductive effort in plants. 2. Does carbon reflect the allocation of other resources? *Amer. Natur.* **129**, 876–906.

Reekie, E.G. & Bazzaz, F.A. 1987c. Reproductive effort in plants. 3. Effect of reproduction on vegetative activity. *Amer. Natur.* **129**, 907–19.

Reeve, H.K., Westneat, D.F., Noon, W.A., Sherman, P.W. & Aquadro, C.F. 1990. DNA 'fingerprinting' reveals high levels of inbreeding in colonies of the eusocial naked mole-rat. *Proc. Natl. Acad. Sci. USA* **87**, 2496–500.

Reitter, E. 1960. *Beetles*. Paul Hamlyn, London.

Reznick, D. 1985. Costs of reproduction: an evaluation of the empirical evidence. *Oikos* **44**, 257–67.

Reznick, D.A., Bryga, H. & Endler, J.A. 1990. Experimentally induced life-history evolution in a natural population. *Nature* **346**, 357–9.

Reznick, D. & Endler, J.A. 1982. The impact of predation on life history evolution in Trinidanian guppies (*Poecilia reticulata*). *Evolution* **36**, 160–77.

Reznick, D., Perry, E. & Travis, J. 1986. Measuring the cost of reproduction: a comment on papers by Bell. *Evolution* **40**, 1338–44.

Rich, P.V., Rich, T.H., Wagstaff, B.E., Mason, J.M., Douthitt, C.B., Gregory, R.T. & Felton, E.A. 1988. Evidence for low temperature and biological diversity in Cretaceous high latitudes of Australia. *Science* **242**, 1403–6.

Richerson, P.J. & Lum, K. 1980. Patterns of plant species diversity in California: relation to weather and topography. *Amer. Natur.* **116**, 504–36.

Richmond, R.C. & Senior, A. 1981. Esterase 6 of *Drosophila melanogaster*. kinetics of transfer to females, decay in females and male recovery. *J. Insect Physiol.* **27**, 849–55.

Richter-Dyn, N. & Goel, N.S. 1972. On the extinction of a colonizing species. *Theor. Popul. Biol.* **3**, 406–33.

Ricklefs, R.E. 1980. Geographical variation in clutch size among passerine birds: Ashmole's hypothesis. *Auk* **97**, 38–49.

Ricklefs, R.E. 1987. Community diversity: relative roles of local and regional processes. *Science* **235**, 167–71.

Ridley, M. 1983. *The Explanation of Organic Diversity: The Comparative Method and Adaptations for Mating.* Clarendon Press, Oxford.

Ridley, M. 1989. The incidence of sperm displacement in insects: four conjectures, one corroboration. *Biol. J. Linn. Soc.* **38**, 349–67.

Riechert, S.E. 1986. Spider fights: a test of evolutionary game theory. *Amer. Sci.* **4**, 604–10.

Riechert, S.E. & Hammerstein, P. 1983. Game theory in an ecological context. *Ann. Rev. Ecol. Syst.* **14**, 377–409.

Riska, B. 1986. Some models for development, growth and morphometric correlation. *Evolution* **40**, 1303–11.

Riska, B. & Atchley, W.R. 1985. Genetics of growth predict brain-size evolution. *Science* **229**, 668–71.

Robbins, C.S., Sauer, J.R., Greenberg, R.S. & Droege, S. 1989. Population declines in North American birds that migrate to the tropics. *Proc. Natl. Acad. Sci. USA* **86**, 7658–62.

Robertson, A. 1952. The effect of inbreeding on the variation due to recessive genes. *Genetics* **37**, 189–207.

Robinson, E.R. & Gibbs Russell, G.E. 1982. Speciation environments and centres of diversity in southern Africa. I. Conceptual framework. *Bothalia* **14**, 83–8.

Roff, D.A. & Mousseau, T.A. 1987. Quantitative genetics and fitness: lessons from *Drosophila*. *Heredity* **58**, 103–18.

Rohde, K. 1979. A critical evaluation of intrinsic and extrinsic factors responsible for niche restriction in parasites. *Amer. Natur.* **114**, 648–71.

Rohde, K. 1982. *Ecology of Marine Parasites*. University of Queensland Press, St Lucia, Queensland.

Room, P.M. 1990. Ecology of a simple plant–herbivore system: biological control of *Salvinia*. *Trends Ecol. Evol.* **5**, 74–9.

Rose, M.R. 1984. The evolution of animal senescence. *Can. J. Zool.* **62**, 1661–7.

Rose, M.R. & Charlesworth, B. 1980. A test of evolutionary theories of senescence. *Nature* **287**, 141–2.

Rose, M.R., Dorey, M.L., Coyle, A.M. & Service, P.M. 1984. The morphology of postponed senescence in *Drosophila melanogaster*. *Can. J. Zool.* **62**, 1576–80.

Rose, M.R., Service, P.M. & Hutchinson, E.W. 1987. Three approaches to trade-offs in life history evolution. In: V. Loeschcke (ed.) *Genetic Constraints on Adaptive Evolution*, pp. 91–105. Springer-Verlag, Berlin.

Rosenzweig, M.L. 1978. Competitive speciation. *Biol. J. Linn. Soc.* **10**, 275–89.

Rosenzweig, M.L., Brown, J.S. & Vincent, T.L. 1987. Red Queens and ESS: the coevolution of evolutionary rates. *Evol. Ecol.* **1**, 59–94.

Ross, M. 1990. Sexual asymmetry in hermaphroditic plants. *Trends Ecol. Evol.* **5**, 43–7.

Roth, J., LeRoith, D., Shiloach, J., Rosenzweig, J.L., Lesniak, M.A. & Havrankova, J. 1982. The evolutionary origins of hormones, neurotransmitters, and other extracellular chemical messengers. *New Engl. J. Med.* **306**, 523–7.

Roughgarden, J. 1979. *Theory of Population Genetics and Evolutionary Ecology: An Introduction*. MacMillan, New York.

Roughgarden, J., Gaines, S. & Possingham, H. 1988. Recruitment dynamics in complex life cycles. *Science* **241**, 1460–6.

Rouhani, S. & Barton, N.H. 1987. The probability of peak shifts in a founder population. *J. theor. Biol.* **126**, 51–62.

Rowell, D. 1987. Complex sex-linked translocation heterozygosity: its genetics and biological significance. *Trends Ecol. Evol.* **2**, 242–6.

Rowley, I. 1968. Communal breeding of Australian birds. *Bonn Zool. Beitr.* **19**, 362–70.

Rummell, J.D. & Roughgarden, J. 1985. A theory of faunal buildup for competition communities. *Evolution* **39**, 1009–33.

Russell, E.M. 1989. Cooperative breeding—a Gondwanan perspective. *Emu* **89**, 61–2.

Ryan, M.J. 1990. Signals, species and sexual selection. *Amer. Sci.* **78**, 46–62.

Ryan, M.J., Fox, J.H., Wilczynski, W. & Rand, A.S. 1990. Sexual selection for sensory exploitation in the frog *Physalaemus pustulosus*. *Nature* **343**, 66–7.

Ryan, M.J. & Rand, A.S. 1990. The sensory basis of sexual selection for complex calls in the Túngara frog, *Physalaemus pustulosus* (sexual selection for sensory exploitation). *Evolution* **44**, 305–14.

Sage, R.D., Heyneman, D., Lim, K.-C. & Wilson, A.C. 1986. Wormy mice in a hybrid zone. *Nature* **324**, 60–3.

Sale, P.F. 1977. Maintenance of high diversity in coral reef fish communities. *Amer. Natur.* **111**, 337–59.

Sale, P.F. 1982. Stock–recruit relationships and regional coexistence in a lottery competitive system: a simulation study. *Amer. Natur.* **120**, 139–59.

Salisbury, E. 1942. *The Reproductive Capacity of Plants*. Bell, London.

Salisbury, E. 1964. *Weeds and Aliens*. Collins, London.

Salvini-Plawen, L.V. & Splechtna, H. (1979) Zur Homologie der Keinblatter. *Z. Systemat. Evolutionsforsch.* **17**, 10–30.

Sander, K. 1983. The evolution of patterning mechanisms: gleanings from insect embryogenesis and spermatogenesis. In: B.C. Goodwin, L. Holder & C.C. Wylie (eds.) *Development and Evolution*, pp. 137–59. Cambridge University Press, Cambridge.

Sanderson, N. 1989. Can gene flow prevent reinforcement? *Evolution* **43**, 1223–5.

Sanson, G. 1982. Evolution of feeding adaptations in fossil and recent macropodids. In: P.V. Rich & E.M. Thompson (eds.) *The Fossil Vertebrate Record of Australia*, pp. 489–506. Monash University Press, Clayton.

Sanson, G. 1989. Morphological adaptations of teeth to diets and feeding in the Macropodoidea. In: G. Grigg, P. Jarman & I. Hume (eds.) *Kangaroos, Wallabies and Rat-Kangaroos*, pp. 151–68. Surrey Beatty & Son, New South Wales.

Sattaur, O. 1989. The shrinking gene pool. *New Scientist* **1675**, 37–41.

Schaffer, W.M. & Schaffer, M.V. 1977. The adaptive significance of variation in the reproductive habit in Agavaceae. In: B. Stonehouse & C.M. Perrins (eds.) *Evolutionary Ecology*, pp. 261–76. University Park Press, Baltimore.

Schaffer, W.M. & Schaffer, M.V. 1979. The adaptive significance of variations in the reproductive habit in the Agavaceae. II: Pollinator foraging behavior and selection for increased reproductive expenditure. *Ecology* **60**, 1051–69.

Schaller, G.B., Jinchu, H., Wenshi, P. & Jing, Z. 1985. *The Giant Pandas of Wolong*. University of Chicago Press, Chicago.

von Schantz, T., Göransson, G., Andersson, G., Fröberg, I., Grahn, M., Helgée, A. & Wittzell, H. 1989. Female choice selects for a viability-based male trait in pheasants. *Nature* **337**, 166–9.

Schemske, D.W. & Lande, R. 1985. The evolution of self-fertilization and inbreeding depression in plants. II. Empirical observations. *Evolution* **39**, 41–52.

Schlichting, C.D. 1986. The evolution of phenotypic plasticity in plants. *Ann. Rev. Ecol. Syst.* **17**, 667–93.

Schlichting, C.D. & Levin, D.A. 1986. Phenotypic plasticity: an evolving plant character. *Biol. J. Linn. Soc.* **29**, 37–47.

Schluter, D. 1988a. Estimating the form of natural selection on a quantitative trait. *Evolution* **42**, 849–61.

Schluter, D. 1988b. Character displacement and the adaptive divergence of finches on islands and continents. *Amer. Natur.* **131**, 799–824.

Schluter, D. 1988c. The evolution of finch communities on islands and continents: Kenya vs. Galapagos. *Ecol. Monogr.* **58**, 229–49.

Schluter, D., Price, T.D. & Grant, P.R. 1985. Ecological character displacement in Darwin's finches. *Science* **227**, 1056–9.

Schmalhausen, I.I. 1949. *Factors of Evolution*. Blakiston, Philadelphia.

Schmitt, J. & Antonovics, J. 1986a. Experimental studies of the evolutionary significance of sexual reproduction. III. Maternal and paternal effects during seedling establishment. *Evolution* **40**, 817–29.

Schmitt, J. & Antonovics, J. 1986b. Experimental studies of the evolutionary significance of sexual reproduction. IV. Effect of neighbour relatedness and aphid infestation on seedling performance. *Evolution* **40**, 830–6.

Schmitt, J. & Erhardt, D.W. 1987. A test of the sib-competition hypothesis for outcrossing advantage in *Impatiens capensis*. *Evolution* **41**, 579–90.

Schmitt, J. & Erhardt, D.W. 1990. Enhancement of inbreeding depression by dominance and suppression in *Impatiens capensis*. *Evolution* **44**, 269–78.

Schoener, T.W. 1983a. Field experiments on interspecific competition. *Amer. Natur.* **122**, 240–85.

Schoener, T.W. 1983b. Rate of species turnover decreases from lower to higher organisms: a review of the data. *Oikos* **41**, 372–7.

Schoener, T.W. 1984. Size differences among sympatric bird-eating hawks: a world-wide survey. In: D.R. Strong, D. Simberloff, L.G. Abele & A.B. Thistle (eds.) *Ecological Communities: Conceptual Issues and the Evidence*, pp. 254–81. Princeton University Press, Princeton.

Schoener, T.W. 1985. Some comments on Connell's and my reviews of field experiments on interspecific competition. *Amer. Natur.* **125**, 730–40.

Schoener, T.W. 1987. The geographical distribution of rarity. *Oecologia* **74**, 161–73.

Schoener, T.W. 1989. Food webs from the small to the large. *Ecology* **70**, 1559–89.

Schoener, T.W. & Spiller, D.A. 1987. High population persistence in a system with high turnover. *Nature* **330**, 474–7.

Schopf, T.J.M. 1974. Permo-Triassic extinctions: relation to sea-floor spreading. *J. Geol.* **82**, 129–43.

Schulz, D.L. 1989. The evolution of phenotypic variance with iteroparity. *Evolution* **43**, 473–5.

Scott, D. 1986. Inhibition of female *Drosophila melanogaster* remating by a seminal protein (esterase 6). *Evolution* **40**, 1084–91.

Scott, D.K. & Clutton-Brock, T.H. 1990. Mating systems, parasites and plumage dimorphism in waterfowl. *Behav. Ecol. Sociobiol.* **26**, 261–76.

Seger, J. 1985. Unifying genetic models for the evolution of female choice. *Evolution* **39**, 1185–93.

Seger, J. & Brockmann, H.J. 1987. What is bet-hedging? *Oxford Surv. Evol. Biol.* **4**, 182–211.

Seger, J. & Hamilton, W.D. 1988. Parasites and sex. In: R.E. Michod & B.R. Levin (eds.) *The Evolution of Sex*, pp. 176–93. Sinauer, Sunderland, Massachusetts.

Semlitsch, R.D. 1987. Density-dependent growth and fecundity in the paedomorphic salamander *Ambystoma talpoideum. Ecology* **68**, 1003–8.

Semlitsch, R.D., Scott, D.E. & Pechmann, J.H.K. 1988. Time and size at metamorphosis related to adult fitness in *Ambystoma talpoideum. Ecology* **69**, 184–92.

Sepkoski, J.J. 1986. Phanerozoic overview of mass extinction. In: D.M. Raup & D. Jablonski (eds.) *Patterns and Processes in the History of Life*, pp. 277–95. Springer-Verlag, Berlin.

Sepkoski, J.J. 1987. Environmental trends in extinction during the Paleozoic. *Science* **235**, 64–6.

Sepkoski, J.J. 1989. Periodicity in mass extinction and the problem of catastrophism in the history of life. *J. Geol. Soc. London* **146**, 7–19.

Service, P.M., Hutchinson, E.W. & Rose, M.R. 1988. Multiple genetic mechanisms for the evolution of senescence in *Drosophila melanogaster. Evolution* **42**, 708–16.

Service, P.M. & Rose, M.R. 1985. Genetic covariation among life-history components: the effects of novel environments. *Evolution* **39**, 943–45.

Sessions, S.K. & Larson, A. 1987. Developmental correlates of genome size in plethodontid salamanders and their implications for genome evolution. *Evolution* **41**, 1239–51.

Sharp, P.L. & Hayman, D.L. 1988. An examination of the role of chiasma frequency in the genetic system of marsupials. *Heredity* **60**, 77–85.

Sharp, P.M. & Li, W.-H. 1987. Molecular evolution of ubiquitin genes. *Trends Ecol. Evolution* **2**, 328–32.

Shaw, R.F. & Moehler, J.D. 1983. The selective advantage of the sex ratio. *Amer. Natur.* **87**, 337–42.

Sheldon, P.R. 1987. Parallel gradualistic evolution of Ordovician trilobites. *Nature* **330**, 561–3.

Shine, R. 1988. The evolution of large body size in females: a critique of Darwin's 'fecundity advantage' model. *Amer. Natur.* **131**, 124–31.

Sibley, C.G., Ahlquist, J.E. & Monroe, B.L. 1988. A classification of the living birds of the world based on the DNA–DNA hybridization studies. *Auk* **105**, 409–23.

Sibly, R. & Calow, P. 1985. Classification of habitats by selection pressures: a synthesis of life-cycle and r/K theory. In: R. Sibly & R. Smith (eds.) *Behavioural Ecology: Ecological Consequences of Adaptive Behaviour*, pp. 75–90. Blackwell Scientific Publications, Oxford.

Sibly, R. & Calow, P. 1986. Why breeding earlier is always worthwhile. *J. theor. Biol.* **123**, 311–19.

Silvertown, J. 1984. Phenotypic variety in germination behavior: the ontogeny and evolution of somatic polymorphism in seeds. *Amer. Natur.* **124**, 1–16.

Silvertown, J. 1985. When plants play the field. In: P.J. Greenwood, P.H. Harvey & M. Slatkin (eds.) *Evolution: Essays in Honour of John Maynard Smith*, pp. 143–53. Cambridge University Press, Cambridge.

Silvertown, J. 1988. The demographic and evolutionary consequences of seed dormancy. In: A.J. Davy, M.J. Hutchings & A.R. Watkinson (eds.) *Plant Population Ecology*, pp. 205–19. Blackwell Scientific Publications, Oxford.

Silvertown, J. 1989. The paradox of seed size and adaptation. *Trends Ecol. Evol.* **4**, 24–6.

Simberloff, D.S. 1981. Community effects of introduced species. In: M.H. Nitecki (ed.) *Biotic Crises in Ecological and Evolutionary Time*, pp. 53–81. Academic Press, New York.

Simberloff, D.S. 1986. The proximate causes of extinction. In: D.M. Raup & D. Jablonski (eds.) *Patterns and Processes in the History of Life*, pp. 259–76. Springer-Verlag, Berlin.

Simberloff, D.S. 1988. The contribution of population and community biology to conservation science. *Ann. Rev. Ecol. Syst.* **19**, 473–511.

Simmonds, N.W. 1981. Genotype (G), environment (E) and GE components of crop yields. *Exp. Agric.* **17**, 355–62.

Simon, C. 1987. Hawaiian evolutionary biology: an introduction. *Trends Ecol. Evol.* **2**, 175–8.

Simonsen, L. & Levin, B.R. 1988. Evaluating the risk of releasing genetically engineered organisms. *Trends Ecol. Evol.* **3**, S27–30.

Simpson, G.G. 1952. How many species? *Evolution* **6**, 342.

Simpson, G.G. 1953. *The Major Features of Evolution*. Columbia University Press, New York.

Sinclair, A.R.E. & Fryxell, J.M. 1985. The Sahel of Africa: ecology of a disaster. *Can. J. Zool.* **63**, 987–94.

Singh, G. & Geissler, E.A. 1985. Late Cainozoic history of vegetation, fire, lake levels and climate, at Lake George, New South Wales, Australia. *Phil. Trans. R. Soc. Lond.* **B311**, 379–447.

Skaife, S.H. 1955. *Dwellers in Darkness*. Longmans Green, London.

Slatkin, M. 1974. Hedging one's evolutionary bets. *Nature* **250**, 704–5.

Slatkin, M. 1981. Populational heritability. *Evolution* **35**, 859–71.

Slatkin, M. 1985a. Somatic mutations as an evolutionary force. In: P.J. Greenwood, P.H. Harvey & M. Slatkin (eds.). *Evolution: Essays in Honour of John Maynard Smith*, pp. 19–30. Cambridge University Press, Cambridge.

Slatkin, M. 1985b. Gene flow in natural populations. *Ann. Rev. Ecol. Syst.* **16**, 393–430.

Slatkin, M. 1987. Gene flow and the geographic structure of natural populations. *Science* **236**, 787–92.

Slobodkin, L.B., Smith, F.E. & Hairston, N.G. 1967. Regulation in terrestrial ecosystems and the implied balance of nature. *Amer. Natur.* **101**, 109–24.

Smith, A.B. & Patterson, C. 1988. The influence of taxonomic method on the perception of patterns in evolution. *Evol. Biol.* **23**, 127–216.

Smith, D.C. 1988. Heritable divergence of *Rhagoletis pomonella* host races by seasonal asynchrony. *Nature* **336**, 66–7.

Smith, J.N.M. 1981. Does high fecundity reduce survival in song sparrows? *Evolution* **35**, 1142–61.

Smith, T.B. 1987. Bill size polymorphism and intraspecific niche utilization in an African finch. *Nature* **329**, 717–19.

Smith, T.B. 1990a. Resource use by bill morphs of an African finch: evidence for intraspecific competition. *Ecology* **71**, 1246–57.

Smith, T.B. 1990b. Natural selection on bill characters in the two bill morphs of the African finch *Pyrenestes ostrinus*. *Evolution* **44**, 832–42.

Smith-Gill, S.J. 1983. Developmental plasticity: developmental conversion versus phenotypic modulation. *Amer. Zool.* **24**, 47–55.

Snyder, M. & Doolittle, W.F. 1988. P

elements in *Drosophila*: selection at many levels. *Trends Genet* **4**, 147–9.

Sober, E. 1984. *The Nature of Selection: Evolutionary Theory in Philosophical Focus*. MIT Press, Cambridge, MA.

Sokal, R.R. 1970. Senescence and genetic load: evidence from *Tribolium*. *Science* **167**, 1733–4.

Soler, M. & Møller, A.P. 1990. Duration of sympatry and coevolution between the great spotted cuckoo and its magpie host. *Nature* **343**, 748–50.

Solomon, B.P. 1985. Environmentally influenced changes in sex expression in an andromonoecious plant. *Ecology* **65**, 1321–32.

Soltis, D.E. & Soltis, P.S. 1987. Polyploidy and breeding systems in homosporous Pteridophyta: a reevaluation. *Amer. Natur.* **130**, 219–32.

Soulé, M. (ed.) 1986. *Conservation Biology: the Science of Scarcity and Diversity*. Sinauer, Sunderland, Massachussets.

Soulé, M. (ed.) 1987. *Viable Populations for Conservation*. Cambridge University Press, Cambridge.

Soulé, M. & Simberloff, D. 1986. What do genetics and ecology tell us about the design of nature reserves? *Biol. Conserv.* **35**, 19–40.

Southwood, T.R.E. 1977. Habitats, the templet for ecological strategies. *J. Anim. Ecol.* **46**, 337–65.

Southwood, T.R.E. 1978. The components of diversity. In: L.A. Mound & N.Waloff (eds.) *Diversity of Insect Faunas*. Symposia of the Royal Entomological Society of London, No. 9 pp. 19–40. Blackwell Scientific Publications, London.

Southwood, T.R.E. 1987. The concept and nature of the community. In: J.H.R. Gee & P.S. Giller (eds.) *Organization of Communities: Past and Present*, pp. 3–27. Blackwell Scientific Publications, Oxford.

Southwood, T.R.E. 1988. Tactics, strategies and templets. *Oikos* **52**, 3–18.

Spencer, H.G., McArdle, B.H. & Lambert, D.M. 1986. A theoretical investigation of speciation by reinforcement. *Amer. Natur.* **128**, 241–62.

Spiller, D.A. & Schoener, T.W. 1990. A terrestrial field experiment showing the impact of eliminating top predators on foliage damage. *Nature* **347**, 469–72.

Stanley, S.M. & Yang, X. 1987. Approximate evolutionary stasis for bivalve morphology over millions of years: a multivariate, multilineage study. *Paleobiology* **13**, 113–39.

Stanton, M.L. 1984. Short-term learning and the searching accuracy of egg-laying butterflies. *Anim. Behav.* **32**, 33–40.

Stearns, S.C. 1976. Life-history tactics: A review of the ideas. *Q. Rev. Biol.* **51**, 3–47.

Stearns, S.C. 1977. The evolution of life history traits: A critique of the theory and a review of the data. *Ann. Rev. Ecol. Syst.* **8**, 145–71.

Stearns, S.C. 1982. The role of development in the evolution of life histories. In: J.T. Bonner (ed.) *Evolution and Development*, pp. 237–58. Springer-Verlag, Berlin.

Stearns, S.C. 1983. The influence of size and phylogeny on patterns of covariation among life-history traits in mammals. *Oikos* **41**, 173–87.

Stearns, S.C. 1986. Natural selection and fitness, adaptation and constraint. In: D.M. Raup & D. Jablonski (eds.) *Patterns and Processes in the History of Life*, pp. 23–44. Springer-Verlag, Berlin.

Stearns, S.C. 1987. Introduction. In: S.C. Stearns (ed.) *The Evolution of Sex and its Consequences*, pp. 15–31. Birkhäuser Verlag, Basel.

Stearns, S.C. 1989. Trade-offs in life-history evolution. *Functional Ecol.* **3**, 259–68.

Stearns, S.C. & Crandall, R.E. 1981. Quantitative predictions of delayed maturity. *Evolution* **35**, 455–63.

Stebbins, R.C. 1966. *A Field Guide to the Western Reptiles and Amphibians*. Houghton Mifflin, Boston.

Stebbins, G.L. 1950. *Variation and Evolution in Plants*. Columbia University Press, New York.

Stebbins, G.L. 1970. Adaptive radiation of reproductive characters in angiosperms. I. Pollination mechanisms. *Ann. Rev. Ecol. Syst.* **1**, 307–26.

Stebbins, G.L. 1977. In defense of evolution: tautology or theory? *Amer. Natur.* **111**, 386–90.

Stebbins, G.L. & Ayala, F.J. 1981. Is a new evolutionary synthesis necessary?

Science **213**, 967−71.

Stebbins, G.L. & Hartl, D.L. 1988. Comparative evolution: latent potentials for anagenetic advance. *Proc. Natl. Acad. Sci. USA* **85**, 5141−5.

Stehli, F.G., Douglas, R.G. & Newell, N.D. 1969. Generation and maintenance of latitudinal gradients in taxonomic diversity. *Science* **164**, 947−9.

Stenseth, N.C. & Maynard Smith, J. 1984. Coevolution in ecosystems: Red Queen evolution or stasis? *Evolution* **38**, 870−80.

Stephens, D.W. & Krebs, J.R. 1986. *Foraging Theory*. Princeton University Press, Princeton.

Stephenson, A.G. & Bertin, R.I. 1983. Male competition, female choice, and sexual selection in plants. In: L. Real (ed.) *Pollination Biology*, pp. 109−49. Academic Press, New York.

Stevens, G.C. 1989. The latitudinal gradient in geographical range: how so many species coexist in the tropics. *Amer. Natur.* **133**, 240−56.

Stewart, C.-B., Schilling, J.W. & Wilson, A.C. 1987. Adaptive evolution in the stomach lysozymes of foregut fermenters. *Nature* **330**, 401−4.

Stigler, S.M. & Wagner, M.J. 1987. A substantial bias in nonparametric tests for periodicity in geophysical data. *Science* **238**, 940−1.

Stork, N.E. 1988. Insect diversity: facts, fiction and speculation. *Biol. J. Linn. Soc.* **35**, 321−37.

Strahan, R. (ed.) 1983. *The Complete Book of Australian Mammals*. Angus & Robertson, Sydney.

Strathmann, R.R. 1978. Progressive vacating of adaptive types through the Phanerozoic. *Evolution* **32**, 907−14.

Strathmann, R.R. 1985. Feeding and non-feeding larval development and life-history evolution in marine invertebrates. *Ann. Rev. Ecol. Syst.* **16**, 339−61.

Strathmann, R.R. & Slatkin, M. 1983. The improbability of animal phyla with few species. *Paleobiology* **9**, 97−106.

Stresemann, E. 1954. Die Entdeckungsgeschichte der Paradiesvögel. *J. Ornithologie* **95**, 263−91.

Strong, D.R., Lawton, J.H. & Southwood, R. 1984. *Insects on Plants: Community Patterns and Mechanisms*. Blackwell Scientific Publications, Oxford.

Subramanyam, K. 1962. *Aquatic Angiosperms*. Botanical Monographs No. 3. Council of Scientific and Industrial Research, New Delhi.

Sulston, J.E., Shierenberg, E., White, J.G. & Thomson, J.N. 1983. The embryonic cell lineage of the nematode *Caenorhabditis elegans*. *Dev. Biol.* **100**, 64−119.

Sultan, S.E. 1987. Evolutionary implications of phenotypic plasticity in plants. *Evol. Biol.* **21**, 127−78.

Sutherland, W.J. & Parker, G.A. 1985. Distribution of unequal competitors. In: R.M. Sibly & R.H. Smith (eds.) *Behavioural Ecology: Ecological Consequences of Adaptive Behaviour*, pp. 255−73. Blackwell Scientific Publications, Oxford.

Szidat, L. 1956. Geschichte, Anwendung und einige Folgerungen aus den parasitogenetischen. *Z. Parasitenk.* **17**, 237−68.

Taper, M.L. & Case, T.J. 1985. Quantitative genetic models for the coevolution of character displacement. *Ecology* **66**, 355−71.

Tasch, P. 1973. *Paleobiology of the Invertebrates*. John Wiley & Sons, New York.

Tauber, M.J., Tauber, C.A. & Masaki, S. 1986. *Seasonal Adaptations of Insects*. Oxford University Press, New York.

Temple, S.A. 1977. Plant−animal mutualism: coevolution with dodo leads to near extinction of plants. *Science* **197**, 885−6.

Templeton, A.R. 1980. The theory of speciation via the founder principle. *Genetics* **94**, 1011−38.

Templeton, A.R. 1981. Mechanisms of speciation − a population genetic approach. *Ann. Rev. Ecol. Syst.* **12**, 23−48.

Templeton, A.R. 1986. Coadaptation and outbreeding depression. In: M.E. Soulé (ed.) *Conservation Biology: the Science of Scarcity and Diversity*, pp. 105−16. Sinauer, Sunderland, Massachusetts.

Templeton, A.R. & Levin, D.A. 1979. Evolutionary consequences of seed pools. *Amer. Natur.* **114**, 232−49.

Templeton, A.R. & Read, B. 1983.

The elimination of inbreeding depression in a captive herd of Speke's gazelle. In: C.M. Schonewald-Cox, S. Chambers, B. MacBryde & W.L. Thomas (eds.) *Genetics and Conservation: a Reference for Managing Wild Plant and Animal Populations*, pp. 241–61. Benjamin/Cummings, Menlo Park, California.

Terborgh, J. 1973. On the notion of favourableness in plant ecology. *Amer. Natur.* **107**, 481–501.

Terborgh, J. 1986. Keystone plant resources in the tropical forest. In: M. Soulé (ed.) *Conservation Biology: The Science of Scarcity and Diversity*, pp. 330–44. Sinauer, Sunderland, Massachussets.

Thomas, C.D. 1990. Fewer species. *Nature* **347**, 237.

Thompson, J.N. 1982. *Interaction and Coevolution*. Wiley, New York.

Thompson, J.N. 1986. Constraints on arms races in evolution. *Trends Ecol. Evol.* **1**, 105–7.

Thompson, J.N. 1987. Symbiont-induced speciation. *Biol. J. Linn. Soc.* **32**, 385–93.

Thompson, J.N. 1988. Variation in species interactions. *Ann. Rev. Ecol. Syst.* **19**, 65–87.

Thompson, J.N. 1989. Concepts of coevolution. *Trends Ecol. Evol.* **4**, 179–83.

Thompson, J.N. & Barrett, S.C.H. 1981. Temporal variation of gender in *Aralia hispida* (Araliaceae). *Evolution* **35**, 1094–1107.

Thompson, J.N. & Brunet, J. 1990. Hypotheses for the evolution of dioecy in seed plants. *Trends Ecol. Evol.* **5**, 11–16.

Thornhill, R. 1980. Mate choice in *Hylobittacus apicalis* (Insecta: Mecoptera) and its relation to models of female choice. *Evolution* **34**, 519–38.

Thornhill, R. & Alcock, J. 1983. *The Evolution of Insect Mating Systems*. Harvard University Press, Cambridge.

Thornton, I.W.B., Zann, R.A., Rawlinson, P.A., Tidemann, C.R., Adikerana, A.S. & Widjoya, A.H.T. 1988. Colonization of the Krakatau Islands by vertebrates: equilibrium, succession, and possible delayed extinction. *Proc. Natl. Acad. Sci. USA* **85**, 515–18.

Thorpe, A. & Duve, H. 1984. Insulin-and glucagon like peptides in insects and molluscs. *Mole. Physiol.* **5**, 139–68.

Tiedje, J.M., Colwell, R.K., Grossman, Y.L., Hodson, R.E., Lenski, R.E., Mack, R.N. & Regal, P.J. 1989. The planned introduction of genetically engineered organisms: ecological considerations and recommendations. *Ecology* **70**, 298–351.

Tomlinson, I.P.M. 1988. Diploid models of the handicap principle. *Heredity* **60**, 283–93.

Trevelyan, R., Harvey, P.H. & Pagel, M.D. 1990. Metabolic rates and life histories in birds. *Funct. Ecol.* **4**, 135–41.

Trivers, R.L. 1972. Parental investment and sexual selection. In: B. Campbell (ed.) *Sexual Selection and the Descent of Man 1871–1971*. pp. 136–79. Heinemann, London.

Trivers, R.L. 1988. Sex differences in rates of recombination and sexual selection. In: R.E. Michod & B.R. Levin (eds.) *The Evolution of Sex*, pp. 270–86. Sinauer, Sunderland, Massachusetts.

Trivers, R.L. & Willard, D.E. 1973. Natural selection of parental ability to vary the sex ratio of offspring. *Science* **191**, 249–63.

Tudge, C. 1989. Variety in vogue. *New Scientist* **1656**, 50–3.

Tuljapurkar, S. 1989. An uncertain life: demography in random environments. *Theor. Popul. Biol.* **35**, 227–94.

Tuljapurkar, S. 1990. Delayed reproduction and fitness in variable environments. *Proc. Natl. Acad. Sci. USA* **87**, 1139–43.

Turelli, M. 1984. Heritable genetic variation via mutation–selection balance: Lerch's zeta meets the abdominal bristle. *Theoret. Popul. Biol.* **25**, 138–93.

Turner, J.R.G. 1981. Adaptation and evolution in *Heliconius*: a defense of neo-Darwinism. *Ann. Rev. Ecol. Syst.* **12**, 99–121.

Turner, J.R.G. 1986. The genetics of adaptive radiation: a neo-Darwinian theory of punctuational evolution. In: D.M. Raup & D. Jablonski (eds.) *Patterns and Processes in the History of Life*, pp. 183–207. Springer-Verlag, Berlin.

Turner, J.R.G., Gatehouse, C.M. & Corey, C.A. 1987. Does solar energy control organic diversity? Butterflies, moths and British climate. *Oikos* **48**, 195–205.

Turner, J.R.G., Lennon, J.J. & Lawrenson, J.A. 1988. British bird species distributions and the energy theory. *Nature* **335**, 539–41.

Valentine, J.W. & Jablonski, D. 1986. Mass extinctions: sensitivity of marine larval types. *Proc. natl. Acad. Sci. USA* **83**, 6912–14.

Van Valen, L. 1973a. A new evolutionary law. *Evol. Theory* **1**, 1–30.

Van Valen, L. 1973b. Body size and numbers of plants and animals. *Evolution* **27**, 27–35.

Venable, D.L. 1985. The evolutionary ecology of seed heteromorphism. *Amer. Natur.* **126**, 577–95.

Venable, D.L. & Levin, D.A. 1985a. Ecology of achene dimorphism in *Heterotheca latifolia*. I. Achene structure, germination and dispersal. *J. Ecol.* **73**, 133–45.

Venable, D.L. & Levin, D.A. 1985b. Ecology of achene dimorphism in *Heterotheca latifolia*. II. Demographic variation within populations. *J. Ecol.* **73**, 743–55.

Vermeij, G.J. 1973a. Adaptation, versatility and evolution. *Syst. Zool.* **22**, 466–77.

Vermeij, G.J. 1973b. Biological versatility and earth history. *Proc. Natl. Acad. Sci. USA* **70**, 1936–8.

Via, S. 1986. Genetic covariance between oviposition preference and larval performance in an insect herbivore. *Evolution* **40**, 778–85.

Via, S. 1987. Genetic constraints on the evolution of phenotypic plasticity. In: V. Loeschcke (ed.) *Genetic Constraints on Adaptive Evolution*, pp. 47–71. Springer-Verlag, Berlin.

Vitousek, P.M., Ehrlich, P.R., Ehrlich, A.H. & Matson, P.A. 1986. Human appropriation of the products of photosynthesis. *Bio Science* **36**, 368–73.

Vrba, E.S. 1984. Evolutionary pattern and process in the sister-group Alcephalini–Aepycerotini (Mammalia: Bovidae). In: N. Eldredge & S.M. Stanley (eds.) *Living Fossils*, pp. 62–79. Springer-Verlag, Berlin.

Vrba, E.S. & Eldredge, N. 1984. Individuals, hierarchies and process: towards a more complete evolutionary theory. *Paleobiology* **10**, 146–71.

Waage, J. 1979. Duel function of the damselfly penis: sperm removal and transfer. *Science* **203**, 916–18.

Waddington, C.H. 1952. Canalization of the development of quantitative characters. In: E.C.R. Reeve & C.H. Waddington (eds.) *Quantitative Inheritance*, pp. 43–6. HMSO, London.

Waddington, K.D., Allen, T. & Heinrich, B. 1981. Floral preferences of bumblebees (*Bombus edwardsii*) in relation to intermittent versus continuous rewards. *Anim. Behav.* **29**, 779–84.

Wade, M.J. 1977. An experimental study of group selection. *Evolution* **31**, 134–53.

Wade, M.J. 1982. Group selection: migration and the differentiation of small populations. *Evolution* **36**, 949–61.

Wade, M.J. & Arnold, S.J. 1980. The intensity of sexual selection in relation to male behaviour, female choice and sperm precedence. *Anim. Behav.* **28**, 446–61.

Wade, M.J. & McCauley, D.E. 1980. Group selection: the phenotypic and genotypic differentiation of small populations. *Evolution* **34**, 799–812.

Wagner, G.P. 1986. The systems approach: an interface between developmental and population genetic aspects of evolution. In: D.M. Raup & D. Jablonski (eds.) *Patterns and Processes in the History of Life*, pp. 149–65. Springer-Verlag, Berlin.

Wainwright, P.C. 1986. Motor correlates of learning behaviour: feeding on novel prey by pumpkinseed sunfish (*Lepomis gibbosus*). *J. exp. Biol.* **126**, 237–47.

Wake, D.B. & Larson, A. 1987. Multidimensional analysis of an evolving lineage. *Science* **238**, 42–8.

Wake, D.B., Roth, G. & Wake, M.H. 1983. On the problem of stasis in organismal evolution. *J. theor. Biol.* **101**, 211–24.

Wake, D.B. & Yanev, K.P. 1986. Geographic variation in allozymes in a ring species, the plethodontid salamander *Ensatina eschscholtzii* of western North America. *Evolution* **40**, 702–15.

Waldman, B. 1989. Do anuran larvae retain kin recognition abilities following metamorphosis? *Anim. Behav.* **37**, 1055–8.

Walker, W.F. 1980. Sperm utilization strategies in nonsocial insects. *Amer. Natur.* **115**, 780–99.

Wallace, B. 1981. *Basic Population Genetics*. Columbia University Press, New York.

Walls, S.C. 1990. Interference competition in postmetamorphic salamanders: interspecific differences in aggression by coexisting species. *Ecology* **71**, 307–14.

Walls, S.C. & Jaeger, R.G. 1987. Aggression and exploitation as mechanisms of competition in larval salamanders. *Can. J. Zool.* **65**, 2938–44.

Warner, R.R. 1984. Deferred reproduction as a response to sexual selection in a coral reef fish: a test of the life historical consequences. *Evolution* **38**, 148–62.

Warner, R.R. & Chesson, P.L. 1985. Coexistence mediated by recruitment fluctuations: a field guide to the storage effect. *Amer. Natur.* **125**, 769–87

Warner, R.R., Robertson, D.R. & Leigh, E.G. 1975. Sex change and sexual selection. *Science* **190**, 633–8.

Waser, N.M., Price, M.V., Montalvo, A.M. & Gray, N.M. 1987. Female mate choice in a perennial herbaceous wildflower, *Delphinium nelsoni*. *Evol. Trends Plants* **1**, 29–33.

Waser, P.M. & Jones, W.T. 1983. Natal philopatry among solitary mammals. *Q. Rev. Biol.* **58**, 355–90.

Watkinson, A.R. 1988. On the growth and reproductive schedules of plants: a modular viewpoint. *Acta Œcologica/Œcologia Plant.* **9**, 67–81.

Watkinson, A.R. & White, J. 1986. Some life-history consequences of modular construction in plants. *Phil. Trans. R. Soc. Lond.* **B313**, 31–51.

Watson, J.D., Hopkins, N.H., Roberts, J.W., Steitz, J.A. & Weiner, A.M. 1987. *Molecular Biology of the Gene.* 4th edn. Benjamin/Cummings, Menlo Park, California.

Watson, M.A. 1984. Developmental constraints: effect on population growth and patterns of resource allocation in a clonal plant. *Amer. Natur.* **123**, 411–26.

Watson, M.A. & Casper, B.B. 1984. Morphogenetic constraints on patterns

of carbon distribution in plants. *Ann. Rev. Ecol. Syst.* **15**, 233–58.

Watt, K.E.F. 1964. Comments on fluctuations of animal populations and measures of community stability. *Canad. Entomol.* **96**, 1434–42.

Watt, W.B., Carter, P.A. & Donohue, K. 1986. Females' choice of 'good genotypes' as mates is promoted by an insect mating system. *Science* **233**, 1187–90.

Wei, K.-Y. & Kennett, J.P. 1988. Phyletic gradualism and punctuated equilibrium in the late Neogene planktonic foraminiferal clade *Globoconella*. *Paleobiology* **14**, 345–63.

Weider, L.J., Beaton, M.J. & Hebert, P.D.N. 1987. Clonal diversity in high-arctic populations of *Daphnia pulex*, a polyploid apomictic complex. *Evolution* **41**, 1335–46.

Weismann, A. 1893. *The Germ-Plasm: a Theory of Heredity*. Scott Publishing, London.

Wellington, A.B. & Noble, I.R. 1985. Seed dynamics and factors limiting recruitment of the mallee *Eucalyptus incrassata* in semi-arid, south-eastern Australia. *J. Ecol.* **73**, 657–66.

Werner, E.E. 1986. Amphibian metamorphosis: growth rate, predation risk, and optimal size at transformation. *Amer. Natur.* **128**, 319–41.

Werner, E.E. 1988. Size, scaling, and the evolution of complex life cycles. In: B. Ebenmann & L. Persson (eds.) *Size-Structured Populations*, pp. 60–81. Springer-Verlag, Berlin & Heidelberg.

Werner, E.E. & Gilliam, J.F. 1984. The ontogenetic niche and species interactions in size-structured populations. *Ann. Rev. Ecol. Syst.* **15**, 393–425.

Werner, P.A. 1979. Competition and coexistence of similar species. In: O.T. Solbrig, S. Jain, G.B. Johnson & P.A. Raven (eds.) *Topics in Plant Population Biology*, pp. 287–310. Columbia University Press, Columbia.

Werner, P.A. & Platt, W.J. 1976. Ecological relationship of co-occurring goldenrods (*Solidago*: Compositae). *Amer. Natur.* **110**, 959–71.

Werren, J.H. 1983. Sex ratio evolution under local mate competition in a parasitic wasp. *Evolution* **37**, 116–24.

Werren, J., Nur, U. & Wu, C.-I. 1988. Selfish genetic elements. *Trends Ecol.*

Evol. **3**, 297–302.

West-Eberhard, M.J. 1986. Alternative adaptations, speciation, and phylogeny (a review). *Proc. Natl. Acad. Sci. USA* **83**, 1388–92.

West-Eberhard, M.J. 1989. Phenotypic plasticity and the origins of diversity. *Ann. Rev. Ecol. Syst.* **20**, 249–78.

White, M.E. 1986. *The Greening of Gondwana.* Reed, French's Forest, New South Wales.

White, G. 1770. *The Natural History of Selbourne* (reprinted 1977). Penguin, Harmondsworth.

White, M.J.D. 1978. *Modes of Speciation.* Freeman, San Francisco.

White, M.M., Mane, S.D. & Richmond, R.C. 1988. Studies of esterase 6 in *Drosophila melanogaster.* XVIII. Biochemical differences between the fast and slow allozymes. *Mol. Biol. Evol.* **5**, 41–62.

Whitham, T.G. 1989. Plant hybrid zones as sinks for pests. *Science* **244**, 1490–93.

Whitham, T.G. & Slobodchikoff, C.N. 1981. Evolution by individuals, plant–herbivore interactions, and mosaics of genetic variability: the adaptive significance of somatic mutations in plants. *Oecologia* **49**, 287–92.

Whittaker, R.H. 1972. Evolution and measurement of species diversity. *Taxon* **21**, 213–51.

Wiklund, C. & Ahrberg, C. 1978. Hostplants, nectar source plants, and habitat selection of males and females of *Anthocharis cardaneus* L. (Lepidoptera). *Oikos* **31**, 169–83.

Wilbur, H.M. 1980. Complex life cycles. *Ann. Rev. Ecol. Syst.* **11**, 67–93.

Wiens, D. 1978. Mimicry in plants. *Evol. Biol.* **11**, 365–403.

Wilcove, D.S., McLellan, C.H. & Dobson, A.P. 1986. Habitat fragmentation in the temperate zone. In: M.E. Soulé (ed.) *Conservation Biology: the Science of Scarcity and Diversity,* pp. 237–56. Sinauer, Sunderland, Massachussets.

Wilcox, B.A. 1980. Insular ecology and conservation. In: M.E. Soulé & B.A. Wilcox (eds.) *Conservation Biology: an Evolutionary–Ecological Perspective,* pp. 95–117. Sinauer, Sunderland, Massachussets.

Wildt, D.E., Bush, M., Goodrowe, K.L., Packer, C., Pusey, A.E., Brown, J.L.,

Joslin, P. & O'Brien, S.J. 1987. Reproductive and genetic consequences of founding isolated lion populations. *Nature* **329**, 328–31.

Williams, E.E. 1972. The origin of faunas. Evolution of lizard congeners in a complex island faunas: a trial analysis. *Evol. Biol.* **6**, 47–79.

Williams, G.C. 1957. Pleiotropy, natural selection, and the evolution of senescence. *Evolution* **11**, 398–411.

Williams, G.C. 1966a. *Adaptation and Natural Selection.* Princeton University Press, Princeton.

Williams, G.C. 1966b. Natural selection, the costs of reproduction, and a refinement of Lack's principle. *Amer. Natur.* **100**. 687–90.

Williams, G.C. 1975. *Sex and Evolution.* Princeton University Press, Princeton.

Williams, G.C. 1979. The question of adaptive sex ratio in outcrossed vertebrates. *Proc. R. Soc. Lond.* **B205**, 567–80.

Williamson, M. 1988. Potential effects of recombinant DNA organisms on ecosystems and their components. *Trends Ecol. Evol.* **3**, S32–5.

Williamson, P.G. 1981. Palaeontological documentation of speciation in Cenozoic molluscs from Turkana Basin. *Nature* **293**, 437–43.

Williamson, P.G. 1987. Selection or constraint?: a proposal for the mechanism of stasis. In: K.S.W. Campbell & M.F. Day (eds.) *Rates of Evolution,* pp. 129–42. Allen & Unwin, London.

Willson, M.F. 1982. Sexual selection and dicliny in angiosperms. *Amer. Natur.* **119**, 579–83.

Willson, M.F. & Burley, N. 1983. *Mate Choice in Plants.* Princeton University Press, Princeton.

Willson, M.F., Hoppes, W.G., Goldman, D.A., Thomas, P.A., Katusic-Malmborg, P.L. & Bothwell, J.L. 1987. Sibling competition in plants: an experimental study. *Amer. Natur.* **129**, 304–11.

Wilson, D.S. 1977. Structured demes and the evolution of group-advantageous traits. *Amer. Natur.* **111**, 157–85.

Wilson, D.S. 1980. *The Natural Selection of Populations and Communities.* Benjamin/Cummings, Menlo Park, California.

Wilson, D.S. 1983. The group selection

controversy: history and current status. *Ann. Rev. Ecol. Syst.* **14**, 159–87.

Wilson, S.D. & Keddie, P.A. 1986a. Measuring diffuse competition along an environmental gradient: results from a shoreline plant community. *Amer. Natur.* **127**, 862–9.

Wilson, S.D. & Keddie, P.A. 1986b. Species competitive ability and position along a natural environmental gradient. *Ecology* **67**, 1236–42.

Wilson, D.S. & Turelli, M. 1986. Stable underdominance and the evolutionary invasion of empty niches. *Amer. Natur.* **127**, 835–50.

Wilson, E.O. 1975. *Sociobiology: The New Synthesis.* Harvard University Press, Cambridge, Massachussets.

Wilson, E.O. 1987. Causes of ecological success: the case of the ants. *J. Anim. Ecol.* **56**, 1–9.

Wilson, E.O. (ed.) 1988. *Biodiversity.* National Academy Press, Washington, DC.

Wilson, E.O. & Bossert, W.H. 1971. *A Primer of Population Biology.* Sinauer, Sunderland, Massachussets.

Winn, A.A. 1985. Effects of seed size and microsite on seedling emergence of *Prunella vulgaris* in four habitats. *J. Ecol.* **73**, 831–40.

Wright, D.H. 1983. Species–energy theory: an extension of species–area theory. *Oikos* **41**, 496–506.

Wright, S. 1931. Evolution in Mendelian populations. *Genetics* **16**, 97–159.

Wright, S. 1977. *Evolution and the Genetics of Populations: a Treatise. III. Experimental Results and Evolutionary Deductions.* University of Chicago Press, Chicago.

Wright, S. 1978. *Evolution and the Gen-*

etics of Populations: a Treatise. IV. Variation Within and Among Natural Populations. University of Chicago Press, Chicago.

Wright, S. 1982. Character change, speciation, and the higher taxa. *Evolution* **36**, 427–43.

Wright, S. 1988. Surfaces of selective value revisited. *Amer. Natur.* **131**, 115–23.

Yamazaki, K., Beauchamp, G.K., Wysocki, C.J., Bard, J., Thomas, L. & Boyse, E.A. 1983. Recognition of H-2 types in relation to the blocking of pregnancy in mice. *Science* **221**, 186–8.

Young, H.M. & Stanton, M.L. 1990. Influence of environmental quality on pollen competitive ability in wild radish. *Science* **248**, 1631–3.

Young, T.P. 1981. A general model of comparative fecundity for semelparous and iteroparous life histories. *Amer. Natur.* **118**, 27–36.

Zahavi, A. 1975. Mate selection—a selection for a handicap. *J. theor. Biol.* **53**, 205–214.

Zahavi, A. 1977. The cost of honesty (further remarks on the handicap principle). *J. theor. Biol.* **67**, 603–5.

Zohary, D. & Spiegel-Roy, P. 1975. Beginnings of fruit growing in the Old World. *Science* **187**, 319–27.

Zuk, M. 1987. The effects of gregarine parasites, body size, and time of day on spermatophore production and sexual selection in field crickets. *Behav. Ecol. Sociobiol.* **21**, 65–72.

Zuk, M. 1988. Parasite load, body size, and age of wild-caught male field crickets (Orthoptera: Gryllidae): effects on sexual selection. *Evolution* **42**, 969–76.

Index

An Introduction to Evolutionary Ecology

Why was 1983 so bad for giant pandas? Why have sex in the sea? Why are there so many species of beetle? These and other problems are addressed in this beautifully illustrated textbook. Using a wealth of real examples, from both the animal and plant kingdoms and often in areas of current controversy, Andrew Cockburn explores the questions at the heart of evolutionary ecology: the origin and maintenance of the diversity of organisms; the pressures that determine their form and shape their behaviour; and the way in which they interact. In addition, a final chapter considers the application of evolutionary ecology, covering the management of endangered species, the causes of extinction, the importance of genetic diversity and the release of genetically engineered organisms.

The book includes mathematical approaches to the study of evolution and ecology, but no great mathematical competence is assumed – emphasis is placed on the testing of theory rather than its algebraic development. In contrast, the text includes a much more detailed discussion of genetics than is usual in an ecology text, and breaks new ground in surveying developments in molecular genetics, palaeontological theory, and the application of quantitative genetics to evolution.

Aimed at senior undergraduate and postgraduate students this well-structured and lucid text provides an up-to-date, critical evaluation of many of the dilemmas at the heart of evolutionary ecology.

Professor John Jaenike of the University of Rochester says:
'This is a terrific book. It is extraordinarily up-to-date and covers virtually all of what I consider to be most interesting in the field of evolutionary ecology. There is a nice balance between examples drawn from the plant and animal literature, and it is not heavily weighted towards any one geographical region.'

ISBN 0-632-02729-0

9 780632 027293

BLACKWELL SCIENTIFIC PUBLICATIONS
Oxford London Edinburgh Boston
Melbourne Paris Berlin Vienna